Pathobiology of Alzheimer's Disease

NEUROSCIENCE PERSPECTIVES

Series Editor
Peter Jenner
Pharmacology Group
Biomedical Sciences Division
King's College London
Manresa Road
London SW3 6LX

Titles in this series
Roger Horton and Cornelius Katona (eds), Biological Aspects of Affective Disorders
Trevor Stone (ed), Adenosine in the Nervous System
Judith Pratt (ed), The Biological Bases of Drug Tolerance and Dependence
Michel Hamon (ed), Central and Peripheral 5-HT_3 Receptors
John Waddington (ed), D_1 : D_2 Dopamine Receptor Interactions
Thomas Barnes (ed), Antipsychotic Drugs and Their Side-Effects
Eva Giesen-Crouse (ed), Peripheral Benzodiazepine Receptors
David Nicholson (ed), Anti-Dementia Agents
Alan Harvey (ed), Natural and Synthetic Neurotoxins
John Wood (ed), Capsaicin in the Study of Pain
Yossef Itzhak (ed), Sigma Receptors

Pathobiology of Alzheimer's Disease

edited by

Alison M. Goate and Frank Ashall
Department of Psychiatry,
St Louis University School of
Medicine, St Louis, USA

ACADEMIC PRESS
Harcourt Brace & Company, Publishers
London San Diego New York
Boston Sydney Tokyo Toronto

ACADEMIC PRESS LIMITED
24/28 Oval Road,
London NW1 7DX

United States Edition published by
ACADEMIC PRESS INC.
San Diego, CA 92101

This book is printed on acid free paper

A catalogue record for this book
is available from the British Library

ISBN 0-12-286965-6

Typeset by Servis Filmsetting Ltd, Manchester, UK
Printed and bound in Great Britain
by T J Press, Padstow, Cornwall.

Contents

The Color Plates are located between pages 18 and 19.

Contributors

Frank Ashall Saint Louis University School of Medicine, 1402 South Grand Boulevard, St Louis, MO 63104, USA

Phillipe Bertrand McGill Centre for Studies in Aging, 6875 LaSalle Boulevard, Verdun, Quebec, Canada H4H 1R7

David R. Borchelt Neuropathology Laboratory, The Johns Hopkins University School of Medicine, 720 Rutland Avenue, 558 Ross Research Building, Baltimore, MD 21205–2196, USA

Frances Busfield Department of Psychiatry, Washington University School of Medicine, 4940 Children's Place, St Louis, MO 63110, USA

Joseph D. Buxbaum Laboratory of Molecular and Cellular Neuroscience, The Rockefeller University, New York, NY 10021, USA

Patricia Edwards Cambridge Brain Bank Laboratory, University of Cambridge Clinical School, Department of Psychiatry, MRC Centre, Hills Road, Cambridge CB2 2QH, UK

Alison M. Goate Department of Psychiatry, Washington University School of Medicine, 4940 Children's Place, St Louis, MO 63110, USA

Todd E. Golde Division of Neuropathology, Institute of Pathology, Case Western Reserve University, Cleveland, OH 44106, USA

Dmitry Goldgaber Department of Psychiatry and Behavioral Science, State University of New York at Stony Brook, Stony Brook, NY 11794–8101, USA

Daniel Guillaume McGill Centre for Studies in Aging, 6875 LaSalle Boulevard, Verdun, Quebec, Canada H4H 1R7

Charles R. Harrington Cambridge Brain Bank Laboratory, University of Cambridge Clinical School, Department of Psychiatry, MRC Centre, Hills Road, Cambridge CB2 2QH, UK

Robert Lai Cambridge Brain Bank Laboratory, University of Cambridge Clinical School, Department of Psychiatry, MRC Centre, Hills Road, Cambridge CB2 2QH, UK

Virginia M.-Y. Lee Department of Pathology and Laboratory Medicine, Division of Anatomic Pathology, University of Pennsylvania School of Medicine, Philadelphia, PA 19104, USA

Raul Mena Cambridge Brain Bank Laboratory, University of Cambridge Clinical School, Department of Psychiatry, MRC Centre, Hills Road, Cambridge CB2 2QH, UK

John C. Morris Washington University School of Medicine, 660 South Euclid Avenue, St Louis, MO 63110, USA

Elizabeth Mukaetova-Ladinska Cambridge Brain Bank Laboratory, University of Cambridge Clinical School, Department of Psychiatry, MRC Centre, Hills Road, Cambridge CB2 2QH, UK

Josephine Nalbantoglu McGill Centre for Studies in Aging, 6875 LaSalle Boulevard, Verdun, Quebec, Canada H4H 1R7

Judes Poirier McGill Centre for Studies in Aging, 6875 LaSalle Boulevard, Verdun, Quebec, Canada H4H 1R7

Donald L. Price Neuropathology Laboratory, The Johns Hopkins University School of Medicine, 720 Rutland Avenue, 558 Ross Research Building, Baltimore, MD 21205–2196, USA

Wolfgang W. Quitschke Department of Psychiatry and Behavioral Science, State University of New York at Stony Brook, Stony Brook, NY 11794–8101, USA

Martin Roth Cambridge Brain Bank Laboratory, University of Cambridge Clinical School, Department of Psychiatry, MRC Centre, Hills Road, Cambridge CB2 2QH, UK

Sangram S. Sisodia Neuropathology Laboratory, The Johns Hopkins University School of Medicine, 720 Rutland Avenue, 558 Ross Research Building, Baltimore, MD 21205–2196, USA

John Q. Trojanowski Department of Pathology and Laboratory Medicine, Division of Anatomic Pathology, University of Pennsylvania School of Medicine, Philadelphia, PA 19104, USA

Steven L. Wagner SIBIA, 505 Coast Boulevard South, La Jolla, CA 92037, USA

Claude M. Wischik Cambridge Brain Bank Laboratory, University of Cambridge Clinical School, Department of Psychiatry, MRC Centre, Hills Road, Cambridge CB2 2QH, UK

John Xuereb Cambridge Brain Bank Laboratory, University of Cambridge Clinical School, Department of Psychiatry, MRC Centre, Hills Road, Cambridge CB2 2QH, UK

Kazuaki Yoshikawa Department of Regulation of Macromolecular Functions, Institute for Protein Research, Osaka University, 3–2 Yamadaoka, Suita, Osaka 565, Japan

Steven G. Younkin Division of Neuropathology, Institute of Pathology, Case Western Reserve University, Cleveland, OH 44106, USA

Series Preface

The neurosciences are one of the most diverse and rapidly changing areas in the biological sphere. The need to understand the workings of the nervous system pervades a vast array of different research areas. By definition research in the neurosciences encompasses anatomy, pathology, biochemistry, physiology, pharmacology, molecular biology, genetics and therapeutics. Ultimately, we are striving to determine how the human brain functions under normal circumstances and perhaps more importantly how function changes in organic disease and in altered states of mind. The key to many of these illnesses will unlock one of the major therapeutic challenges remaining in this era.

The difficulty lies in the vastness of the subject matter. However I try, I find it almost impossible to keep abreast of the changes occurring in my immediate sphere of interest, let alone those exciting advances being made in other areas. The array of journals dealing with neurosciences is enormous and the flow of reprints needed to keep me updated is daunting. Inevitably piles of papers accumulate on my desk and in my briefcase. Many lie there unread until sufficient time has passed for their content to be overtaken by yet more of the ever rising tide of publications.

There are various approaches that one can take to deal with this problem. There is the blinkered approach in which you pretend that literature outside your area does not exist. There is the ignore it totally option. Indeed, one colleague of mine has ceased to read the literature in the belief that, if there is a publication of critical importance to his research, someone will tell him about it. I am not that brave and instead I arrived at what I thought was the ideal solution. I started to read critical reviews of areas of current interest. But I soon came unstuck as I realized that, for many subjects of importance to the neurosciences, such authoritative works did not exist.

Instead of simply moaning that the literature was incomplete, I conceived the idea of *Neuroscience Perspectives*. For purely selfish reasons I wanted to have available a series of individually edited monographs dealing in depth with issues of current interest to those working in the neuroscience area. Already a number of volumes have been published which have been well received and the series is thriving with books on a range of topics in preparation or in production. Each volume is designed to bring a multidisciplinary approach to the subject matter by pursuing the topic from the laboratory to the clinic. The editors of the individual volumes are producing balanced critiques of each topic to provide the reader with an up-to-date, clear and comprehensive view of the state of the art.

As with all ventures of this kind, I am simply the individual who initiates a chain of events leading to the production of the series. In reality, it is key individuals at Academic Press who are really responsible for the success of *Neuroscience Perspectives*. In particular Léona Daw had the uneviable task of recruiting editors and authors and keeping the ship on an even keel.

Finally, I hope that *Neuroscience Perspectives* will continue to be enjoyed by my colleagues in the neurosciences. Already the series is being read, understood and enjoyed by a wide audiences and it is fast becoming a reference series in the field.

Peter Jenner

Preface

Improvements in the diagnosis and treatment of the major diseases of childhood and early to mid-adulthood has led to a major increase in mean life span in the populations of the developed world during the last hundred years. This has obviously been accompanied by an increase in the number of people suffering from diseases of old age. Indeed, Alzheimer's disease is now the fourth leading cause of death in the United States after heart disease, cancer and stroke. This disease not only causes an irreversible reduction in the quality of life of the sufferer but also places a tremendous burden on care-givers and welfare systems. During the next 40 years with the ageing of the baby-boom generation it is predicted that the number of people affected with Alzheimer's disease will more than double. Understanding the causes of Alzheimer's disease and developing treatments is therefore one of the major scientific and medical challenges of the late 20th century.

The major aim of this book is to outline the various approaches being taken to understand the pathobiology of Alzheimer's disease. Two striking features of the neuropathology of Alzheimer's disease have received most attention concerning the pathogenesis of Alzheimer's disease and provide the major focus for this volume. They are the senile plaque and the neurofibrillary tangle. Neither is unique to Alzheimer's disease but in combination their characteristic distribution is pathognomonic for Alzheimer's disease.

The recent use of molecular genetics to study early-onset Alzheimer's disease has led to the identification of the first cause of Alzheimer's disease – mutations within the amyloid β-protein precursor (APP) gene. Indeed, the demonstration that only a very few cases of Alzheimer's disease are caused by this mechanism shows that Alzheimer's disease is not a single disease entity but probably has many causes manifesting with a similar phenotype. Subsequently much research has focused upon an understanding of the expression and function of APP in normal and disease states. From these studies a common pattern seems to be emerging in neurodegenerative diseases: that the selective neuronal vulnerability observed in each disease is not a direct result of selective expression of the disease gene, rather other, as yet unknown, downstream factors must bring about this selective vulnerability.

The study of the genetics of late-onset Alzheimer's disease has also led to the identification of the first putative risk factor for this common form of the disease. Alleles of the apolipoprotein E (ApoE) gene appear to influence risk for disease – the ApoE-ε4 allele increases the risk of developing disease, whilst ApoE-ε2 appears to decrease the risk, at least in the Caucasian population. Although ApoE is found deposited in senile plaques and neurofibrillary tangles, the biology of ApoE with regard to Alzheimer's disease had not previously been well studied. One of our present challenges is to understand how ApoE alleles influence risk for disease. Several hypotheses have been proposed involving both β-amyloid and tau, the major constituent of neurofibrillary tangles. However, experimental evidence is not conclusive in favour of any hypothesis.

While we have progressed a long way in our understanding of Alzheimer's disease, we still have a considerable way to go before rational drug design can be used to develop treatments. The research outlined in the following pages presents some of the major ideas and fields contributing to our present understanding of Alzheimer's disease.

Common Abbreviations
AD=Alzheimer's disease
Aβ=Amyloid β-protein
ApoE=Apolipoprotein E
APP=Amyloid β-protein precursor
FAD=Familial Alzheimer's disease
NFT=Neurofibrillary tangle
PHF=Paired helical filament

CHAPTER 1

PLAQUES AND TANGLES IN ALZHEIMER'S DISEASE: A HISTORICAL INTRODUCTION

Frank Ashall

Saint Louis University School of Medicine, 1402 South Grand Boulevard, St Louis, Missouri 63104, USA

Table of Contents

1.1 The Alzheimer century

Progress in understanding the biochemical, cellular and molecular genetic basis of Alzheimer's disease has been so rapid in recent years that it is almost impossible for texts such as this one to keep abreast of developments. It has been almost 90 years since Alois Alzheimer first reported the disease that bears his name, and it is not unrealistic to believe that rational prevention or treatment will be available before the year 2007, when the 'Alzheimer century' comes to an end.

Many significant contributions from diverse fields have helped to produce the ever-clearer picture we have of Alzheimer's disease. Even Alzheimer's own work could not have been done without new techniques in tissue staining and fixation, application of cell theory, improvements in psychiatric definitions and diagnoses, and theories concerning the organic basis of psychiatric diseases. As the 20th century progressed, epidemiological and family studies contributed to an understanding of the impact of Alzheimer's disease on human society and led to the realization that there are genetic causes of the disease. More sophisticated tissue studies of brains allowed clearer definition of the lesions that characterize Alzheimer's disease and provided information on how these lesions are related to the progress of the disease. Biochemistry provided the tools for isolation and molecular characterization of the major components of plaques and tangles, as well as identification of other molecular defects that occur during the

Pathobiology of Alzheimer's Disease
ISBN 0-12-286965-5

course of the disease. Use of cultured cells has allowed dynamic studies of some of the molecular and cellular processes that are associated with the disease. Thanks to breathtaking advances in recombinant DNA technology, gene mutations predisposing to Alzheimer's disease have been identified and valuable animal models have now been developed. Even as this book was in preparation, a major step forward was made in the production of a mouse model of the disease. Using microinjection technology, transgenic mice were generated that overexpress a human amyloid precursor protein (APP) gene carrying a mutation ($APP_{717V \rightarrow F}$) found in familial cases of Alzheimer's disease (Games *et al.*, 1995). Careful neuropathological examination of the brains from these transgenic mice reveals the presence of large numbers of extracellular thioflavin S-positive Aβ deposits, neuritic plaques, synaptic loss, astrocytosis and microgliosis. Preliminary attempts to identify neurofibrillary tangles with tau antibodies were negative. However, the animals are still only about a year old and the life span of a laboratory mouse can be as high as 24 months. As in humans, the plaque density increases with age in these transgenic mice. These mice support a primary role for APP/Aβ in the genesis of Alzheimer's disease and may provide a preclinical model for testing potentially therapeutic drugs in Alzheimer's disease.

Many questions remain to be answered but research progress is clearly such that many puzzles, which were seemingly unanswerable a decade ago, are not far from being explained.

1.2 Early history of dementia and Alzheimer's disease

Refined definitions of psychiatric disorders, and particularly of dementia, were crucial for our understanding of different forms of dementias. Reviews of the history of dementia and Alzheimer's disease have been published (Torack, 1978, 1983; Amaducci *et al.*, 1986; Berrios, 1987, 1990) and much of the following comes from these articles.

Prior to the late 19th century, the definition of dementia was vague and the term usually referred to general mental deterioration associated with chronic brain disease. By the end of that century, dementia was largely synonymous with memory loss. Indeed, as early as 1838, Esquirol had defined several types of dementia, including a senile form, 'demence senile', which he said was associated with 'the progress of age' and involved 'loss of memory, especially recent memory'.

From the mid-19th century onwards, not only were efforts made to find pathological abnormalities associated with dementia (and other brain disorders), but also improved diagnostic techniques were available for both identifying and quantifying the severity of dementia.

Until histological procedures were developed that allowed brain sections to be preserved and appropriately stained, attempts at discovering pathological changes in brain-associated diseases were based chiefly on gross anatomical changes, such as size, colour, mass, and relative amounts of fluid and solid material. In 1822, for example,

Bayle reported changes in arachnoid in patients with motor and psychiatric symptoms, and numerous studies suggested that senile dementia was associated with cerebral cortical atrophy, enlarged ventricles and changes in the physical consistency of brain tissue.

With the advent of improved tissue storage, preservation and staining methods, combined with microscopy, scientists had at their disposal techniques for examining the brain with unprecedented detail, and many of them recognized that here was a chance to find pathological changes in mental disorders that may be too small to be detected using gross anatomical examination. During the last few decades of the 19th century, microscopic and histological studies had allowed the visualization of necrotic brain cells, senile plaques and neurofibrillary tangles, even before Alzheimer described them in his first patient.

In 1882, Blocq and Marinesco described plaques (often called 'miliary foci' at the time) in the brains of an epileptic who had no dementia. This is considered by many science historians to be the first ever description of neuritic plaques. Subsequently, similar plaques were found in the brains of patients with other diseases. For example, in 1898, Redlich observed plaques in brains of two senile patients with memory loss and confusion; both brains showed cerebral atrophy. In 1907, Fischer noted 'miliary necrosis' in the cerebral cortices of 12 out of 16 patients with senile dementia, but failed to find significant numbers of plaques in any of 45 patients diagnosed with progressive paralysis but having no dementia.

Neurofibrillary tangles had also been detected in brain specimens before Alzheimer described them. Fragnito, in 1904, observed neurofibril degeneration and damage in cerebral cortical cells of brains from patients with senile dementia, and three years later Fuller also described neurofibrillary accumulations in senile dementia.

Alzheimer himself contributed significantly to the advances in preservation and staining of brain tissue and, by 1910, many methods were available for staining brain tissue; more than ten of them could detect neurofibrils.

By the early 1900s, it was clear that at least some mental diseases were caused by pathological changes in the brain. Indeed, syphilis was one of the few mental diseases for which a specific physical cause was known. Dementia due to cerebral infarcts caused by arteriosclerosis was also known to occur in elderly people, although other patients with dementia were found at autopsy to have general cerebral atrophy without arteriosclerotic lesions. It should be borne in mind that dementia was not as serious a public health problem in those days as it is nowadays: the life expectancy at the turn of the century was well below 65 years, so senile dementia was not so common. However, it was known that dementia could occur at any age and 'presenile dementia' was a recognized disease entity.

Freud's psychoanalysis movement was also powerful at the beginning of this century, and his ideas that mental illness was due mainly to disturbances of the subconscious mind were at odds with those who believed that such disorders were caused by specific pathological, cellular and chemical lesions. The great psychiatrist, Kraepelin, was one of the main proponents of the organic theory of mental illness and he was highly influential in the eventual definition of Alzheimer's disease.

1.3 Alois Alzheimer and his disease

Kraepelin was the head of the Munich Psychiatric Clinic where Alois Alzheimer made his 'discovery'. Kraepelin had already made important contributions to our understanding of the different forms of dementia, particularly those associated with the psychoses, cerebral arteriosclerosis and senile dementia. He became convinced that diseases of the mind were caused by abnormalities of tissues, cells and biochemicals, and the institute he headed represented this hypothesis. Alzheimer held the same convictions and sought, using up-to-date histological techniques, to contribute to the neuroanatomical basis of psychiatric disease.

More than 10 years prior to Alzheimer's description of the 'first' patient with Alzheimer's disease, he and Kraepelin detected arteriosclerotic lesions and cerebral atrophy in the brains of 50–70-year-olds with dementia ('arteriosclerotic dementia'). One of the staining methods that was valuable for Alzheimer's work, and the one he used to detect lesions in his 'first patient', was a modified silver staining procedure developed by Bielschowsky in 1903, which allowed neurofibrils within neurons to be stained.

On 3 November 1906, Alzheimer attended a meeting of South Western German Alienists (psychiatrists) in Tubingen, where he presented the now famous case history that led to the baptism of Alzheimer's disease. A year later he published the findings in a paper entitled 'Uber eine eigenartige Erkrankung der Hirnrinde' ['On a peculiar disease of the cerebral cortex'] (Alzheimer, 1907). The patient was a 51-year-old woman who was jealous of her husband, showed signs of paranoia and loss of memory. Her short-term memory loss was particularly defective and her symptoms were progressive. Four-and-a-half years after she first visited Alzheimer, the patient died as a result of chronic deterioration of her symptoms. Alzheimer could find no obvious macroscopic abnormalities of her brain, although when he examined brain sections microscopically after Bielschowsky staining, he noticed cellular atrophy and neurofibrillary tangles. 'Miliary foci' (plaques) were abundant in the cerebral cortex, particularly in the upper layers. Alzheimer also noticed arteriosclerotic lesions.

By the time Kraepelin published the eighth edition of his famous handbook of psychiatry (*Psychiatrie: Ein Lehrbuch für Studierende und Arzte*) in 1910, several more brains from cases of presenile dementia had been studied, and plaques and tangles were found in most of these. This led Kraepelin to coin the term 'Alzheimer's disease'. Although he gave the name to the presenile forms of dementia characterized by the pathology described by Alzheimer, it appears that he did consider it to be essentially the same as senile dementia, and that it differed from the senile form mainly insofar as its age of onset was concerned. Arguments as to whether or not presenile dementia is simply an early-onset form of senile dementia lasted from that time until recently. Nowadays, it is generally accepted that the two forms are pathologically identical, and in any case, both types are referred to as Alzheimer's disease.

Many arguments have been put forward to explain why Kraepelin so eagerly named Alzheimer's disease on the basis of studies of only a handful of brains. For example, it has been suggested that he was eager to justify funding of research at his clinic or that he wanted a member of his own institute to be credited with the discovery. It has also been proposed that he wished to strengthen the case for the organic cause of psychiatric illness in order to counter Freud's theories. Torack (1978) even suggests that 'in reality, neither Alzheimer nor Kraepelin established Alzheimer's disease; it was Freud who did so'.

Alzheimer published a study of another case of the disease in 1911 (Alzheimer, 1911; Tonkogony and Moak, 1988), although this time he did not find neurofibrillary tangles. By 1912, nearly 50 articles had been published on plaques and tangles in dementia, and it was recognized that tangles and senile plaques were not significantly elevated in syphilis, affective disorders and arteriosclerotic dementia.

1.4 Further progress with plaques and tangles

The biochemical nature of senile plaques goes back further than the discovery of plaques in human brains. When they were first discovered, many scientists saw similarities between the plaques and other deposits known generally as 'amyloid', from which the name of the peptide deposited in senile plaques derives its name. Such deposits were first detected in 1842 following iodine staining of tissues by Rokitansky and, in the 1850s, Virchow coined the name 'amyloid' to describe these apparently starch-like deposits. The diseases in which these deposits formed were generally named the 'amyloidoses'. In 1859, Friedrich and Kekule obtained evidence that 'amyloid' deposits, far from being starchy, were mainly proteinaceous.

By the early 1970s, plaques had been isolated from brains of Alzheimer's disease patients and subjected to biochemical analysis. Nikaido *et al.*, for example, showed that their major constituent was proteinaceous: amino-acid analyses demonstrated that isolated plaques were rich in glutamate, glycine, leucine and alanine, while plaque cores were rich in glycine, glutamate, aspartate, serine and leucine (Nikaido *et al.*, 1970, 1971). These and other studies were the beginnings of serious biochemical investigations of Alzheimer plaques and they coincided with important improvements in biochemical methods for characterizing cellular molecules. Another great advance was the isolation and sequencing of the major protein constituent, the β-amyloid peptide (Aβ), from cerebrovascular amyloid deposits of Alzheimer brains (Glenner and Wong, 1984). Masters *et al.* (1985) subsequently confirmed this finding when they isolated a similar peptide from neuritic plaques. Sequencing of the Aβ soon led to cloning of the gene encoding it, which turned out to be a larger precursor protein, the β-amyloid precursor protein (APP).

Major advances in the study of neurofibrillary tangles also coincided with the appearance of advanced cellular and biochemical techniques. Electron microscopic studies (Kidd, 1963; Terry, 1963) provided ultrastructural information about them, and Kidd was the first to refer to the 'paired helical filaments' that comprised the tangles found in Alzheimer's disease. Once tangles were isolated biochemically (Crowther and Wischik, 1985; Wischik *et al.*, 1985), their molecular characteristics could be more readily characterized, and Brion *et al.* (1985), followed by other groups, subsequently showed that the microtubule-association protein, tau, was a major constituent of tangles.

Genetic components of Alzheimer's disease have also contributed a great deal to our understanding of Alzheimer's disease in general and to the specific involvement of β-amyloid. Several early studies (Meggendorfer, 1925; Schottky, 1932; Lowenberg and Waggoner, 1934) described familial clustering particularly in early-onset cases of Alzheimer's disease, suggesting that at least some cases of Alzheimer's disease may be heritable. In a classic study published in 1952, Sjorgren *et al.* (1952) studied 36 cases of Alzheimer's disease and more than 1200 of their relatives, and provided convincing evidence that predisposition to the disease may be heritable. The identification of families in which an early-onset form of the disease behaves as a classic autosomal dominant trait led eventually to the discovery of the first mutation known to be associated with Alzheimer's disease (Goate *et al.*, 1991).

The current status of plaques and tangles in Alzheimer's disease, the genetics of Alzheimer's disease and the recently demonstrated role of apolipoprotein E in the disease are described in the remaining chapters of this volume.

Where is Alzheimer's disease research leading us? Certainly, it is producing a clearer understanding of the cellular and biochemical changes that underlie the pathology, and this will inevitably lead to rational approaches to treatment or prevention of the disease. Such clinical applications will be timely, for Alzheimer's disease is most frequent in the age group over 80, which also happens to be the most rapidly growing age group in the USA and other developed countries. It will not be too long before the disease also becomes a major problem in underdeveloped countries, where the elderly portion of the population is also increasing.

References

Alzheimer, A. (1907) *Allg. Z. Psychiatrie Psychisch-Gerichtliche Med.* **64**, 146–148. [English translation (1987) in *The Early Story of Alzheimer's Disease* (eds Buck, K., Amaducci, L. & Pepeu, G.), pp. 1–3. New York, Rana Press.]

Alzheimer, A. (1911) *Gesanipte Neurol. Psychiatr.* **4**, 356–385.

Amaducci, L.A., Rocca, W.A. & Schoenberg, B.S. (1986) *Neurology* **36**, 1497–1499.

Berrios, G.E. (1987) *Psychol. Med.* **17**, 829–837.

Berrios, G.E. (1990) *Int. J. Geriatr. Psychiatr.* **5**, 355–365.

Brion, J.P., Passareiro, H., Nunez, J. & Flament-Durand, J. (1985) *Arch. Biol.* **95**, 229–235.

Crowther, R.A. & Wischik, C.M. (1985) *EMBO J.* **4**, 3661–3665.

Games, D., Adams, D., Alessandrini, R., Barbour, R., Berthelette, P., Blackwell, C., Carr, T., Clemens, J., Donaldson, T., Gillespie, F., Guido, T., Hagopian, S., Johnson-Wood, K., Khan, K., Lee, M., Leibowitz, P., Lieberburg, I., Little, S., Masliah, E., McConlogue, I., Montoya-Zavala, M., Mucke, L., Paganini, L., Penniman, E., Power, M., Schenk, D., Scubert, P., Snyder, B., Soriano, F., Tan, H., Vitale, J., Wadsworth, S., Wolozin, B. & Zhao, J. (1995) *Nature* **373**, 523–527.

Glenner, G.G. & Wong, C.W. (1984) *Biochem. Biophys. Res. Commun.* **120**, 885–890.

Goate, A., Chartier Harlin, M.C., Mullan, M., Brown, J., Crawford, F., Fidani, I., Giuffra, L.,

Haynes, A., Irving, N., James, L., Mant, R., Newton, P., Rooke, K., Roques, P., Talbot, C., Pericak-Vance, M., Roses, A., Williamson, R., Rossor, M., Owen, M. & Hardy, J. (1991) *Nature* **349**, 704–706.

Kidd, M. (1963) *Nature* **197**, 192–193.

Lowenberg, K. & Waggoner, R. (1934) *Arch. Neurol. Psychiatr.* **31**, 737–754.

Masters, C., Simms, G., Weinman, N. *et al.* (1985) *Proc. Natl Acad. Sci. USA* **82**, 4245–4249.

Meggendorfer, F. (1925) *Zentralbl. Neurol. Psychiatr.* **40**, 359–364.

Nikaido, T., Austin, J., Trueb, L., Hutchinson, J., Rinehart, R., Stuckenbrok, H. & Miles, B. (1970) *Trans. Am. Neurol. Assoc.* **95**, 47–50.

Nikaido, T., Austin, J., Rinehart, R., Trueb, L., Hitchinson, J., Stukenbrok, H. & Miles, B. (1971) *Arch. Neurol.* **25**, 198–211.

Schottky, J. (1932) *Z. ges. Neurol. Psychiatr.* **140**, 333–397.

Sjorgren, T., Sjorgren, H. & Lindgren, H. (1952) *Acta Psychiatr. Scand.* **82** (Suppl. 9).

Terry, R.D. (1963) *J. Neuropathol. Exp. Neurol.* **22**, 269–292.

Tonkonogy, J. & Moak, G.S. (1988) *J. Geriatr. Psychiatr. Neurol.* **1**, 199–206.

Torack, R.M. (1978) *The Pathologic Physiology of Dementia.* Heidelberg, Springer-Verlag.

Torack, R.M. (1983) In *Alzheimer's Disease: The Standard Reference* (ed. Reisberg, B.), pp. 23–28. New York, The Free Press.

Wischik, C.M., Crowther, R.A., Stewart, M. & Roth, M. (1985) *J. Cell. Biol.* **100**, 1905–1912.

CHAPTER 2

STRUCTURE, BIOCHEMISTRY AND MOLECULAR PATHOGENESIS OF PAIRED HELICAL FILAMENTS IN ALZHEIMER'S DISEASE

Claude M. Wischik, Robert Lai, Charles R. Harrington, Elizabeta B. Mukaetova-Ladinska, John H. Xuereb, Raul Mena, Patricia C. Edwards and Martin Roth

Cambridge Brain Bank Laboratory, University of Cambridge Clinical School, Department of Psychiatry, MRC Centre, Hills Road, Cambridge CB2 2QH, UK

Table of Contents

2.1 The anatomy of neurofibrillary degeneration

The German psychiatrist, Alois Alzheimer, based his initial description of the disease that bears his name on a 51-year-old female patient whose clinical presentation began with a paranoid delusional disorder of late onset, progressing rapidly to severe cognitive deterioration and death 4½ years later. Post-mortem examination of her brain showed generalized atrophy with no macroscopic lesions. Using a silver staining method developed by Bielschowsky, Alzheimer found extracellular 'multiple miliary

Pathobiology of Alzheimer's Disease
ISBN 0-12-286965-5

lesions' (plaques) and, in more than a quarter of the cortical neurons, flame-shaped fibrillary structures (neurofibrillary tangles). He also reported severe neuronal loss, particularly in the outer layers of the neocortex (Alzheimer, 1907).

Neurofibrillary tangles are generally found within the cell bodies of large pyramidal neurons, from where they extend into the proximal apical dendrite, but not into the axonal hillock. As the tangle comes to occupy more of the cytoplasm, the nucleus is displaced to an eccentric position and the dendritic arborization is pruned. The morphology of the tangle conforms to the outlines of the pyramidal cell while it remains intracellular (Plate I), but, after cell death, the tangle swells and disperses in the extracellular space as it is invaded and progressively degraded by astrocytes. The loss of pyramical cells is highly correlated with the accumulation of neurofibrillary tangles ($r = 0.76$, $p < 0.001$; Bondareff *et al.*, 1993).

The appearance of neurofibrillary pathology, in the form of neurofibrillary tangles and dystrophic neurites, follows a stereotyped pattern recently described by Braak and Braak (1991) on the basis of cross-sectional analysis of unselected cases coming to post-mortem. Neurofibrillary tangles are first found in the pre-α layer (layer II) of the transentorhinal cortex and the entorhinal cortex (Stages I and II). By the time neurofibrillary pathology spreads to layer I of Ammon's horn in the hippocampus, extracellular tangles have appeared in the entorhinal cortical regions (Stage III). Next, tangles appear in the deeper pre-α layer (layer IV) of the entorhinal cortex and in CA1 of the hippocampus (Stage IV). Pathology then spreads into sectors CA2, 3 and 4 of the hippocampus, the subiculum, and into isocortical areas, particularly the temporal and parietal cortices (Stage V). The final Stage VI is characterized by more extensive spread in neocortical areas, with relative sparing of primary motor and sensory areas, and the appearance of neurofibrillary tangles in the granule cells of the fascia dentata.

Van Hoesen and Damasio (1990) have discussed the neurophysiological implications of this pattern in terms of the progressive disconnection of the hippocampal formation from the rest of the neocortex. Multimodal cortico-cortical projection pathways converge on the layer II pyramidal cell islands of the entorhinal cortex. These are the cells of origin of the perforant pathway, projecting to dendrites of the granule cells of the fascia dentata and directly on to pyramidal cell dendrites of CA1 of the hippocampus. The granule cell axons project to dendrites of hippocampal and subicular pyramidal cells, which provide the major hippocampal output back to layer IV of the entorhinal cortex, which in turn has wide projections in the cerebrum (Figure 1). The striking feature of the pathology of Alzheimer's disease is the progressive and systematic disconnection of both the inputs and the outputs of this circuit.

The cognitive neurophysiology of this circuit has been investigated in terms of habituation, or its converse, the detection of novelty (Gray, 1982) and/or episodic memory (Treves and Roles, 1994). Habituation to multimodal stimuli is accompanied by inhibition of firing in the subicular output of this circuit, whereas stimuli coded for attentive processing gain access via CA1 and subicular pyramidal cell projections to the 'feedback' isocortical projection system. The relationship between these phenomena of attentional coding and memory is not well understood, but it is clear that bilateral hippocampal disconnection is associated with profound deficits in episodic and autobiographical update memory.

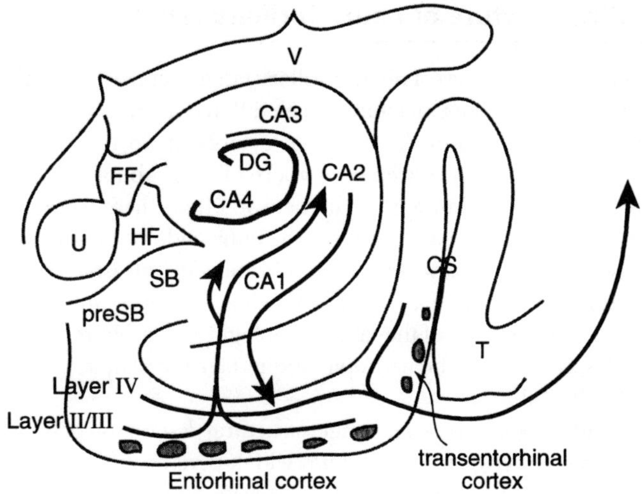

Figure 1 Schematic representation of the input and output of the hippocampus. Multimodal cortico-cortical projections terminate on the layer II islands of the entorhinal cortex. Pyramidal cells of this layer are the cells of origin of the perforant pathway, which projects to dendrites of granule cells of the dentate gyrus and also to dendrites of pyramidal cells of stratum pyramidale of the hippocampus (CA3) and subiculum. Axons from these cells project to dendrites of CA1 pyramidal cells and subiculum, which send projections back to layer IV of the entorhinal cortex. This represents the major hippocampal output, projecting to a variety of telencephalic and diencephalic structures. The striking feature of neurofibrillary pathology of Alzheimer's disease is the graded destruction of pyramidal cells at each of the points of this circuit, excepting the granule cells of the dentate gyrus. The latter are prominently affected in another neurodegenerative disorder involving changes in tau protein processing, Pick's disease. CS, collateral sulcus; DG, dentate gyrus; FF, fimbria-fornix; HF, hippocampal fissure; HP, hippocampus; PROSB, prosubiculum; SB, subiculum; T, temporal isocortex; U, uncal hippocampus; V, lateral ventricele.

Although there is as yet no precise quantitative metric for the Braak staging system, qualitative staging is broadly correlated with degree of dementia determined prospectively by the Blessed scale for dementia (Bancher *et al.*, 1993; Braak *et al.*, 1993). Furthermore, the fundamental axiom of the system, that the greater the density of neurofibrillary pathology in any given region, the greater the number of regions in which neurofibrillary pathology can be detected, has been borne out in a recent prospective study of unselected cases coming to post-mortem at early stages of pathology ($r = 0.85$; Gertz *et al.*, submitted).

Thus, the anatomically ordered progression of neurofibrillary degeneration is closely linked to neuropsychological deficit in ways that appear to have meaning in terms of our understanding of the neurophysiology of higher cortical function. There can be little doubt that the destruction of specific cortico-cortical projection systems, which is linked quantitatively with neurofibrillary degeneration of the Alzheimer type, represents neuronal damage, which must at some point underlie irreversible dementia.

2.2 The ultrastructure of neurofibrillary tangles

Electron microscopical studies of neurofibrillary tangles showed that their main structural constituent was an abnormal fibre, which differed in appearance from any of the normal fibrous constituents of the neuronal cytoplasm. In his initial report, Kidd (1963) observed that there appeared to be more of these abnormal fibres inside dystrophic neurites scattered throughout the neuropil than there were in tangles. This morphological impression has since been confirmed in quantitative studies of advanced cases comparing biochemical and histological measures of neurofibrillary pathology (Mukaetova-Ladinska *et al.*, 1993). Thus, an important unifying feature of this pathology is the presence of the same pathological 'neurofibres' in tangles, in dystrophic neurites throughout the neuropil, and in neuritic plaques (Braak *et al.*, 1986; Braak and Braak, 1988).

There was initial debate as to the correct structural description of these fibres, which alternate in diameter between 8 and 20 nm with a periodicity of 80 nm. Terry's view that these fibres were 'twisted tubules' (Terry, 1963) for some time gave rise to the erroneous belief that the pathological 'neurofibres' were abnormal microtubules. This impression arose from certain electron micrograph images in which the filament could be observed to have a clear centre. So strong was this expectation that biochemical isolates were produced, which duly reported the presence of a 50 kDa band thought to be tubulin as a major constituent of paired helical filaments (PHFs) (Iqbal *et al.*, 1974, 1975).

Kidd used the term 'paired helical filaments' because he observed midline splitting of the fibres in some images. The structural model of two strands twisted helically around each other cannot be reconciled with the twisted tubule model. One way of explaining the unstained centre could be in terms of peripheral distribution of stain in axial images. Since a normal thin section for electron microscopy has a thickness generally in excess of 70 nm, an axial image of a PHF would include an almost complete helical rotation, and so would appear circular. A preferential distribution of stain at the periphery would produce an image with a dense periphery and an 'empty' centre, hence giving the impression of a tubule.

The PHF model produced a different series of biochemical and immunohistochemical interpretations. Careful analysis of well-prepared biopsy specimens suggested to Metuzals that continuity could be demonstrated between the two filaments of the PHF and the neurofilamentous network, which he interpreted as implying that the PHF was an altered neurofilament (Metuzals *et al.*, 1984, 1987). This impression received immunohistochemical support from the work of Anderton, in which he showed that monoclonal antibodies raised against bovine neurofilament preparations, which labelled neurofilament proteins, also labelled neurofibrillary tangles, concluding that neurofilament proteins make a structural contribution to PHFs (Anderton *et al.*, 1982). Selkoe, attempting to purify PHFs on the basis of sodium dodecyl sulphate (SDS) and urea insolubility, proposed that the PHF was composed of altered neurofilament proteins with pathological ε-amino linkages (Selkoe *et al.*, 1982). This view also had good theoretical support from basic cell biology, since the collapsed neurofilament networks seen in cells after depolymerization of microtubules have a strong resemblance to neurofibrillary tangles.

The work on PHF structure undertaken in Cambridge was based on analysis of isolated PHFs visualized by negative stain (Crowther and Wischik, 1985; Wischik *et al.*, 1985; Figure 2). This produced superior resolution of ultrastructural detail and permitted the use of image reconstruction techniques. Images obtained from such preparations confirmed that filaments split in the midline, implying that the paired helical model was correct. Characteristic clean transverse truncations without fraying implied the absence of the axial protofilaments seen in neurofilaments. Metal shadowing suggested that the marked wide-to-narrow modulation was the product of a twisted ribbon structure with a high aspect ratio, i.e. a structure with greater breadth than depth. Thus, the PHF could be characterized structurally as a double helical assembly of transverse subunits twisted into a left-handed helical ribbon. Further substructural information could be deduced from regularly repeating 3/4 longitudinal striations. This could be explained by the presence of two C-shaped subunits each containing three domains. This domain substructure could also be demonstrated directly in computed cross-sectional density maps of isolated PHFs (Figure 3). A similar organization was demonstrated by Crowther (1991) in the 15 nm straight filaments, which copurify with PHFs and which tend to be located in axons. The 'straight filament' appearance is caused by a lower aspect ratio, i.e. the transverse breadth is comparable to the depth of the filament. Although the same paired C-shaped subunit organization is present, the subunits are arranged back to back rather than base to base, producing a more nearly rounded cross-section. Thus the straight filament is a structural variant of the PHF.

This structural description of the PHF made no predictions regarding its molecular composition. The main implication of this regular structure is that there are specific sites of molecular association, which are reproduced throughout the filament. Many proteins have been found to have the capacity to form such paracrystalline filamentous arrays. However, since PHFs are a constant feature of neurofibrillary pathology, regardless of histological appearance as tangles, dystrophic neurites or neuritic plaques, the important theoretical implication at the molecular level is that a characteristic modification of one or a small set of proteins is responsible for a process of *de novo* polymerization, which becomes extremely widespread throughout the brain in Alzheimer's disease.

2.3 Molecular analysis of the PHF core

Immunohistochemical studies demonstrated that neurofibrillary tangles contain neurofilament proteins (Anderton *et al.*, 1982; Perry *et al.*, 1985; Sternberger *et al.*, 1985; Gambetti *et al.*, 1986; Miller *et al.*, 1986), vimentin (Yen *et al.*, 1983), actin (Vogelsang *et al.*, 1988), ubiquitin (Mori *et al.*, 1987; Perry *et al.*, 1987; Shaw and Chau, 1988), MAP2 (Yen *et al.*, 1987), tau (Brion *et al.*, 1985; Delacourte and Defossez, 1986; Grundke-Iqbal *et al.*, 1986; Ihara *et al.*, 1986; Kosik *et al.*, 1986; Nukina and Ihara, 1986; Wood *et al.*, 1986) and amyloid β-protein (Aβ) (Kidd *et al.*, 1985; Masters *et al.*,

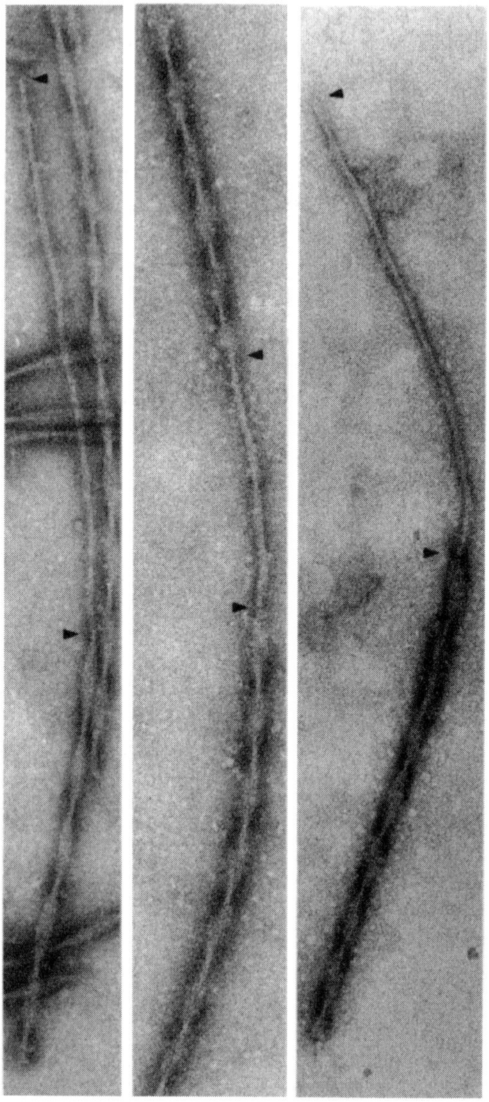

Figure 2 Electron micrographs of isolated paired helical filaments (PHFs), showing splitting of filaments into single strands, confirming the paired helical organization of the filament. Longitudinal striation patterns are also readily seen, which are characteristically three-fold or four-fold. This is caused by the superimposition of domains of the underlying C-shaped subunit organization of the PHF.

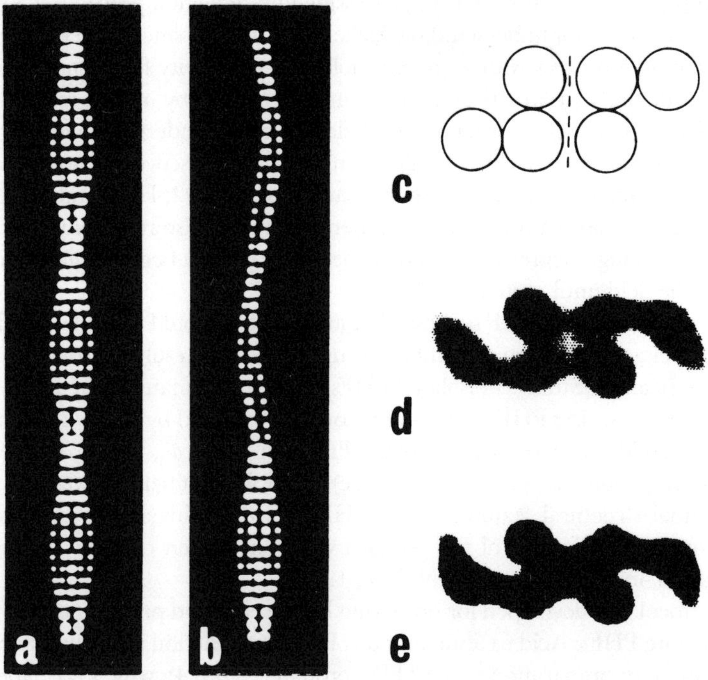

Figure 3 Schematic representations of the subunit organization of the PHF (a, b, c) and computed cross-sectional density maps (d, e). In the course of the helical rotation of the basic C-shaped subunits, superimposition produces an impression of either three or four elements. 'Straight filaments' are structural variants of this basic C-shaped organization, except that the subunits are base to base rather than back to back. The actual packaging of molecules within this overall structure is unknown. From mass data, each of the domains could correspond to a single 12 kDa core tau unit, since there are probably six of these per nanometre in the core PHF.

1985). Isolated PHFs could be immunodecorated with antibodies against MAP2, neurofilament, ubiquitin and tau. Antibodies raised against PHF preparations included those with reactivity against tau, MAP2 and Aβ. These results suggested a theoretical view at odds with the structural view described above. Whereas the structure of the PHF suggested that neurofibrillary degeneration should be understood in terms of *de novo* polymerization of a characteristically modified subunit, the immunohistochemical data suggested rather that neurofibrillary degeneration should be viewed as a nonspecific disorder of cytoskeletal proteins, an end-stage process of cytoskeletal aggregation that might be initiated in many different pathological settings, including head trauma, viral diseases, the Guam–Parkinsonism–dementia complex and Down's syndrome (Shankar *et al.*, 1989). The link between neurofibrillary degeneration and cell death did not appear particularly surprising, if this simply represented an agonal precipitation of cytoplasmic/cytoskeletal proteins.

In this context, the discovery of a genetic association between abnormalities in amyloid β-protein precursor (APP) processing and dementia proved to be extremely attractive as a basis for understanding Alzheimer's disease, since it offered the prospect of elucidating a process with apparent molecular specificity for Alzheimer's disease. More particularly, a practical judgement was made by many researchers that abnormalities in the processing of APP held the key to understanding early stages in molecular pathogenesis, which would open the way to discovering novel therapeutic approaches (Goate *et al.*, 1991; Crawford and Goate, 1992; Haas and Selkoe, 1993). The possibility that neurofibrillary degeneration might also represent a pathological process with a high degree of molecular specificity has not begun to gain wide acceptance in the field until quite recently.

The structure of the PHF suggests that its assembly should be understood in terms of specific changes in a small number of molecules that result in a relentless process of polymerization. Studies of isolated PHFs showed that many of the immunological ambiguities regarding PHF composition could be resolved by removing a fuzzy coat, which surrounds the core structure of the PHF (Wischik *et al.*, 1988a). It was observed that after digestion with proteases (Pronase) a core was left behind, which retained all the essential structural features of the PHF. However, this core was found to lack immunoreactivity for any of the proteins which had been claimed as tangle/PHF constituents, including tau (Wischik *et al.*, 1988a).

A protocol was developed for preparing highly enriched preparations of protease-resistant core PHFs. Acid treatment of this PHF core fraction was found to release an essentially pure preparation of a 12 kDa protein (Figure 4), which was labelled by a monoclonal antibody (mAb) raised against the PHF core fraction, which also labelled PHFs (mAb 423). Sequence analysis of this protein showed that it was derived from the tandem repeat region of the microtubule-associated protein tau. This established that tau protein represents at least one integral constituent of the core structure of the PHF (Wischik *et al.*, 1988b).

Tau protein is known from sequencing of cDNA libraries prepared from human and other mammalian sources to exist in six isoforms, ranging from 352 to 441 amino acids derived by alternative RNA splicing from a single gene on chromosome 17 (Figure 5). Each isoform contains three or four tandem repeats of 31 or 32 amino acids located in the C-terminal half of the molecule. Each repeat contains a characteristic Pro-Gly-Gly-Gly motif. The repeats are rich in basic amino acids and are involved in the binding of tau to an acidic domain of tubulin. The extra repeat in the four-repeat isoform is inserted within the first repeat in a way that preserves the periodic pattern. Similar microtubule binding repeats are found in the carboxy-terminal domain of high molecular weight microtubule-associated proteins MAP2 and MAP-U (Goedert *et al.*, 1988, 1989a, b).

The 12 kDa tau peptides derived from the PHF core originate from both three- and four-repeat isoforms, but are restricted to three repeats in length (Jakes *et al.*, 1991). The N-termini of these fragments are aligned at homologous positions with respect to the maximum homology alignment of the repeat region, but are 14–16 residues out of phase. Information regarding the C-terminal extent of the PHF-core tau fragments was determined by detailed analysis of the epitope of antibody mAb 6.423, which was

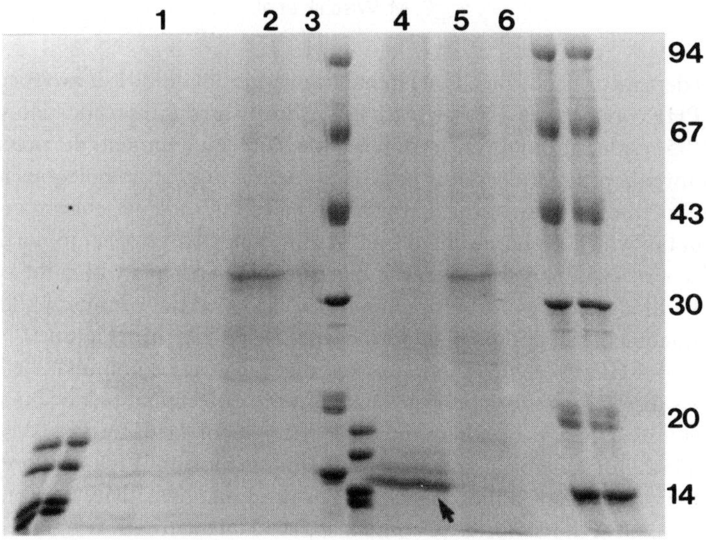

Figure 4 Coomassie-stained gel of PHF-core preparation from control elderly brain (lanes 1–3) and from a case of Alzheimer's disease (lanes 4–6). The first lane in each set (lanes 1 and 4) represents an acid supernatant fraction ('F5.5') obtained after sonicating the preparation at pH 5.5. There is a prominent 12 kDa band only in the AD preparation (arrow). The second pair of lanes in each set contains the material insoluble at pH 5.5 brought into solution using a protocol developed in this laboratory for solubilizing PHF-core preparations completely. Apart from the 12 kDa band, no other major difference between the control and AD preparations is apparent. Thus, the 12 kDa material is the major constituent of the PHF core preparation.

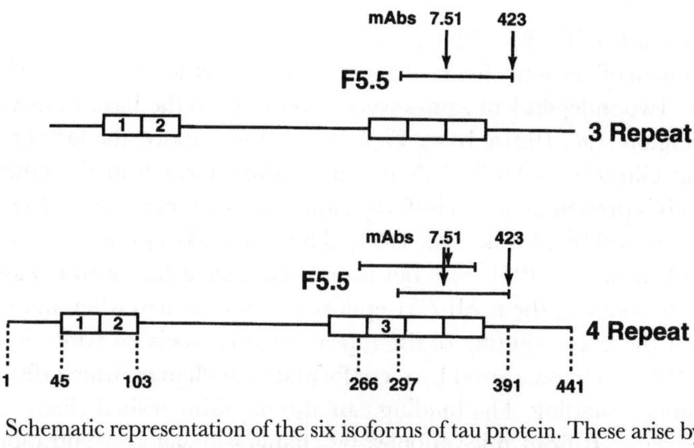

Figure 5 Schematic representation of the six isoforms of tau protein. These arise by alternative splicing to include inserts shown as '1' and '2' in the N-terminal half of the molecule, and insert'3' in the tandem repeat region, giving rise to six possible combinations. The segments denoted 'F5.5', which span precisely three repeats of both three- and four-repeat isoforms, correspond to the species found to comigrate as the 12 kDa band shown in Figure 4. The fragments are phase-shifted by half a repeat with respect to the maximum homology alignment structure of the tandem repeat region, and terminate at position Glu-391 at the C-terminus, or a postion homologous to this in the 4-repeat isoform. The epitopes of mAb's 7.51 and 423 are shown.

found to depend critically on a C-terminal truncation at Glu-391 in two species of the 12 kDa PHF core peptides (Novak *et al.*, 1993). Surprisingly, this truncation site is situated 93/95 residues from the N-terminus of the core tau fragment, the precise equivalent of three repeats. Indeed it is possible to demonstrate a homologous alignment structure for the core tau species, which preserves the three-repeat organization of intact tau but which is phase-shifted by 14/16 residues with respect to normal. Thus the limits of susceptibility to proteolytic digestion of tau protein within the core structure of the PHF are very precisely defined. It is likely that these limits are determined by the attachment site of the tau molecule within the PHF core (Figure 6).

Although PHFs isolated without protease digestion can be immunolabelled by tau antibodies directed against epitopes located in the N-terminal half of the molecule, this immunoreactivity is lost after proteolytic stripping of the fuzzy coat (Wischik *et al.*, 1988a). Further biochemical and immunochemical studies of PHF core preparations have failed to reveal any amino acid sequence or immunoreactivity derived from the N-terminal half of the molecule (Caputo *et al.*, 1992). Thus, the N-terminal half of the tau molecule is located entirely in the fuzzy outer coat of the PHF and makes no structural contribution to the regular paracrystalline structure of the core. Furthermore, the N-terminal half of the tau molecule can be removed proteolytically without affecting the structural integrity of the PHF.

2.4 Development and application of quantitative methods for measuring PHF-bound tau

The relative purity of the core PHF preparation has made it a useful starting point for the development of immunochemical assays for measurement of tau, which is integral to the PHF. Two independent assays were developed on the basis of mAbs 423 and 7.51 (Harrington *et al.*, 1991a, b; see Figures). As noted above, the former detects tau truncated at Glu-391. mAb 7.51 detects an epitope located in the tandem repeat region, which is present in all tau isoforms, and is also present in the 12 kDa tau fragment of the core PHF (Novak *et al.*, 1991). The mAb 7.51 epitope is occluded when tau is bound within the PHF core but is exposed after it has been released by acid treatment. By contrast, the mAb 7.51 epitope is not occluded when tau is free. This suggests that the phase-shifting of the repeat region associated with binding of tau within the PHF is characterized by a conformational change, which distinguishes it from tau-tubulin binding. The binding can also be distinguished chemically, in that tau can be released from microtubules by changes in salt concentration, whereas much harsher treatments are required to release tau from PHFs.

Acid-dependent mAb 7.51 immunoreactivity has been found to be highly correlated with mAb 423 immunoreactivity in PHF-core preparations ($r = 0.95$ Wischik *et al.*, 1995). Since mAb 7.51 is able to detect all tau species that contribute to the PHF core irrespective of C-terminal truncation and, since not all of these terminate at Glu-391 (for example, the fragment encompassing the first three repeats from the four-

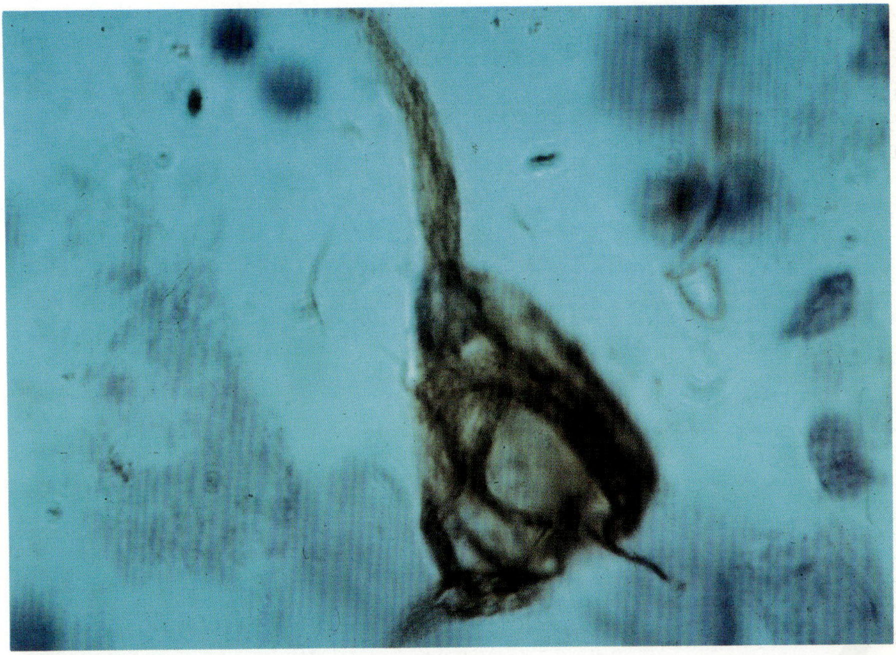

Plate I Intracellular, Type I neurofibrillary tangle visualized immunohistochemically using mAb 423, a monoclonal antibody raised against PHF-core preparations which has been shown to recognize tau C-terminally truncated at Glu-391. Adjacent is a normal pyramidal cell which lacks neurofibrillary pathology.

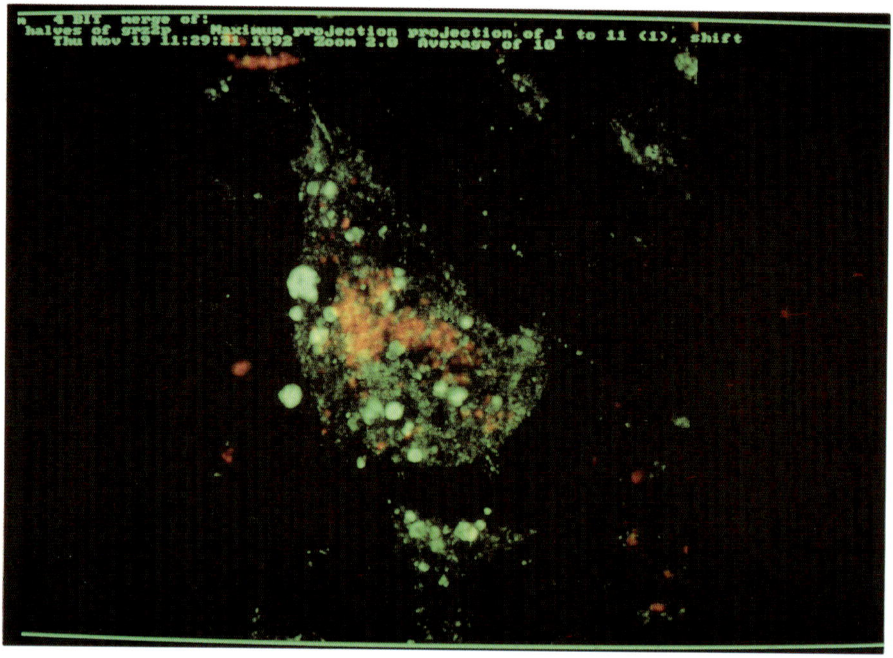

Plate II An early stage of neurofibrillary degeneration involves the accumulation in the somatodendritic compartment of pyramidal cells of amorphous deposits of tau truncated at Glu-391 and recognized by mAb 423, as shown in this figure. Other structures that are recognized by mAb 423 include the larger granules of the GVD complex. The red deposits are autofluorescent unlabelled lipofuscin granules.

A [R1,R3,R4]

D L K N V K S K I G S T E N

↓L K↓H Q P G G G K V Q I V Y K P V D L S K V T S K C G S L G N

I H H K P G G G Q V E V K S E K L D F KDR V Q S K I G S L D N

I T H V P G G G N K K I E T H K L t f ren a k a k t d h g a e↓

B [R1,R2,R3]

D L K N V K S K I G S T E N

↓L K↓H Q P G G G K V Q I I N K K L D L S N V Q S K C G S K D N

I K H V P G G G S V Q I V Y K P V D L S K V T S K C G S L G N

I H H K P G G G Q V E V K S E K L D F KDR V Q S K I G S L D N?

C [R2,R3,R4]

D L S N V Q S K C G S K D N

↓I K↓H V P G G G S V Q I V Y K P V D L S K V T S K C G S L G N

I H H K P G G G Q V E V K S E K L D F KDR V Q S K I G S L D N

I T H V P G G G N K K I E T H K L t f ren a k a k t d h g a e↓

Figure 6 N-Terminal sequence analysis of the 12 kDa F5.5 fragments released from the PHF core shows the existence of six fragments beginning at the positions indicated by the arrows. These positions are located half-way through tubulin-binding domains (shown underlined). The C-termini of some of the fragments have been defined by epitope mapping of mAb 423 used in Plate I. Using recombinant tau fragments, all and only those terminating at Glu-391 were found to be recognized by mAb 423.

repeat isoform), the correlation with mAb 423 immunoreactivity implies that the tau isoforms which contribute to the core structure of the PHF do so in a regular stoichiometric relationship. The precise stoichiometry is not yet known.

These assays have been used in an extensive series of studies of cases coming to post-mortem with and without a diagnosis of Alzheimer's disease. Alzheimer's disease

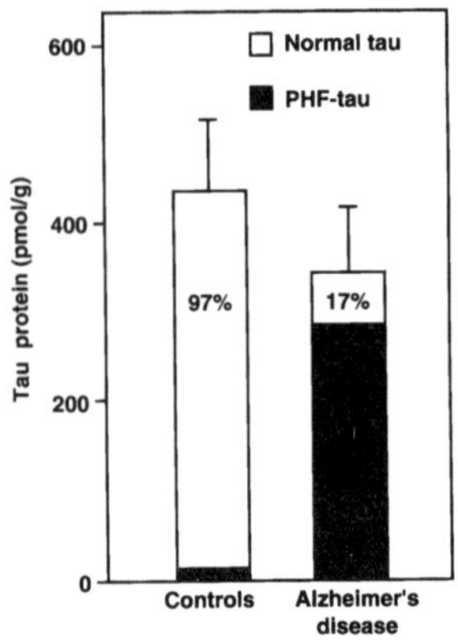

Figure 7 Total tau protein content in controls and Alzheimer's disease, as measured by mAb 7.51 immunoreactivity. This antibody recognizes an epitope located in the 12 kDa tau fragment of the core PHF, and also normal tau protein. PHF-bound tau can be distinguished by the absence of mAb 7.51 immunoreactivity until after the core tau fragment has been released by acid treatment (see Figure 4). Although total tau protein is somewhat reduced in AD, this is not statistically significant. The major change appears in the redistribution of the tau protein pool from the soluble phase (white) to the PHF-bound phase (black).

cases were shown to be characterized biochemically by a major redistribution of the tau protein pool from the soluble form to a form which is assembled into paired helical filaments. In frontal cortex, for example, there is a 95% loss of soluble tau and, in tempero-parietal cortices, there is a 45-fold increase in tau protein incorporated into PHFs relative to controls (Mukaetova-Ladinska *et al.*, 1992, 1993; Figure 7). This redistribution of a protein, which is essential for maintaining axonal microtubules in a polymerized state, may underlie the loss of synaptic connectivity in cortico-cortical association circuits (Terry *et al.*, 1991), which is though to be associated with clinical dementia (Figure 7).

The contribution made by various histological forms of neurofibrillary pathology to the measured protease-resistant PHF pool has also been examined. PHF content is most closely related to the amount of dystrophic neurite pathology and only weakly predicted by counts of neurofibrillary tangles in case of advanced Alzheimer's disease (Mukaetova-Ladinska *et al.*, 1993). This supports the initial ultrastructural observations made by Kidd, namely that the main bulk of PHFs in the Alzheimer's disease (AD) brain is located in dystrophic neurites. This impression was strongly confirmed by cases in which very high levels of protease-resistant PHFs were observed largely in the absence of neurofibrillary tangles (Mukaetova-Ladinska *et al.*, 1992). The claim that protease-resistant PHFs are derived preferentially from neurofibrillary tangles at advanced stages of pathology (Goedert *et al.*, 1992a) is not supported by this data.

2.5 Quantitative studies of tau protein in PHF preparations

Although the studies discussed above show that at least the minimal protease resistant tau unit contributes to the structure of the PHF, they do not demonstrate that tau is the sole constituent of the PHF. Claims that tau is the sole constituent of PHFs have come from studies based on the analysis of a subpopulation of PHFs, variously known as 'A68-PHFs', 'soluble PHFs', 'dispersed PHFs', which are prepared as a fraction insoluble in 1% sarkosyl from an initial low-speed supernatant from brain homogenate (Rubenstein *et al.*, 1986; Greenberg and Davies, 1990; Lee *et al.*, 1991). Although tau proteins are a minor constituent of the SDS-soluble species present in this material, these can be visualized selectively by immunoblot. Since the major tau species seen in such preparations are those corresponding to a characteristic triplet of bands of gel mobility 60, 64 and 68 kDa, the 'A68-proteins', the inference has been drawn that the PHFs that are present are composed solely and entirely of A68 proteins (Lee *et al.*, 1991; Goedert *et al.*, 1992a).

A68 proteins are full-length abnormally phosphorylated tau species. That at least some of the tau protein associated with some PHFs is abnormally phosphorylated can be deduced from the fact that some PHFs, notably those in the sarkosyl-insoluble preparation, can be labelled by phosphorylation-dependent tau antibodies, which also label the A68-proteins. Since this immunolabelling can be abolished by protease digestion, which leaves the PHF core intact, it follows that the phosphorylated epitopes are located exclusively in the fuzzy coat of the PHF. Indeed, mass-spectroscopic analysis of tau protein released from the protease-resistant PHF core preparations have failed to demonstrate that any sites within this segment of the molecule are phosphorylated (Poulter *et al.*, 1993). However, some of the tau species released by SDS treatment of A68-PHF preparations are phosphorylated in the tandem repeat region (Hasegawa *et al.*, 1992). Presumably such species do not originate from the PHF core.

The proportion of PHFs which are phosphorylated in the fuzzy coat has recently been determined in a study comparing measures of total PHF tau with measures of phosphorylated tau isolated from the same brain tissues. The technical feasibility of determining this parameter derives from the use of a protocol developed for bulk isolation of PHFs from AD brain tissues (Wischik *et al.*, 1988b). Total PHF-tau in this fraction can be determined from acid-dependent mAb 7.51 immunoreactivity (Harrington *et al.*, 1991). Phosphorylated PHF-tau can be determined using an ELISA based on mAb AT8 immunoreactivity (Harrington *et al.*, 1994a). These parameters were recently compared in 90 preparations derived from cases coming to postmortem at early stages of neurofibrillary degeneration as determined by the Braak staging system. The contribution of phosphorylated PHFs was found to be less than 5% of total PHFs. Thus, phosphorylated PHFs represent a minority species even at the earliest stages of AD pathology (Lai *et al.*, 1995).

Another important quantitative question is how much of the tau in A68-PHF preparations is of full length. Scanning transmission electron microscopic (STEM) measurements have shown that the molecular mass of the protease-resistant core of the PHF is 65.2±0.7 kDa/nm (Wischik *et al.*, 1988a). The only tau species identified

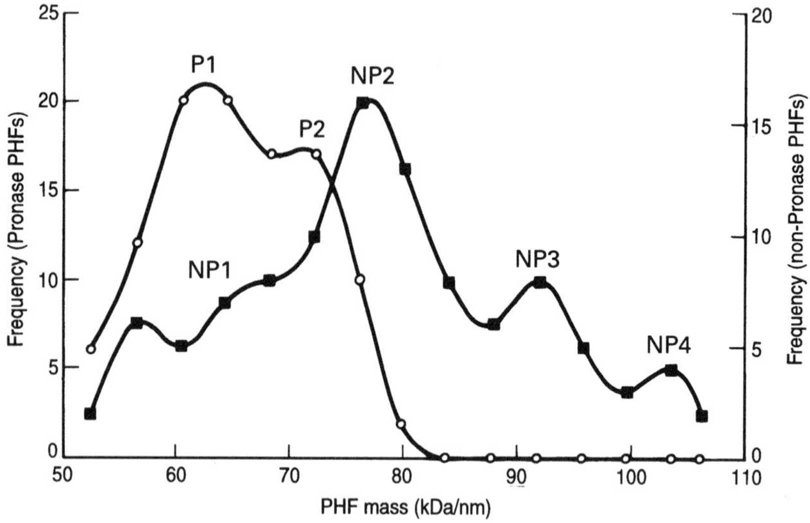

Figure 8 Frequency distribution of PHFs by mass in Pronase-treated [(○)] and non-Pronase-treated [(■)] PHF preparations as determined by scanning transmission electron microscopy. In each case 100 filaments were measured. The non-Pronase PHFs exist in several size classes (designated NP1–4, corresponding to non-Pronase Classes 1–4 in Table 1), of which the top 10% group has a mean mass of 103 kDa/nm. There are two peaks in the Pronase-PHF preparation (designated P1, P2, corresponding to Pronase Classes 1, 2 in Table 1), corresponding to 61.6 kDa/nm and 71.6 kDa/nm (Table 1). These would correspond to six or seven of the 93/95 amino-acid residue core PHF fragments shown in Figures 5 and 6. This sets an upper limit of about one full-length tau molecule per nanometre contributing to the fuzzy outer coat of the PHF, which is lost in the course of Pronase digestion. Thus, of the tau molecules, which are intrinsic to the PHF, about 6/7 are truncated, and about 1/7 is N-terminally intact. A schematic model corresponding to these estimates is shown in Figure 9.

after extensive biochemical (Wischik *et al.*, 1988a) and immunochemical (Caputo *et al.*, 1992) analyses of such core PHF preparations are restricted to a 93/95 amino acid fragment (predicted mass 10.2 kDa) derived from the tandem repeat region, with a variable contribution of fragments from the C-terminal tail of tau. If the core PHF mass is to be accounted for entirely by tau, then there must be six tandem-repeat fragments per nanometre to account for the observed mass. The N-terminal domain of tau, which is only found in the fuzzy outer coat of the PHF and which is removed in the course of Pronase digestion, has a predicted mass in excess of 24 kDa. If each of the core tau fragments were to extend to the N-terminus in the intact PHF, this would add a further 6×24 kDa/nm (144 kDa/nm), and the predicted mass of the intact PHF would therefore exceed 208 kDa/nm. STEM measurements of PHFs isolated without protease digestion have shown that PHFs in the top decile of the frequency distribution (NP4) have a mass of 102.7±0.88 kDa/nm (Wischik *et al.*, 1988a; Figure 8 and Table 1). A similar figure has been reported for STEM measurements of PHFs in the

Table 1. Frequency distribution according to mass of PHFs isolated from AD brain tissues with or without Pronase digestion, with mean ± SEM for mass in each class. In the non-Pronase PHFs, classes 1–4 are designated NP1–NP4 in Figure 8. Class 1 (NP1) includes two sub-populations which are grouped as one. In the Pronase PHFs, there are only two main classes, designated P1 and P2 in Figure 8. The maximum mass loss after Pronase digestion is 37.5 ± 1.11 kD/nm, the mean mass loss is 12.9 ± 1.20 kD/nm.

Non-Pronase PHF Mass		
Class 1	(12%)	62.1 ± 0.98 kD/nm
Class 2	(47%)	77.6 ± 0.58 kD/nm
Class 3	(18%)	91.5 ± 0.70 kD/nm
Class 4	(9%)	102.7 ± 0.88 kD/nm
All	(100%)	78.1 ± 1.30 kD/nm
Pronase PHF Mass		
Class 1	(67%)	59.0 ± 0.51 kD/nm
Class 2	(33%)	73.9 ± 0.46 kD/nm
All	(100%)	65.2 ± 0.68 kD/nm
Maximum Mass Loss After Pronase		37.5 ± 1.11 kD/nm
Mean Mass Loss After Pronase		12.9 ± 1.20 kD/nm

A68-PHF preparation (Ksiezak-Reding and Wall, 1994). This sets an upper limit of about 40 kDa/nm to the mass difference before and after Pronase digestion, the equivalent of about one full-length tau molecule per nanometre. The mass data for intact PHFs therefore imply that, if tau is the sole constituent of the PHF, the majority the tau molecules which contribute to its structure, even in A68-PHF preparations, must have undergone N-terminal truncation, and that only one in six or seven extends to the N-terminus (Figure 9).

If a substantial proportion of the tau molecules which contribute to the structure of the PHF have already undergone truncation in the course of PHF assembly, then A68-PHFs could not be composed solely of full-length tau, whether or not hyperphosphorylated, as has been claimed (Lee *et al.*, 1991; Goedert *et al.*, 1992a). Either tau is not the sole constituent of the PHF, or if it is, then the bulk is not full-length. Either way, the claims generally accepted as defining the molecular composition of the PHF are unlikely to be true.

This then raises the question of the origin of the hyperphosphorylated tau species extracted by SDS from the sarkosyl-insoluble PHF preparation. These

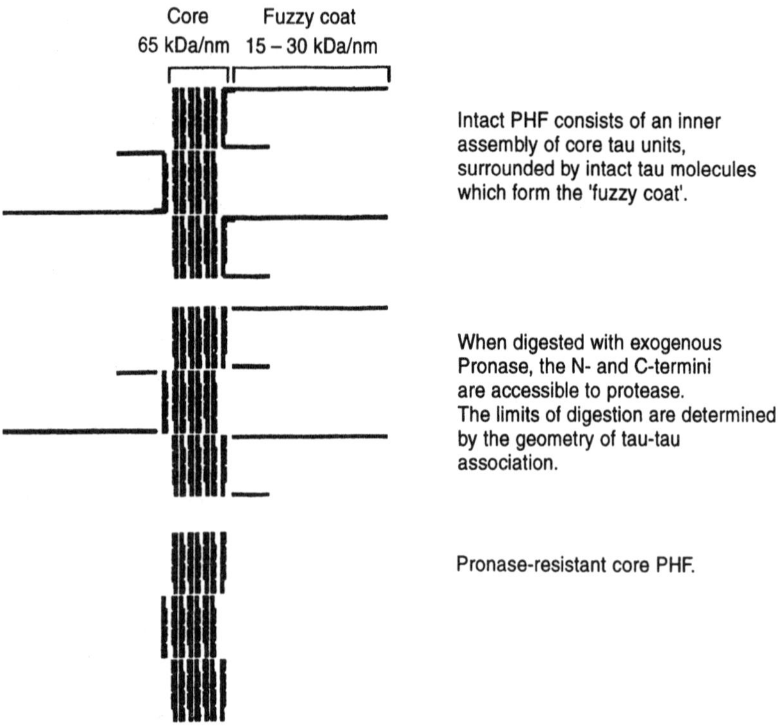

Figure 9 Schematic model of PHF structure based on mass data shown in Figure 8 and biochemical data shown in Figure 4. The N-terminal half of the tau molecule contributes only to the fuzzy outer coat of the PHF, which is lost after Pronase digestion. The core structure of the PHF, which is left after Pronase digestion and which contains the basic C-shaped subunit organization shown in Figure 3, contains only tau species derived from the tandem repeat region, phase-shifted by 14–16 residues as shown in Figures 5 and 6. The contribution of the C-terminal tail of the tau molecule is more complex, and is shown as entirely lost after Pronase digestion. However, a portion, but not all, of the C-terminal tail survives Pronase digestion. The only regions of the tau molecule which are phosphorylated in some PHFs are those located in the fuzzy outer coat.

could derive either from the surface tau molecules which coat the PHF core, or else they might represent tau protein, which copurifies with PHFs, or a combination of both.

In a recent study which compared SDS-soluble A68-protein with total phosphorylated tau content in the sarkosyl-insoluble fraction derived from AD brain tissues, no quantitative relationship could be demonstrated (Table 2). The reason for this is that only 12% of the phosphorylated tau which precipitates in sarkosyl can be found in the bulk PHF fraction, if the sarkosyl precipitation step is omitted. The remainder appears in several low-speed supernatant fractions where soluble or partially assembled tau species, but not PHFs, can be detected. Thus the bulk of phosphorylated tau

Table 2. Correlation matrix for measures (direct or competitive ELISA using mAb AT8, or densitometry of SDS-soluble A68-proteins) of phosphorylated tau prepared via the A68-protocol on the basis of insolubility in 1% sarkosyl. These are compared with each other, and with measures of PHF-bound tau in the Pronase-resistant core PHF preparation obtained from the same brain homogenate by the if-II protocol. Assays were carried out in 24 large scale preparations from 12 cases on neuropathologically confirmed Alzheimer's disease. SDS-soluble A68 proteins are correlated neither with the phosphorylated tau content of the A68-tau fraction as measured by mAb AT8, nor with the PHF-tau content. By contrast, mAb AT8 immunoreactivity in the A68-tau fraction is highly correlated with PHF content. Therefore, the SDS-solubility of tau protein in the A68-tau fraction is non-quantitative. In any case, only 12% of phosphorylated tau in this preparation is in the form of PHFs. Therefore, SDS-gels of the A68-tau fraction do not provide a basis for determining the molecular composition of PHFs. Data from Wischik *et al.*, 1995.

	A68-proteins SDS-soluble	*mAb AT8-direct*	*mAb AT8-competive*
PHF-core (mAb 423)	0.256	0.835***	0.842***
		0.357	0.257
A68-proteins -SDS-soluble			
mAb AT8 -direct			0.788***

*** $p < 0.0001$, otherwise NS

recovered in the sarkosyl-insoluble fraction does not have the sedimentation properties of PHFs in the absence of sarkosyl. Presumably these non-PHF forms of phosphorylated tau contribute disproportionately to the SDS-soluble forms which are visualized an immunoblots (Wischik *et al.*, 1995).

Thus, two post-translational modifications of tau have so far been identified: truncation of tau which is intrinsic to the core structure of the PHF and abnormal phosphorylation of portions of the tau molecule located in the fuzzy outer coat of the PHF. The quantitative relationship between these two processes is balanced very much in favour of truncation, in that only a minority of PHFs in AD brain tissues are abnormally phosphorylated, and only a minority of tau molecules which are incorporated into PHFs are phosphorylated. Thus, although the terms 'PHF-tau' and 'phosphorylated-tau' have tended to become synonymous in the literature, the two are in fact equivalent in less than 5% of the tau molecules which appear in PHFs. This therefore leaves open the question as to the actual pathogenic significance of phosphorylation and indeed at what point it occurs during the events leading to PHF assembly.

2.6 Phosphorylation of tau protein

Significant progress has been made in the elucidation of abnormal phosphorylation sites in tau in AD. Using site-directed mutagenesis and *in vitro* phosphorylation of mutant recombinant tau, Biernat *et al.* (1993) were able to map the phosphorylation-dependent antibody mAb AT8. mAb AT8 recognizes A68-tau proteins before but not after dephosphorylation. The antibody also fails to recognize the bulk of adult tau proteins that are isolated from normal brain tissues. The epitope of mAb AT8 is very close to, if not identical with, the Tau-1 epitope, around residues 189–207 (Kosik *et al.*, 1988), and is dependent on phosphorylation in the vicinity of Ser-202 (Goedert *et al.*, 1993). mAb Tau-1 has a complementary property to mAb AT8 in that it requires dephosphorylation (Binder *et al.*, 1985) of Ser-199 and/or Ser-202 (Liu *et al.*, 1993). Lichtenberg-Kraag *et al.* (1992) have used a series of phosphorylation-dependent neurofilament antibodies (SMI-31, SMI-33 and SMI-34), which also recognize A68-proteins, to show that tau is phosphorylated at Ser-235, Ser-396 and Ser-404, Ser-235 and Ser-396. Ser-46 in one of the N-terminal tau inserts has also been shown to be phosphorylated in the A68-tau proteins of 64 and 68 kDa mobility (Brion *et al.*, 1993). Hasegawa *et al.* (1992) have shown variable phosphorylation at Thr-231 and Ser-262 by mass spectroscopy. The interest of Ser-262 is that it is the only site of abnormal phosphorylation so far identified in the tandem repeat region. However, even this site is outside the minimal protease-resistant tau unit of the core PHF (see above). Most of the sites so far identified are within Ser-Pro or Thr-Pro pairs, of which there are 17 distributed throughout the N- and C-terminal domains of tau.

Numerous enzymes have been found which phosphorylate tau *in vitro*. Mitogen-activated protein kinase (MAP kinase) phosphorylates most of the 17 Ser-Pro and Thr-Pro sites (Drewes *et al.*, 1992; Goedert *et al.*, 1992b; Gustke *et al.*, 1992). Glycogen-synthetase kinase-3β phosphorylates most of the Ser-Pro motifs (Hanger *et al.*, 1992; Mandelkow *et al.*, 1992). Both MAP kinase and GSK-3β are microtubule-associated proteins as defined by the criterion of copurification with microtubules during cycles of assembly and disassembly (Mandelkow *et al.*, 1992). Two further kinases, called tau protein kinase 1 and 2, have been purified from microtubule-associated proteins and found to phosphorylate tau (Ishiguro *et al.*, 1988, 1991, 1992). Tau protein kinase 1 appears to be identical to GSK-3β, and phosphorylates tau at Ser-199, Thr-231, Ser-391 and Ser-413. Tau protein kinase 2 phosphorylates Ser-202, Thr-205, Ser-235 and Ser-404, which is similar to MAP kinase. Rodder and Ingram (1991) isolated two further kinases (PK36 and PK40), which phosphorylate tau at Ser-Pro and Thr-Pro sites, and also phosphorylate intermediate and heavy neurofilament subunits. Other kinases include cAMP-dependent kinase (Pierre and Nunez, 1983), calcium-calmodulin-dependent kinase II (CaM kinase; Yamamoto *et al.*, 1983), casein kinase I (Pierre and Nunez, 1983), casein kinase II (Steiner *et al.*, 1990), proline-directed kinase (Vulliet *et al.*, 1992), cdc2 kinase (Drewes *et al.*, 1992; Ledesma *et al.*, 1992; Mawal-Dewan *et al.*, 1992) and the cyclin-dependent cdk2 and cdk5 kinases (Baumann *et al.*, 1993). Biernat *et al.* (1993) have isolated a 35/41 kDa kinase from rat brain extract that phosphorylates Ser-262, which produces a substantial reduction in tau binding affinity for microtubules. It is not known which, if any, of these kinases are involved in the phosphorylation of tau *in vivo*, whether they are pathological and what their significance is regarding the development of neurofibrillary pathology in AD.

Since phosphorylation may also be caused by a loss of phosphatase activity, candidate phosphatases have also been examined. Protein phosphatase 2A dephosphorylates tau phosphorylated by p42 MAP kinase (also known as ERK2), whereas phosphatase 1 is ineffective (Goedert *et al.*, 1992b; Drewes *et al.*, 1993). Both phosphatases 2A and 2B (calcineurin) dephosphorylate phosphorylated Ser-Pro and Thr-Pro motifs (Drewes *et al.*, 1993; Gong *et al.*, 1994), as well as Ser-262 within the first repeat (Drewes *et al.*, 1993). The significance of these observations is at present unknown.

One argument that has been advanced in favour of attributing a pathogenic significance to states of abnormal phosphorylation of tau has been the observation that phosphorylated tau has a reduced binding affinity for microtubules. Bramblett *et al.* (1992) have shown that tau protein isolated from the sarkosyl-insoluble fraction binds with less affinity to microtubules before dephosphorylation. Recombinant tau protein phosphorylated *in vitro* also has reduced binding affinity for microtubules (Gustke *et al.*, 1992), and phosphorylation at Ser-262 appears to be particularly effective in this regard, although phosphorylation at Ser-Pro and Thr-Pro sites also contributes (Biernat *et al.*, 1993). Similarly, MAP kinase-phosphorylated tau also has a marked reduction in binding affinity for microtubules.

Two hypotheses regarding the pathogenesis of neurofibrillary degeneration in Alzheimer's disease have been advanced on the basis of this data. The first has been that abnormal phosphorylation of tau reduces binding affinity to microtubules and hence cause their destabilization, and so reduces fast axonal transport. The second hypothesis is that abnormal phosphorylation of tau, by neutralizing positive charges in the basic residues just upstream from the repeat region, reduces electrostatic repulsion, and so favours the self-assembly of tau into PHFs (Crowther, 1993).

However, there is as yet no evidence that any abnormal state of phosphorylation of tau favours self-association. Indeed, the only study that has examined the effect of phosphorylation on the polymerization of tau *in vitro* failed to demonstrate any quantitative difference between phosphorylated and unphosphorylated tau (Crowther *et al.*, 1994). We have recently measured the binding affinity of phosphorylated tau in a tau–tau binding assay, and found that phosphorylation in fact inhibits pathological self-association of tau ten-fold (Figure 10). Furthermore, tau which is sequestered in

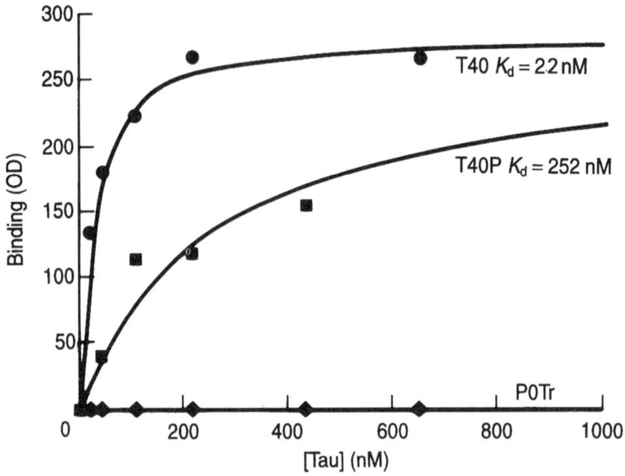

Figure 10 A major argument, which has been advanced in support of phosphorylation of tau as a rate-determining post-translational modification responsible for PHF assembly, is the observation that phosphorylated tau binds with a lower affinity to microtubules than non-phosphorylated tau. However, using a tau–tau binding assay developed in this laboratory, the binding affinity of phosphorylated tau was found to be ten-fold lower than unphosphorylated tau. Thus, phosphorylation inhibits pathological tau–tau association to the same extent as it inhibits tau–tubulin binding. T40, recombinant tau protein; T40P, recombinant tau protein phosphorylated *in vitro* with rat brain kinase extract; POTr, post-natal day 0 rat tau.

the somatodendritic compartment for whatever reason is more phosphorylated relative to axonal tau (Papasozomenos and Binder, 1987; Riederer and Innocenti, 1991), and this appears to be a mechanism for reducing both tau–tubulin association and transport of tau into the axon. Indeed the presence of phosphorylated tau in PHFs may simply be biologically synonymous with their occurrence in the somatodendritic compartment of the cell, and may carry no further pathogenic significance than this.

2.7 Immunohistochemical studies of altered tau processing in Alzheimer's disease

The main evidence which is cited to show that phosphorylation is an early pathological event is histological. Bancher *et al.* (1989) first observed pyramidal cells in regions vulnerable to neurofibrillary degeneration containing tau-immunoreactive granular material in the cytoplasm in the absence of neurofibrillary tangles, and Wolozin *et al.* (1986) also reported Alz-50 immunoreactivity in pyramidal cells without neurofibrillary tangles. The latter has been shown to recognize the N-terminus of tau (Goedert *et al.*, 1991). Further immunohistochemical studies with antibodies such as mAb AT8 have likewise shown that these early deposits contain tau phosphorylated at Ser-199/202, and a variety of other phosphorylation-dependent antibodies that detect other phosphorylation sites have shown similar results. Most interestingly, Braak *et al.* (1994) have recently reported that the earliest changes found in transitional entorhinal cortex, occurring as early as the fourth decade of life, are pyramidal cells immunoreactive with mAb AT8.

However, similar findings have also been reported using mAb 423, which detects tau truncated at Glu-391. Pyramidal cells vulnerable to neurofibrillary degeneration, which do not yet contain neurofibrillary tangles, have also been found to contain fine immunoreactive granular material (Mena *et al.*, 1991; Plate II). This material is also immunoreactive with mAb 7.51 only after formic acid treatment of the section, showing that the tau protein is in a partially assembled state which occludes the tandem repeat region (Mena *et al.*, 1995). When examined by electron microscopy, these deposits are ultrastructurally amorphous (Mena *et al.*, 1995). Two other abnormal cytoplasmic inclusions that contain truncated tau have also been reported. The granule of the granulovacuolar degeneration (GVD) complex, a tertiary lysosomal structure whose appearance in the hippocampus is one of the characteristic neuropathological features of Alzheimer's disease, contains C-terminally truncated tau (Bondareff *et al.*, 1991; Dickson *et al.*, 1993). A second type of granule, the 'PHF core body', has also been reported (Mena *et al.*, 1992). These granules are typically found in smaller numbers in the cell, have a smoother outline than the GVD granule and are typically located at the base of the apical dendrite. Neither of these two latter granular structures is labelled by markers for phosphorylated tau.

By the time typical intracellular neurofibrillary tangles appear within the cell, similar immunohistochemical features are preserved: mAb 423 immunoreactivity and mAb 7.51 immunoreactivity are both present, but in a form which requires formic acid treatment of the histological section for the epitopes to be accessible (Mena *et al.*, 1995). Some, but by no means all, intracellular neurofibrillary tangles are characterized by mAb AT8 immunoreactivity (Bondareff *et al.*, 1995).

Further stages of processing of neurofibrillary structures can be discerned in the extracellular space after cell death. Stage II extracellular tangles are characterized by loss of the very N-terminus of the tau molecule but preservation of regions closer to the tandem repeat region, and recognized by phosphorylation-dependent antibodies. These tangles can be seen to be extracellular, because they are no longer associated with a pyramidal cell nucleus, but remain fusiform and compact. Stage III extracellular tangles are characterized by loss of all immunoreactivity outside the tandem repeat region, although some immunoreactivity for the very C-terminus of tau is preserved. Such tangles are also more dispersed in appearance and swollen beyond the dimensions of the original pyramidal cell, presumably due to invasion by astrocytic processes (Bondareff *et al.*, 1990, 1994).

Thus the immunohistochemical staging of neurofibrillary pathology appears to involve the following stages of abnormal processing of the tau molecule:

(1) Appearance in the somatodendritic compartment of N-terminally intact tau (phosphorylated) and ultrastructurally amorphous tau deposits, characterized by

C-terminal truncation (mAb 423-reactive), tau–tau association (formic acid-dependent mAb 7.51 immunoreactivity).

(2) Appearance of overt fibrillar assemblies, characterized as polymers with a core of truncated tau and superficial coating of full-length tau, some phosphorylated (Stage I tangles).

(3) Stage II extracellular tangles, with loss of N-terminal tau immunoreactivity.

(4) Stage III extracellular tangles, loss of all markers for the N-terminal half of the tau molecule, re-exposure of the marker for C-terminal truncation.

2.8 Aetiology of abnormal tau processing in Alzheimer's disease – insights from other neurodegenerative disorders

Aetiological theories for AD have included debate over the relative importance of neurofibrillary pathology. The formation and deposition of Aβ is regarded by many investigators as the specific aetiological agent in AD. This view has received the strongest direct support from studies of Down's syndrome and from genetic evidence in familial autosomal dominant AD.

2.8.1 Down's syndrome

Senile plaques and neurofibrillary tangles are found in a large proportion of Down's syndrome (DS) subjects (Jervis, 1948; Mann, 1988; Mattiace *et al.*, 1991) irrespective of genotype (Whalley, 1982; Sylvester, 1986) and these increase with age. Trisomy of only the 21q22 band of chromosome 21 has been shown to be sufficient for the development of AD-type pathology (Neibuhr, 1974). The PHFs found in DS brain tissues are identical to those found in AD (Mukaetova-Ladinska *et al.*, 1994). The extent of neurofibrillary pathology has been found to be correlated both with the loss of normal soluble tau and the quantity of tau in PHFs (Mukaetova-Ladinska *et al.*, 1994). The quantity of PHF-tau is comparable in AD and DS cases with clinical evidence of dementia. Although some reports have claimed that there is more phosphorylated tau in DS relative to AD (Goedert *et al.*, 1992a), phosphorylated tau can in fact be detected in DS individuals regardless of clinical or neuropathological evidence of AD (Mukaetova-Ladinska *et al.*, 1994), and neurons reactive with Alz-50 have been found to persist throughout life in DS (Sparks and Hunsaker *et al.*, 1992).

As with AD, there is a positive correlation between the amount of neurofibrillary degeneration and the amount of PHF-bound tau, and a negative correlation between neurofibrillary pathology and the amount of normal soluble tau (Mukaetova-Ladinska *et al.*, 1993, 1994). No statistically significant relationship has yet been found between the amount of Aβ pathology and accumulation of PHF-bound tau. However, a significant positive correlation can be demonstrated in DS cases between the amount of Aβ pathology and the level of normal soluble tau ($r = 0.52$, $p = 0.048$; Mukaetova-Ladinska *et al.*, 1994). This is also consistent with reports that diffuse Aβ plaques are also positively correlated with neuron count (Cummings *et al.*, 1993). Thus the only quantitative relationship which can be demonstrated either in AD or DS between Aβ pathology and tau protein is likely to be an indirect reflection of neuronal density. No relationship has yet been demonstrated that would support the hypothesis that Aβ deposition is linked in a quantitative manner to alterations in tau-protein processing in either AD or DS.

2.8.2 Senile dementia of the Lewy body type

A number of families have been identified with early-onset forms of AD which have been linked to markers on chromosome 21 (St George-Hyslop *et al.*, 1987), and mutations identified in the APP gene itself (Goate *et al.*, 1991; Fidani *et al.*, 1992). However, not all APP mutations are associated with the clinical phenotype of Alzheimer's disease. Thus a dominant mutation at codon 693 of APP_{770} is associated with a hereditary form of cerebrovascular amyloidosis without tau pathology (Levy *et al.*, 1990). One of the cases with a mutation at codon 717 of APP_{770} has been found at postmortem to be characterized by a distinct pathological profile, which includes cortical Lewy bodies, amyloid angiopathy and stroke (Hardy *et al.*, 1991). This has led to the view that deposition of Aβ represents a unifying feature present in a range of neurodegenerative disorders.

There is some neuropathological overlap between AD, senile dementia of the Lewy body type (SDLT) and Parkinson's disease (PD), so that the current nosological status of Lewy body disorders remains controversial (Hansen and Galasko, 1992). Nevertheless, it is possible to select cases of relatively pure SDLT on neuropathological grounds, which can be distinguished also on clinical grounds as presenting an atypical dementia syndrome characterized by a fluctuating confusional state, behavioural disturbance, visual and/or auditory hallucinations and progressive dementia (McKeith *et al.*, 1992). Such cases have recently been studied biochemically, and were shown to have significant accumulation of SDS-insoluble Aβ in the absence of accumulation either of abnormally phosphorylated tau protein or of tau protein incorporated into PHFs. These cases therefore demonstrate that APP dysmetabolism leading to deposition of SDS-insoluble Aβ can occur without inducing the pathological alterations in tau processing, which distinguish normal ageing from AD (Harrington *et al.*, 1994a).

It has been suggested that possession of the apolipoprotein E3 allele (or possibly the ApoE2 allele) may be protective against PHF formation (Strittmatter *et al.*, 1994). The mechanism proposed is that ApoE3 binds to phosphorylated tau with higher affinity than ApoE4 and hence prevents PHF assembly. Quite apart from the question of whether phosphorylation of tau does or does not promote PHF assembly, there is no evidence that the possession of the ApoE3 allele is protective against PHF assembly. When the amount of PHF-bound tau and phosphorylated tau were compared in cases with and without the ApoE4 allele, no difference could be found (Harrington *et al.*, 1994b). Furthermore, when pure SDLT cases were compared with AD, a similar elevation in apoE4 allele frequency was found (0.341 in SDLT, 0.326 in AD, 0.140 in controls). These SDLT cases were characterized by the absence of any tau pathology (Harrington *et al.*, 1994a), therefore, possession of the ApoE4 allele cannot by itself be responsible for pathological alterations in tau protein processing of the kind seen in AD.

2.8.3 Environmental factors – aluminium exposure

Interest in environmental triggers for the pathology of AD arises most notably from the consistent epidemiological observation that prevalence appears to be a simple exponential function of age (Rocca *et al.*, 1991), with an approximate doubling of the rate of inception of new cases for every 5 years beyond the age of 70 (Paykel *et al.*, 1994). A recent study of renal dialysis patients with elevated levels of brain aluminium caused by high therapeutic exposure to aluminium-containing medications showed that changes in tau protein processing of the kind seen in AD were related in a quantitative manner to brain aluminium accumulation (Harrington *et al.*, 1994c). Specifically, the accumulation of tau protein truncated at Glu-391 was found to be correlated with both the level of aluminium in the brain ($r = 0.753$, $p = 0.001$) and the loss of normal soluble tau ($r = -0.573$, $p = 0.025$). The frequency of cases in this series with accumulation of protease-resistant PHFs in frontal cortex (13%) was significantly higher than expected for this age group (0.1%, $p < 0.0001$). Thus, although elevated levels of aluminium in the brain are not normally found in AD, these findings may indicate a pathological pathway whereby environmental toxins are able to induce pathological processing of tau protein without significant Aβ deposition (Harrington *et al.*, 1994c).

2.9 Conclusion

Alzheimer's disease is an aetiologically heterogeneous disorder. Although several genetic factors have been identified that are associated with AD, there are many cases of neuropathologically typical AD in patients who do not have any evident mutation in the APP gene, show no linkage to the AD3 locus on chromosome 14 and do not carry any apolipoprotein E4 alleles.

Therefore an important practical problem for understanding the molecular pathogenesis of dementia of the Alzheimer type is to define those molecular changes which are the critical substrates for clinical dementia. On the evidence currently available, the argument for deposition of SDS-insoluble Aβ as a critical substrate is weak except in the minority of AD cases with APP mutations. Thus, deposition of SDS-insoluble amyloid is seen in normal ageing, in PD, in SDLT and in AD. Although the levels are higher in AD, SDLT cannot be distinguished from normal ageing in terms of insoluble Aβ. Therefore, the accumulation of this material cannot be the explanation of clinical dementia in SDLT and, since the levels are only marginally higher in AD, the same conclusion follows. As a unifying factor for the neurodegenerative disorders of late life, Aβ deposition emerges as a weak discriminator for dementia but a good marker for some kind of neurodegenerative change, which is very common in late life.

It must be said that the changes in tau protein processing which are seen in AD also cannot be used to explain all of the neurodegenerative dementias of late life. Specifically, the redistribution of the tau protein pool from soluble to PHF-bound phases, which strongly discriminates cases with clinical dementia from normal ageing, is not seen in pure SDLT or Parkinson's disease patients with dementia. Thus, the redistribution of the tau protein pool cannot provide the explanation for dementia in these latter categories of dementia, and some other process, which implicates neither tau protein nor APP, must be found. Furthermore, changes in tau protein processing are not only seen in AD. For example, at least at the immunochemical level, changes in tau protein similar to those found in AD (including endogenous truncation of tau at Glu-391) are also found in Pick's disease in the dentate gyrus (Soillieux *et al.*, unpublished observation). Nevertheless there appears to be a reasonably firm body of evidence to support the view that the extent of neurofibrillary pathology, whether measured histologically or biochemically, correlates with the degree of clinical dementia of the Alzheimer type. Furthermore, the same underlying process also correlates with the degree of synapse loss in the frontal cortex, which in turn correlates with the degree of clinical dementia (Terry *et al.*, 1991).

However, the genetic evidence in regard to the mutations in the APP locus must be accepted as proving that, at least in these familial cases, genetically mediated alterations in the processing of APP are a sufficient cause to explain the observed or implied alterations in tau processing. It follows that changes in APP processing are able to trigger the alterations in tau processing, which are quantitatively linked to degree of clinical dementia. For the reasons discussed above, the molecular linkage between these two processes does not appear to be tight. Thus, it is possible to have extensive deposition of Aβ without triggering tau pathology. Likewise, it appears to be possible to trigger alterations in tau protein processing without the mediation of measurable changes in APP processing, for example, as in the case of aluminium exposure.

It follows that therapeutic strategies which aim to block pathological alterations in APP processing need not necessarily prevent the alterations in tau processing which appear to underlie dementia, at least in AD. On the other hand, a successful therapeutic strategy which prevented pathological alterations in tau protein processing would not be expected to provide a treatment for the other neurodegenerative disorders of old age, particularly SDLT, which is the second most significant cause of dementia after AD.

As to the molecular mechanism of redistribution of the tau protein pool into PHFs which occurs in AD, it is widely held that phosphorylation plays a critical role and therefore that, if a link is to be found between 'APP dysmetabolism' and 'tau dysmetabolism', it is via the processes which regulate the pathological phosphorylation of tau. Arguments have been advanced in this chapter which do not support this position. Specifically, although 'phosphorylated tau' and 'PHF tau' have become synonymous in the literature, this is likely to prove a false conjunction. It is probable that, of tau molecules which have been incorporated into PHFs, as few as 5% are phosphorylated, even at the earliest measurable stages of the disease process and that, at any time, the total amount of phosphorylated tau in the brain is less than 10% of the total PHF-bound tau. It is perfectly compatible with the available evidence that phosphorylation of tau is an epiphenomenon in which a small proportion of PHFs become phosphorylated in the course of the events which lead to PHF assembly. There is no evidence presently available which forces one to conclude that phosphorylation of tau is the rate-determining step in the process of PHF assembly. Indeed, there is evidence to suggest that phosphorylation might even represent a normal physiological post-translational modification of tau, which reduces its tendency to bind either to tubulin or to self-polymerize in the somato-dendritic compartment.

The molecular analysis of enriched PHF preparations suggests that the bulk of the tau protein, which contributes to the core structure of the PHF, has undergone extensive N- and C-terminal truncation, to produce a phase-shifted fragment which is restricted to the tandem repeat region. This 'amyloidogenic' processing of tau protein in Alzheimer's disease probably conforms to generic concepts of amyloidogenesis implicating other proteins, notably the prion protein and APP. Specifically, one can visualize the polymerization of tau protein into PHFs as a process which requires a repetitive cycle of high-affinity capture of normal tau by a truncated tau species, and subsequent partial proteolysis of the captured species to reproduce the truncated tau unit which is able then to capture another normal tau molecule, and so on. In the absence of either further tau capture or partial truncation, the process would cease or saturate. The dynamics of this process would be exponential, since the more truncated tau units which are produced, the more tau will be captured, with the reproduction of further tau capture sites. In other words, such a process would be autocatalytic (Figure 11).

As regards the initiation of pathological tau truncation, it is conceivable that many different aetiological processes could operate, including alterations in APP processing, possibly via the endosomal/lysosomal system. The intrinsic amplification of pathological tau processing which would occur via autocatalytic capture, truncation and polymerization of tau would then produce large-scale alterations in the tau protein pool which need bear no quantitative relationship to the underlying initiating processes. Although a large shift in the tau protein pool might be required to explain clinical dementia, quantitatively minor but highly specific changes in other metabolic pathways, which lead to the initial production of truncated tau units, would be adequate to initiate an explosive and irreversible sequence of events. There is good reason to believe that alterations in the processing of APP represent one potential triggering process, but need not represent the only or necessary trigger.

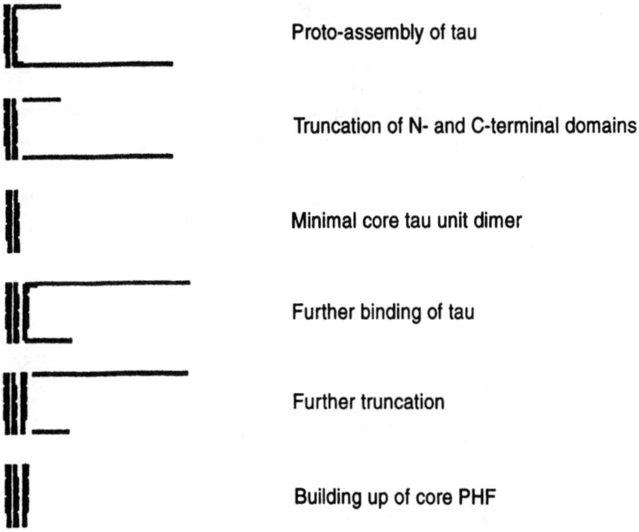

Figure 11 Schematic model of autocatalytic processing of tau protein to produce PHFs in Alzheimer's disease. In this model, initial deposits of truncated tau, of the kind shown in Plate II, represent high-affinity capture sites for normal intact tau (K_d = 22 nM, Figure 10). Partial proteolytic processing of this complex results in removal of N- and C-terminal domains of the tau molecule, leaving a core tau unit with the characteristic C-terminal truncation site at Glu-391 recognized by mAb 423. This is able in turn to capture a further molecule of full-length tau, and so on. This process is envisaged as being exponential and endless, because proteolytic degradation is incomplete, leaving an exposed tandem-repeat fragment able to capture a further molecule of intact tau. However, partial proteolytic processing is seen as essential, since otherwise the capture of full-length tau molecules would become saturable.

References

Alzheimer, A. (1907) *Allg. Z. Psychiatrie Psychisch-Gerichtlich Med.* **64**, 146–148.
Anderton, B.H., Breinburg, D., Biownes, M.J., Green, P.J., Tomlinson, B.E., Ulrich, J., Wood, J.N. & Kahn, J. (1982) *Nature* **298**, 84–86.
Bancher, C., Brunner, C., Lassmann, H., Budka, H., Jellinger, K., Wiche, G., Seitelberger, F., Grundke-Iqbal, I. & Wisniewski, H.M. (1989) *Brain Res.* **477**, 90–99.
Bancher, C., Braak, H., Fischer, P. & Jellinger, K.A. (1993) *Neurosci. Lett.* **162**, 179–182
Baumann, K., Mandelkow, E.M., Biernat, J., Piwnica-Worms, H. & Mandelkow, E. (1993) *FEBS Lett.* **336**, 417–424.
Biernat, J., Mandelkow, E.M., Schroter, C., Lichtenberg-Kraag, B., Steiner, B., Berling, B., Meyer, H., Mercken, M., Vandermeeren, A., Goedert, M. & Mandelkow, E. (1992) *EMBO J.* **11**, 1593–1597.
Biernat, J., Gustke, N., Derwes, G., Mandelkow, E.M. & Mandelkow, E. (1993) *Neuron* **11**, 153–163.
Binder, L.I., Frankfurter, A. & Rebhun, L.I. (1985) *J. Cell Biol.* **101**, 1371–1378.
Bondareff, W., Wischik, C.M., Novak, M., Amos, W.B., Klug, A. & Roth, M. (1990). *Am. J. Pathol.* **137**, 711–723.
Bondareff, W., Wischik, C.M., Novak, M. & Roth, M. (1991). *Am. J. Pathol.* **139**, 641–647.

Bondareff, W., Mountjoy, C.Q., Wischik, C.M., Hauser, D.L., LaBree, L.D. & Roth, M. (1993) *Arch. Gen. Psychiatry* **50**, 350–356.
Braak, H. & Braak, E. (1988) *Neuropathol. Appl. Neurobiol.* **14**, 39–44.
Braak, H. & Braak, E. (1991). *Acta Neuropathol.* **82**, 239–259.
Braak, H., Braak, E., Grundke-Iqbal, I. & Iqbal, K. (1986) *Neurosci. Lett.* **65**, 351–355.
Braak, H., Duyckaerts, C., Braak, E. & Piette, F. (1993) Neuropathological staging of Alzheimer-related changes correlates with psychometrically assessed intellectual status. In *Alzheimer's Disease: Advances in Clinical and Basic Research. Third International Conference on Alzheimer's Disease and Related Disorders*, pp. 131–137. Chichester, John Wiley & Sons.
Braak, E., Braak, H. & Mandelkow, E.M. (1994) *Acta Neuropathol.* **87**, 554–567.
Bramblett, G.T., Trojanowski, J.Q. & Lee, V.M.Y. (1992) *Lab. Invest.* **66**, 212–222.
Bramblett, G.T., Goedert, M., Jakes, R., Merrick, S.E., Trojanowski, J.Q. & Lee, V.M.Y. (1993) *Neuron* **10**, 1089–1099.
Brion, J.P., Passareiro, H., Nunez, J. & Flament-Durand, J. (1985) *Arch. Biol.* **95**, 229–235.
Brion, J.P., Couck, A.M., Robertson, J., Loviny, T.L.F. & Anderton, B.H. (1993) *J. Neurochem.* **60**, 1372–1382.
Caputo, C., Norris, T.E. & Salama, A.I. (1992) *Neurobiol. Aging* **13**, 267–274.
Clark, R.F., Cruts, M., Korenblat, K., He, C., Talbot, C., Van Broeckhoven, C., Goate, A.M. *Human Molecular Genetics* **4** 1347–1354
Cork, L.C., Sternberger, N.H., Sternberger, L.A., Csasnova, M.F., Struble, R.G. & Price, D.I. (1986) *J. Neuropathol. Exp. Neurol.* **45**, 56–64.
Crawford, F. & Goate, A. (1992). *BioEssays* **14.11**: 727–734.
Crowther, R.A. (1991) *Proc. Natl Acad. Sci. USA* **88**, 2288–2292.
Crowther, R.A. (1993) *Curr. Opin. Struct. Biol.* **3**, 202–206
Crowther, R.A. & Wischik, C.M. (1985) *EMBO J.* **4**, 3661–3665.
Crowther, R.A., Oleson, O.F., Smith, M.J., Jakes, R. & Goedert, M. (1994) *FEBS Lett.* **337**, 135–138
Delacourte and Defossez (1986) *J. Neurol. Sci.* **76**, 173–186.
Dickson, D.W., Liu, W.K., Kress, Y., Ku, J., DeJesus, O. & Yen, S.H.C. (1993) *Acta Neuropathol.* **85**, 463–470.
Drewes, G., Lichtenberg-Kraag, B., Doring, F., Mandelkow, E.M., Biernat, J., Doree, M. & Mandelkow, E. (1992) *EMBO J.* **11**, 2131–2138.
Drewes, G., Mandelkow, E.M., Baumann, K., Goris, J., Merlevede, W. & Mandelkow, E. (1993) *FEBS Lett.* **336**, 425–432.
Fidani, L., Rooke, K., Chartier-Harlin, M.C., Huges, D., Tanzi, R., Mullan, M., Roques, P., Rosser, M., Hardy, J. & Goate, A. (1992) *Human Mol. Genet.* **1.3**, 165–168.
Gambetti, P., Velasco, M.E., Dahl, D., Bignani, A., Roessmann, U., Sindley, S.P. (1986) In *Aging of the Brain in Dementia* (eds Amaducci, L., Davison, A.N. & Antuono, P.), pp. 55–63. New York, Raven Press.
Games, D., Adams, D., Alessandrini, R., Barbour, R., Berthelette, P., Blackwell, C., Carr, T., Clemens, J., Donaldson, T., Gillespie, F., Guido, T., Hagopian, S., Johnson-Wood, K., Khan, K., Lee, M., Leibowitz, P., Lieberburg, I., Little, S., Masliah, E., McConlogue, L., Montoya-Zavala, M., Mucke, L., Paganini, L., Penniman, E., Power, M., Schenk, D., Seubert, P., Snyder, B., Soriano, F., Tan, H., Vitale, J., Wadsworth, S., Wolozin, B., Zhao, J. (1995) *Nature* **373**, 523–527.
Goate, A.M., Chartier-Harlin, M.C., Mullan, M., Brown, J., Crawford, F., Fidani, L., Giuffra, L., Haynes, A., Irving, N., James, L. *et al.* (1991) *Nature* **349**, 704–706.
Goedert, M., Wischik, C.M., Crowther, R.A., Walker, J.E. & Klug, A. (1988) *Proc. Natl Acad. Sci. USA* **85**, 4051–4055.
Goedert, M., Spillantini, M.G., Potier, M.C., Ulrich, J. & Crowther, R.A. (1989a) *EMBO J.* **3**, 393–399.
Goedert, M., Spillantini, M.G., Jakes, R., Rutherford, D. & Crowther, R.A. (1989b) *Neuron* **3**, 519–526.
Goedert, M., Spillantini, M.G. & Jakes, R. (1991) *Neurosci. Lett.* **126**, 149–154.

Goedert, M., Spillantini, M.G., Cairns, N.J. & Crowther, R.A. (1992a) *Neuron* **8**, 159–168.
Goedert, M., Cohen, E.S., Jakes, R. & Cohen, P. (1992b) *FEBS Lett.* **312**, 95–99.
Goedert, M., Jakes, R., Crowther, R.A., Six, J., Lubke, U., Vandermeeren, M., Cras, P., Trojanowski, J.Q. & Lee, V.M.Y. (1993) *Proc. Natl Acad. Sci. USA* **90**, 5066–5070.
Gong, C.X., Singh, T.J., Grundke-Iqbal, I. & Iqbal, K. (1994) *J. Neurochem.* **62**, 803–806.
Gray, J.A. (1982) *The Neuropsychology of Anxiety: An Enquiry into the Function of the Septal-Hippocampal System.* Oxford, Clarendon Press.
Greenberg, S.G. & Davies, P. (1990) *Proc. Natl Acad. Sci. USA* **87**, 5827–5831.
Grundke-Iqbal, I., Iqbal, K., Quinlan, M., Tung, Y.-C., Zaidi, M.S. & Wisniewski, H.M. (1986) *J. Biol. Chem.* **261**, 6084–6089.
Gustke, N., Steiner, B., Mandelkow, E.M., Biernat, J., Meyer, H.E., Goedert, M. & Mandelkow, E. (1992). *FEBS Lett.* **307**, 199–205.
Haas, C. & Selkoe, D.J. (1993) *Cell* **75**, 1039–1042.
Hanger, D.P., Hughes, K., Woodgett, J.R., Brion, J.P. & Anderton, B.H. (1992) *Neurosci. Lett.* **147**, 58–62.
Hardy, J., Mullan, M., Chartier-Harlin, M.-C., Brown, J., Goate, A., Rossor, M., Collinge, J., Roberts, G., Luthert, P., Lantos, P., Naruse, S., Janeko, K., Tsuji, S., Miyatake, T., Shimizu, T., Kojima, T., Nakano, I., Yoshioka, K., Sakaki, Y., Miki, T., Katsuya, T., Ogihara, T., Roses, A., Pericak-Vance, M., Haan, J., Roos, R., Lucotte, G. & David, F. (1991) *Lancet* **337**, 1342–1343.
Harrington, C.R., Edwards, P.C. & Wischik, C.M. (1991a) *J. Immunol. Meth.* **134**, 261–271.
Harrington, C.R., Mukaetova-Ladinska, E.B., Hills, R., Edwards, P.C., Montejo de Garcini, E., Novak, N. & Wischik, C.M. (1991b) *Proc. Natl Acad. Sci. USA* **88**, 5842–5846.
Harrington, C.R., Wischik, C.M., McArthur, F.K., Taylor, G.A., Edwardson, J.A. & Candy, J.M. (1994a) *Lancet* **343**, 993–997.
Harrington, C.R., Perry, R.H., Perry, E.K., Hurt, J., McKeith, I.G., Roth, M. & Wischik, C.M. (1994b) *Dementia* **5**, 215–228.
Harrington, C.R., Louwagie, J., Rossau, R., Vanmechelen, E., Perry, R.H., Perry, E.K., Xuereb, J., Roth, M. & Wischik, C.M. (1994c) *Am. J. Pathol.* (in press).
Hasegawa, M., Morishima-Kawashima, M., Takio, K., Suzuki, M., Titani, K. & Ihara, Y. (1992) *J. Biol. Chem.* **267**, 17047–17054.
Higgins, L., Holtzman, D., Rabin, J., Mobley, W., Cordell, B. (1994) *Ann. Neurol.* **35**, 598–607
Ihara, Y., Nukina, N., Sugita, H. & Toyokura, Y. (1981) *Proc. Jap. Acad.* **57**, 152–156.
Ihara, Y., Nukina, N. & Miurar Ogawarra, M. (1986) *J. Biochem.* **99**, 1807–1810.
Iqbal, K., Wisniewsky, H.M., Shelanski, M.L., Brostoff, S., Liwnicz, B.H. & Terry, R.D. (1974) *Brain Res.* **77**, 337–343.
Iqbal, K., Wisniewsky, H.M., Grundke-Iqbal, I., Korthals, J.K. & Terry, R.D. (1975) *J. Histochem. Cytochem.* **23**, 563–569.
Ishiguro, K., Ihara, Y., Uchida, T. & Imahori, K. (1988) *J. Biochem.* **104**, 319–321.
Ishiguro, K., Omori, A., Kato, K., Tomizawa, K., Imahori, K. & Uchida, T. (1991) *Neurosci. Lett.* **128**, 195–198
Ishiguro, K., Takamatsu, M., Tomizawa, K., Omori, A., Takashashi, M., Arioka, M., Uchida, T. & Imahori, K. (1992) *J. Biol. Chem.* **267**, 10897–10901.
Jakes, R., Novak, M., Davison, M. & Wischik, C.M. (1991) *EMBO J.* **10**, 2725–2729.
Jervis, G.A. (1948). *Am. J. Psychiatry* **105**, 102–106.
Kidd, M. (1963). *Nature* **197**, 192–193.
Kidd, M., Allsop, D. & Landon, M. (1985) *Lancet* **i**, 278.
Kopke, E., Tung, Y.C., Shaikh, S., Alonso, A.C., Iqbal, K. & Grundke-Iqbal, I. (1993) *J. Biol. Chem.* **268**, 24374–24384.
Kosik, K.S., Joachim, C.L. & Selkoe, D.J. (1986) *Proc. Natl Acad. Sci. USA* **83**, 4044–4048.
Kosik, K.S., Orecchio, L.D., Binder, L., Trojanowski, J.Q., Lee, V.M.Y. & Lee, G. (1988) *Neuron* **1**, 817–825.
Ksiezak-Reding, H. & Wall, J.S. (1994) *Neurobiol. Aging* **15**, 11–19.

Lamb, B., Sangram, S., Lawler, A., Slunt, H., Kitt, C., Kearns, W., Pearson, P., Price, D., Gearhart, J. (1993) *Nature Genetics* **5**, 22–30.
Ledesma, M.D., Correas, I., Avila, J. & Diaz-Nido, J. (1992) *FEBS Lett.* **308**, 218–224.
Lee, V.M.Y., Balin, B.J., Otvos, L. & Trojanowski, J.Q. (1991) *Science* **251**, 675–678.
Levy, E., Carman, M.D., Fernandez-Madrid, I.J. *et al.* (1990) *Science* **248**, 1124–1126.
Lichtenberg-Kraag, B., Mandelkow, E.M., Biernat, J., Steiner, B., Schroter, C., Gustke, N., Meyer, H.E. & Mandelkow, E. (1992) *Proc. Natl Acad. Sci. USA* **89**, 5384–5388.
Liu, W.K., Ksiezak-Reding, H. & Yen, S.H. (1991) *J. Biol. Chem.* **266**, 21723–21727.
Liu, W.K., Moore, W.T., Williams, R.T., Hall, F.L. & Yen, S.H. (1993) *J. Neurosci. Res.* **34**, 371–376.
Mandelkow, E.M., Derwes, G., Biernat, J., Gustke, N., Van Lint J., Vandenheede, J.R. & Mandelkow, E. (1992) *FEBS Lett.* **314**, 315–321.
Mann, D.M.A. (1988) *Histopathology* **13**, 125–137.
Masters, C.L., Multhaup, G., Simms, G., Pottgiesser, J., Martins, R.N. & Beyreuther, K. (1985) *EMBO J.* **4**, 2757–2763.
Mattiace, L.A., Kress, Y., Davies, P., Ksiezak-Reding, H., Yen, S.H. & Dickson, D.W. (1991). *J. Neuropathol. Exp. Neurol.* **50**, 547–559.
Mawal-Dewan, M., Sen, P.C., Abdel-Ghany, M., Shalloway, D. & Racker, F. (1992) *J. Biol. Chem.* **267**, 19705–19709.
McKeith, I.G., Perry, R.H., Fairbairn, A.F., Jabeen, S. & Perry, E.K. (1992) *Psychol. Med.* **22**, 911–922.
Mena, R., Wischik, C.M., Novak, M., Milstein, C. & Cuello, A.C. (1991) *J. Neuropathol. Exp. Neurol.* **50.4**, 474–490.
Mena, R., Robitaille, Y. & Cuello, C. (1992) *J. Geriat. Psych. Neurol.* **5**, 132–141.
Mena, R., Edwards, P., Pérez-Olivera, O. & Wischik, C.M. (1994). *Acta Neuropathol.* **89**, 50–56.
Metuzals, J., Clapin, D.F. & Montpetit, V. (1984) *Proc. EMSA.* **42**, 302–303.
Metuzals, J., Robitaille, Y., Houghton, S., Gauthier, S. & Leblanc, R. (1987) In *Alterations in the Neuronal Cytoskeleton in Alzheimer Disease* (ed. G. Perry), Advances in Behavioural Biology, Vol. 34, pp. 75–95, New York, Plenum.
Miller, C.C.J., Brion, J.-P., Calvert, R., Chin, T.K., Eagles, P.A.M., Downes, M.J., Flament-Durant, J., Haugh, M., Kahn, J., Probst, A., Ulrich, J. & Anderton, B.H. (1986) *EMBO J.* **5**, 269–276.
Mori, H., Kondo, J. & Ihara, Y. (1987) *Science* **235**, 1641–1644.
Mukaetova-Ladinska, E.B., Harrington, C.R., Hills, R., O'Sullivan, A., Roth, M. & Wischik, C.M. (1992) *Dementia* **3**, 61–69.
Mukaetova-Ladinska, E.B., Harrington, C.R., Roth, M. & Wischik, C.M. (1993) *Am. J. Pathol.* **143**, 565–578.
Mukaetova-Ladinska, E.B., Harrington, C.R., Roth, M. & Wischik, C.M. (1994) *Develop. Brain Dys.* **7**, 311–329
Neibuhr, E. (1974) *Hum. Genet.* **21**, 99–101.
Novak, M., Jakes, R., Edwards, P.C., Milstein, C. & Wischik, C.M. (1991) *Proc. Natl Acad. Sci. USA* **88**, 5837–5841.
Novak, M., Kabat, J. & Wischik, C.M. (1993) *EMBO J.* **12**, 365–370.
Nukina, N. & Ihara, Y. (1986) *J. Biochem.* **99**, 1541–1544.
Papasozomenos, SCh. & Binder L.I. (1987) *Cell Motil. Cytoskeleton* **8**, 210–226.
Paykel, E.S., Brayne, C., Huppert, F.A., Gill, C., Barkley, C., Gehlhaar, E., Beardsall, L., Girling, D.M., Pollitt, P. & O'Connor, D. (1994) *Arch. Gen. Psychiatry* **51**, 325–332.
Pearson, B., Choi, T. *Proc. Natn. Acad. Sci. USA* (1993) **90**, 10578–10582
Perry, G., Rizzuto, N., Autilio-Gambetti, L. & Gambetti, P. (1985) *Proc. Natl Acad. Sci. USA* **82**, 3916–3920.
Perry, G., Friedman, R., Shaw, G. & Chau, V. (1987) *Proc. Natl Acad. Sci. USA* **84**, 3033–3036.
Pierre, M. & Nunez, J. (1983). *Biochem. Biophys. Res. Commun.* **115**, 212–219.
Poulter, L., Barratt, D., Scott, C.W. & Caputo, C.B. (1993) *J. Biol. Chem.* **268**: 9636–9644.

Quon, D., Wang, Y., Catalano, R., Marian Scardina, J., Murakami, K., Cordell, B. (1991) *Nature* **352**, 239–241
Rasool, C.G., Abraham, C., Anderton, B.H., Haugh, M.C., Kahn, J. & Selkoe, D.J. (1984) *Brain Res.* **310**, 249–260.
Riederer, B.M. & Innocenti, G.M. (1991) *Eur. J. Neurosci.* **3**, 1134–1145.
Rodder, H.M. & Ingram, V.M. (1991) *J. Neurosci.* **11**, 3325–3343.
Rubenstein, R., Kascsak, R.J., Merz, P.A., Wisniewski, H.M., Carp, R.I. & Iqbal, K. (1986) *Brain Res.* **372**, 80–88.
Selkoe, D.J., Ihara, Y. & Salazar, F.J. (1982) *Science* **215**, 1243–1245.
Shankar, S.Y., Yanagihara, R., Garruto, R.M., Grundke-Iqbal, I., Kosik, K.S. & Gajdusek, D.C. (1989) *Ann. Neurol.* **25**, 146–151.
Sherrington, R., Rogaev, E., Liang, Y., Rogaeva, E., Levesque, G., Ikeda, M., Chi, H., Lin, C., Holman, K., Tsuda, T., Mar, L., Foncin, J-F., Bruni, A., Montesi, M., Sorbi S., Rainero, I., Pinessi, L., Nee, L., Chumakov, I., Pollen, D., Brookes, A., Sanseau, P., Polinsky, R., Wasco., W., da Silva, H., Haines, J., Pericak-Vance, M., Tanzi, R., Roses, A., Fraser, P., Rommens, J., St. George-Hyslop, P. (1995) *Nature* **375**, 754–760.
St George-Hyslop, P.H., Tanzi, R.E., Polinsky R.J., Haines, J.L., Nee, L., Watkins, P.C., Myers, R.H., Feldman, R.G., Pollen, D., Drachman, D., Growdon, J., Bruni, A., Foncin, J.-F., Salmon, D., Frommelt, P., Amaducci, L., Sorbi, S., Piacentini, S., Stewart, G.D., Hobbs, W.J., Conneally, P.M. & Gusella, J.F. (1987) *Science* **135**, 885–890.
Steiner, B., Mandelkow, E.M., Biernat, J., Gustke, N., Meyer, H.E., Schmidt, B., Mieskes, G., Soling, H.D., Drechsel, D., Kirschner, M.W., Goedert, M. & Mandelkow, E. (1990) *EMBO J.* **9**, 3539–3544.
Sternberger, N.H., Sternberger, L.A. & Ulrich, J. (1985). *Proc. Natl Acad. Sci. USA* **82**, 4272–4276.
Strittmatter, W.J., Saunders, A.M., Schmechel, D., Pericak-Vance, M., Enghild, J., Salvesen, G.S. & Roses, A.D. (1994) *Proc. Natl Acad. Sci. USA* **90**, 1977–1981.
Suzuki, N., Cheung, T., Cai, X-D., Odaka, A., Otvos, L., Eckman, C., Golde, T., Younkin, S. (1994) *Science* **264**, 1336–1340.
Terry, R.D. (1963) *J. Neuropathol. Exp. Neurol.* **22**, 629–641.
Terry, R.D., Masliah, E., Salmon, D.P. *et al.* (1992) *Ann. Neurol.* **30**, 572–580.
Van Hoesen, G.W. & Damasio, A.R. (1990) In *Handbook of Physiology. The Nervous System V.* pp. 871–898.
Vogelsang, G.D., Zemlan, F.P. & Dean, G.E. (1988) In *The Molecular Biology of Alzheimer's Disease* (eds. Finch, C.E. & Davies, T.), Current Communications in Molecular Biology, pp. 150–158, Cold Spring Harbor, New York, Cold Spring Harbor Publications.
Vulliet, R., Halbran, S.M., Braun, R.K., Smith, A.J. & Lee, G. (1992) *J. Biol. Chem.* **267**, 22570–22574.
Whalley, L.J. (1982) *Ann. N.Y. Acad. Sci.* **396**, 39–53.
Wischik, C.M., Crowther, R.A., Stewart, M. & Roth, M. (1985) *J. Cell Biol.* **100**, 1905–1912.
Wischik, C.M., Novak, M., Thogersen, H.C., Edwards, P.C., Runswick, M.J., Jakes, R., Walker, J.E., Milstein, C., Roth, M. & Klug, A. (1988a) *Proc. Natl Acad. Sci. USA* **85**, 4506–4510.
Wischik, C.M., Novak, M., Edwards, P.C., Klug, A., Tichelaar, W. & Crowther, R.A. (1988b) *Proc. Natl Acad. Sci. USA* **85**, 4884–4888.
Wischik, C.M., Harrington, C.R., Mukaetova-Ladinska, E.B., Lai, R., Edwards, P. & Roth, M. (1995) *Neurobiol. Aging* **16**, 409–431.
Wolozin, B.L., Pruchnicki, A., Dickson, D.W. & Davies, P. (1986) *Science* **232**, 648–650.
Wood, J.N. & Kahn, J. (1982) *Nature* **298**, 84–86.
Wood, J.G., Mirra, S.S., Pollock, N.J. & Binder, L.I. (1986) *Proc. Natl Acad. Sci. USA* **83**, 4040–4043.
Yamamoto, H., Fukunaga, K., Tanaka, E. & Miyamoto, E. (1983) *J. Neurochem.* **41**, 1119–1126.
Yen, S.-H., Gaskin, F. & Fu, S.M. (1983) *Am. J. Pathol.* **113**, 373–381.
Yen, S.-H., Dickson, D.W., Crowe, A. *et al.* (1987) *Am. J. Pathol.* **126**, 81–91.

CHAPTER 3

TAU PROTEINS AND THEIR SIGNIFICANCE IN THE PATHOBIOLOGY OF ALZHEIMER'S DISEASE

Virginia M.-Y. Lee and John Q. Trojanowski

Department of Pathology and Laboratory Medicine, Division of Anatomic Pathology, University of Pennsylvania School of Medicine, Philadelphia, Pennsylvania 19104, USA

Table of Contents

Summary

Alzheimer's disease (AD), the most common cause of dementia in the elderly, is characterized by several different types of abnormalities that affect selectively vulnerable regions of the telencephalon. The major abnormalities include the deposition of amyloid β-protein (Aβ) in the extracellular space, the massive loss of cortical neurons and the accumulation of paired helical filaments (PHFs) in neurofibrillary tangles, dystrophic neurites and neuropil threads. The subunit proteins of PHFs are derivatized forms of central nervous system (CNS) τ proteins known collectively as PHFτ. Relative

Pathobiology of Alzheimer's Disease
ISBN 0-12-286965-5

to adult CNS τ proteins, PHFτ is excessively phosphorylated and far more resistant to proteolysis than its normal counterpart although CNS τ is phosphorylated at sites that are similar to those found in PHFτ. Based on these and other differences between PHFτ and normal CNS τ proteins, the formation of neurofibrillary lesions in AD may be the result of the abnormal or altered regulation of the phosphorylation state of adult CNS τ in the AD brain. In this review, we summarize current understanding of the pathogenesis of neurofibrillary lesions in AD. We then consider how the incorporation of abnormally phosphorylated CNS τ into PHFs could disrupt the microtubule network in neurons, impair axonal transport and compromise the function and viability of neurons, thereby contributing to the development and progression of AD.

3.1 Introduction

Over the past 10 years, several different lines of research have provided increasing support for the hypothesis that Alzheimer's disease (AD) is not a unitary disorder, but rather a heterogeneous group of dementing illnesses that share several common clinical and pathological hallmarks. As illustrated in Figure 1, this notion has been reinforced by a number of recent studies of the molecular genetics of familial (FAD) and sporadic AD that implicate genes on at least three different chromosomes (i.e. 14, 19 and 21) in the aetiology of FAD (for reviews and additional citations, see Ashall and Goate, 1993; Hardy, 1993; Mullan and Crawford, 1993; Price *et al.*, 1993; Selkoe, 1993). Additionally, apolipoprotein E (ApoE) has been identified as a risk factor for familial and sporadic AD (Saunders *et al.*, 1993; Strittmatter *et al.*, 1994). For example, individuals who are homozygous for the ApoE4 allele have a far greater susceptibility to develop sporadic AD than individuals who are homozygous for the ApoE2 or ApoE3 allele. Further, mutations in the amyloid-β protein precursor (APP) have been detected in a small number of FAD kindreds (Ashall and Goate, 1993; Hardy, 1993; Mullan and Crawford, 1993; Selkoe, 1993). These important findings implicate mutations in the APP gene in the aetiology of FAD. Nonetheless, missense mutations in the APP gene do not account for the vast majority (i.e. >99%) of sporadic and familial AD cases. Thus, AD exhibits considerable phenotypic and genotypic heterogeneity.

If AD indeed represents a group of chronic progressive neurodegenerative conditions each of which is the result of different aetiologies, this concept will have important implications for research efforts to develop effective therapies for this common dementing illness of the elderly. For example, it might be necessary to elucidate the aetiology and pathogenesis of each variant of AD in order to design effective treatments that will prevent or block the progression of each form of this dementia of late life. Accordingly, AD may turn out to be like cancer or infectious diseases, which require different or individualized therapies for each type of cancer or infectious agent.

While the development of individualized therapies for different variants of AD may be a daunting but unavoidable undertaking, it also might be possible to focus success-

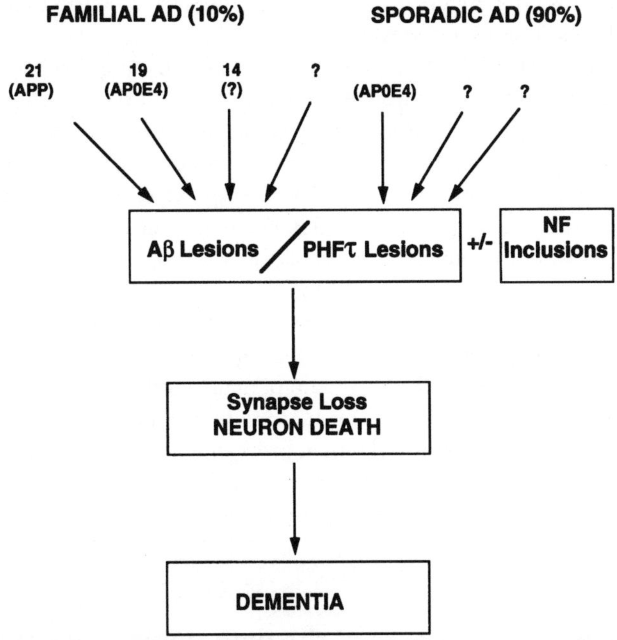

Figure 1 This figure summarizes putative risk factors or aetiologies for AD that converge to generate characteristic AD brain lesions caused by the accumulation of Aβ and PHF. Aβ deposits and PHFτ-rich neurofibrillary lesions may be a final common pathway leading to neuron death and dementia in AD. Lewy bodies comprised of intraneuronal accumulations of neurofilament (NF) proteins (designated NF inclusions here) occur in some cases of AD known as the Lewy body variant of AD. The sequence of events shown here leading from putative risk factor or aetiology to the loss of synpases and neurons followed by dementia is hypothetical, but familial AD accounts for about 10% of cases and is linked to chromosome 21 (locus of the APP gene), 19 (locus of the ApoE gene) and 14. The ApoE4 allele (ApoE4) on chromosome 19 is a risk factor for FAD and sporadic AD.

ful and effective drug discovery on alternative strategies that target neurodegenerative disease mechanisms common to all forms of AD even though they may be invoked or activated 'downstream' from a given aetiological event. For example, since AD is defined operationally as a diagnostic entity if an elderly patient exhibits clinical evidence of a chronic progressive dementia and post-mortem examination reveals abundant senile plaques (SPs) and neurofibrillary tangles (NFTs) in the brain (Khachaturian, 1985), these AD lesions may well represent 'final common pathways' that could serve as therapeutic 'targets'. This would be a plausible strategy if the formation of plaques and tangles is the site of convergence of different aetiological events that cause AD, and if plaques and tangles induce a cascade of cellular perturbations that inexorably lead to the degeneration of neurons and the emergence of the dementia in AD. Although the pathogenesis of plaques and tangles is far from well understood at this time, Figure 2 provides a hypothetical scenario to account for the formation of these lesions in the AD brain.

The deposition of Aβ peptides in diffuse and neuritic amyloid plaques as well as the accumulation of NFTs are not restricted to AD, and deposits of Aβ may be as profuse in cognitively intact elderly individuals as they are in AD patients (Dickson *et al.*, 1991; Arriagada *et al.*, 1992; Goedert, 1993; Trojanowski *et al.*, 1993b). However, there is a close correlation between the burden of NFTs and the dementia in AD (Dickson *et al.*, 1991; Arriagada *et al.*, 1992). Further, while neuronal cytoskeletal proteins known as τ are the building blocks of NFTs, related neurofibrillary lesions are also formed from τ proteins and they too are abundant in AD brains (for recent reviews, see Goedert, 1993; Trojanowski *et al.*, 1993a, b; Trojanowski and Lee, 1994b). Indeed, accumulating evidence suggests that alterations in the phosphorylation of CNS τ proteins could lead to the formation of neurofibrillary lesions in AD. Here we review recent insights into mechanisms that might account for the disruption of the normal metabolism and post-translational processing of CNS τ in AD. Additionally, we also discuss how neurons might become dysfunctional and degenerate as a result of the abnormal metabolism of τ proteins in CNS neurons.

3.2 Molecular components of the pathological signatures of Alzheimer's disease

Paired helical filaments are found aggregated together (in addition to straight filaments and other granular or amorphous cellular debris or material) into NFTs in the perikarya of selectively vulnerable populations of neurons in AD. Despite some similarities between the amyloid fibrils of SPs and PHFs in NFTs, amyloid fibrils and PHFs are composed of different proteins. Specifically, amyloid fibrils are formed from peptides 39–43 amino acid long (designated Aβ) that are normal cleavage products of one or more of the alternatively spliced APPs (reviewed in Ashall and Goate, 1993; Hardy, 1993; Mullan and Crawford, 1993; Selkoe, 1993) while PHFs are composed of derivatized forms of each of the normal adult CNS τ proteins referred to here as PHFτ (reviewed in Goedert, 1993; Trojanowski *et al.*, 1993a, b; Trojanowski and Lee, 1994b).

Figure 2 This two-part figure illustrates hypothetical mechanisms for the generation of neurofibrillary lesions (a) and the deposition of Aβ in amyloid plaques (b).

(a) The conversion of normal brain τ (rectangles overlying 2 MTs on the left) into PHFτ (A68). PHFτ is generated in neuronal perikarya and their processes as a consequence of the overactivity or inappropriate activation of kinase(s) and/or the hypoactivity or loss of phosphatase(s) that regulate the phosphorylation state of CNS τ. PHFτ then accumulates in neuronal processes as NTs (top), in neuronal perikaryal as NFTs (middle) and in plaque-associated dystrophic neurites (bottom). PHFτ loses the ability to bind MTs, and this could lead to the depolymerization of MTs, the disruption of axonal transport and the dysfunction and/or

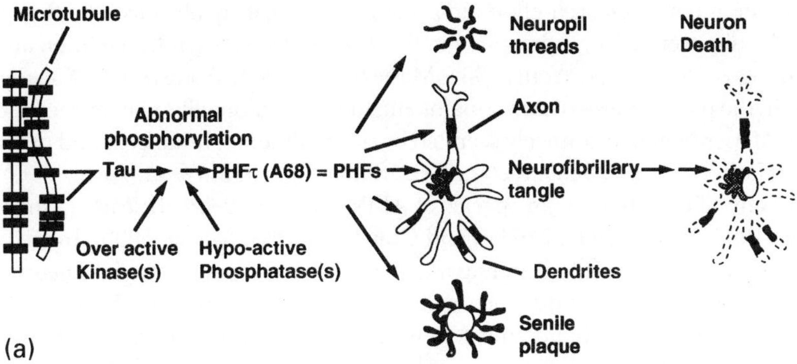

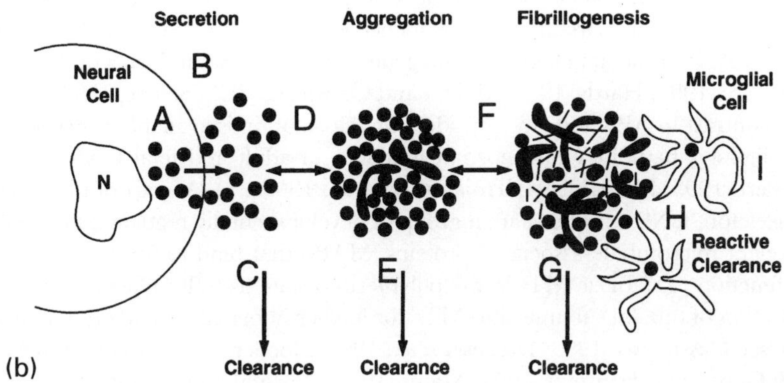

degeneration of neurons in AD. The accumulation of PHFs in neurons could exacerbate this process by sequestering normal τ and by physically blocking the transport of proteins and organelles in neuronal perikarya, axons and dendrites. The death of the neuron (on the right) would be the ultimate outcome of all of these events.

(b) Soluble Aβ (s-Aβ) peptides (depicted by the filled circles here) are produced in neural cells (e.g. neurons) in the brain (N=nucleus of a cell) and secreted (A) into the extracellular space (B) from whence they are cleared in the brains of normal younger individuals (C). With advancing age, s-Aβ aggregates (D) into so-called diffuse amyloid plaques (closely packed filled circles to the right of the D). Some aggregated s-Aβ also may be cleared from these plaques (E). In AD, s-Aβ in diffuse plaques may come into contact with PHFτ proteins (filled curvilinear profiles) that have been released from degenerating neurons or their processes, as well as with other factors (e.g. proteoglycans) in the extracellular space (filled curvilinear profiles). As a consequence of these interactions, s-Aβ may be induced to form fibrillar Aβ (f-Aβ) aggregates (F) that lead to the formation of SPs (to the right of the F), which accumulate or 'trap' many additional components including more s-Aβ (filled circles), PHFτ (curvilinear profiles) and f-Aβ (short linear profiles), in addition to other fibrillogenic factors (curvilinear profiles), microglial cells (I) and reactive astrocytes (H). Nonfibrillar or s-Aβ (and possibly f-Aβ) might be cleared from SPs (G), but the continued presence of factors that induce the fibrillogenesis and insolubility of s-Aβ would favour the persistence of SPs in the AD brain.

Despite intense research efforts into the pathological significance of Aβ deposits since the discovery of the Aβ peptide in the AD and Down's syndrome brain about a decade ago (Glenner and Wong, 1984; Masters *et al.*, 1985; Wong *et al.*, 1985), the role of Aβ in the pathogenesis of AD remains enigmatic. Additionally, neither the function of the Aβ peptides constitutively secreted by normal human brain cells and other cell types (Haas *et al.*, 1992; Seubert *et al.*, 1992; Shoji *et al.*, 1992; Wertkin *et al.*, 1993), nor that of any of the three major species of APPs, has been defined clearly (Ashall and Goate, 1993; Hardy, 1993; Mullan and Crawford, 1993; Selkoe, 1993). Indeed, one of the conundrums in understanding the role of Aβ in the pathogenesis of neuron loss or dysfunction in AD is the presence of abundant deposits of Aβ in the brains of elderly individuals who show no ante-mortem evidence of cognitive deficits (Dickson *et al.*, 1991). On the other hand, the early and excessive accumulation of AD pathology in Down's syndrome patients with a trisomy of chromosome 21, and the presence of mutations in the APP gene within or flanking the Aβ domain in a subset of FAD kindreds, provide strong circumstantial evidence for the involvement of Aβ and altered forms of APP in the aetiology or pathogenesis of at least some forms of AD (Ashall and Goate, 1993; Hardy, 1993; Mullan and Crawford, 1993; Selkoe, 1993).

In contrast to APP and Aβ, considerable information is available concerning the basic biology, pathology and normal functions of adult and fetal CNS τ proteins (Goedert, 1993; Matus, 1994; Trojanowski *et al.*, 1993a, b). As part of the neuronal cytoskeleton, CNS τ proteins are a group of developmentally regulated low molecular weight microtubule-associated proteins (MAPs) that bind to microtubules (MTs) and function to stabilize MTs in the polymerized state as well to facilitate the polymerization of tubulin subunits into MTs (for further information and additional citations, see Cassimeris, 1993; Drechsel *et al.*, 1992; Goedert, 1993; Ludin and Matus, 1993; Goode and Feinstein, 1994; Matus, 1994). Normal adult human brain τ consists of six alternatively spliced proteins encoded by the same gene, and each τ isoform contains either three or four consecutive MT binding motifs that are imperfect repeats of 31 or 32 amino acids (Goedert *et al.*, 1989a, b). Further, human τ isoforms differ with respect to the presence or absence of inserted sequences in the amino-terminal third of τ that are 29 or 58 amino acids in length and of unknown function. Accordingly, the largest τ isoform is 441 amino acids long (with four tandemly repeated MT binding motifs and a 58 amino acid long amino terminal insert), while the smallest τ isoform is 352 amino acids long (with three consecutive MT binding repeats and no amino terminal insert). So-called 'fetal' τ is the shortest τ isoform, and it is expressed early in the developing human nervous system, while all six alternatively spliced τ isoforms (including 'fetal' CNS τ) are expressed in the adult human brain (Bramblett *et al.*, 1993; Goedert *et al.*, 1989a, b; Goedert, 1993; Goedert *et al.*, 1993; Lee *et al.*, 1993). These observations on the delayed expression of five of the six alternatively spliced CNS τ isoforms until post-natal life suggest that the developmental regulation of the number of CNS τ isoforms and their phosphorylation state (see below) may allow for greater flexibility in establishing an appropriate equilibrium between polymerized and depolymerized MTs during the complex series of molecular events that transforms the immature CNS into the mature CNS.

3.3 Formation of neurofibrillary lesions from derivatized forms of CNS τ (PHFτ) in Alzheimer's disease

A number of studies have demonstrated that PHFs are abundant in neuronal perikarya where they are the major structural elements of NFTs, but PHFs also have been shown to accumulate in dystrophic processes (Goedert, 1993; Trojanowski *et al.*, 1993a; Trojanowski and Lee, 1994b). These processes are known as neuropil threads (NTs) when they are dispersed throughout the CNS grey matter. Similar dystrophic neurites are also intermingled with Aβ fibrils in neuritic and diffuse amyloid plaques as well as in the olfactory epithelium (Braak *et al.*, 1986; Ihara, 1988; Shin *et al.*, 1989; Talamo *et al.*, 1989; Arai *et al.*, 1990, 1992; Perry *et al.*, 1990; Schmidt *et al.*, 1990, 1991; Tabaton *et al.*, 1991; Trojanowski *et al.*, 1991; Masliah *et al.*, 1992; Lee *et al.*, 1993). After the initial detection of immunoreactive τ in AD NFTs and in PHFs (Brion *et al.*, 1985; Delacourte and Defossez, 1986; Grundke-Iqbal *et al.*, 1986; Kosik *et al.*, 1986, 1988; Wood *et al.*, 1986), biochemical studies showed that fragments of τ could be isolated from PHFs purified from the AD brain (Goedert *et al.*, 1988; Kondo *et al.*, 1988; Wischik *et al.*, 1988). These studies rapidly confirmed and further refined the emerging notion that τ was involved in the formation of neurofibrillary lesions. For example, they provided compelling evidence that PHFs themselves were formed from polymers of partially proteolysed or full-length τ proteins. Further, they showed that PHFτ was abnormally phosphorylated relative to normal isoforms of human τ obtained from post-mortem samples of adult CNS tissues.

However, controversy about the identity of the subunit proteins of PHFs persisted until sodium dodecyl sulphate (SDS)-soluble PHFs were purified from AD brains and shown to consist of CNS τ proteins that were abnormally phosphorylated at specific residues (e.g. at Ser^{396}) in PHFτ but not in normal τ proteins isolated from the adult post-mortem brain (Lee *et al.*, 1991). Validation of these results came rapidly from a number of different laboratories (Bramblett *et al.*, 1992, 1993; Brion *et al.*, 1991; Ksiezak-Reding and Yen, 1991; Greenberg *et al.*, 1992; Shin *et al.*, 1992; Goedert *et al.*, 1993). Furthermore, AD PHFs were shown to contain variable amounts of all six alternatively spliced human CNS τ proteins (Goedert *et al.*, 1992c). Subsequent studies of recombinant fragments of human τ that were comprised of non-phosphorylated MT binding repeats demonstrated that these fragments were able to polymerize into PHF-like structures under *in vitro* conditions that required an acidic pH (Crowther *et al.*, 1992; Wille *et al.*, 1992). More recently, full-length recombinant human τ with four MT binding repeats (clone htau40, which corresponds to the 441 amino acid long adult CNS τ isoform) was shown to polymerize into PHF-like structures *in vitro* at neutral pH (Crowther *et al.*, 1994). Notably, when these proteins were phosphorylated *in vitro* with brain extracts to resemble PHFτ biochemically and immunochemically, the PHFs formed from these proteins more closely resembled authentic PHFs *in situ* in the AD brain than the filaments formed from the non-phosphorylated recombinant τ proteins.

Thus, by the end of 1992, all of the alternatively spliced CNS τ proteins had been convincingly shown to be the major, if not the sole, building blocks of the PHFs found in the major neurofibrillary lesions of the AD brain including NFTs and dystrophic neurites. However, the highest molecular weight isoform of τ (which contains a unique amino-acid insert in its amino-terminal region) is found in peripheral nervous system neurons, and it does not appear to be converted into PHFτ (Couchie *et al.*, 1992; Goedert *et al.*, 1992b). Taken together, these studies suggest that the generation of PHFτ results from the aberrant or excessive phosphorylation of normal adult CNS τ (see Figure 2). To understand the significance of these events for the pathogenesis of AD, it became apparent that the mechanisms regulating the phosphorylation states of τ need to be elucidated (Lee, 1991; Lee and Trojanowski, 1992; Goedert, 1993; Trojanowski *et al.*, 1993b; Trojanowski and Lee, 1994b).

To begin to understand the mechanisms whereby normal adult CNS τ is converted into PHFτ in AD, a number of laboratories have sought to identify the sites of aberrant phosphorylation that distinguish PHFτ from normal adult CNS τ (Biernat *et al.*, 1992; Lee *et al.*, 1991; Drewes *et al.*, 1992; Hasegawa *et al.*, 1992; Bramblett *et al.*, 1993; Goedert *et al.*, 1993). Several approaches have been used to address this question. For example, two Ser residues found in all CNS τ isoforms that are abnormally phosphorylated in PHFτ relative to adult human CNS τ were identified (i.e. Ser^{202} and Ser^{396}) using immunological methods, synthetic τ phospho-peptides and recombinant human τ subjected to site-directed mutagenesis and *in vitro* phosphorylation (Lee *et al.*, 1991; Lang *et al.*, 1992; Bramblett *et al.*, 1993; Goedert *et al.*, 1993). The numbering system of Goedert *et al.* (1989b) for the longest human CNS τ isoform was used in these studies, which showed that the abnormally phosphorylated Ser^{202} and Ser^{396} residues in PHFτ reside within the epitopes recognized by two well-characterized PHFτ-specific mAbs known as AT8 and PHF1, respectively. Both of these antibodies have been extremely useful in mapping phosphorylated residues in PHFτ because the epitope detected by each of them has been identified. For example, AT8 recognizes an epitope containing phosphorylated Ser^{202} in PHFτ (Goedert *et al.*, 1993). Additionally, the PHF1 mAb has been shown to recognize phosphorylated Ser^{396} in PHFτ just like the T3P antibody (Lee *et al.*, 1991; Lang *et al.*, 1992; Bramblett *et al.*, 1993), but PHF1 also appears to recognize phosphorylated Ser^{404} in PHFτ (Otvos *et al.*, 1994). Although the AT8 mAb was thought to be complementary to the Tau 1 mAb, recent epitope mapping studies suggest that the Tau 1 epitope differs somewhat from the AT8 epitope (Szendrei *et al.*, 1993). Specifically, Tau 1 recognizes dephosphorylated PHFτ and normal adult CNS τ, but not native PHFτ (Goedert *et al.*, 1993), and Tau 1 binds to an epitope contained within amino acids 189–215 only when none of the Ser residues within this region are phosphorylated (Szendrei *et al.*, 1993).

Another approach used to identify normal and abnormal phosphorylation sites in PHFτ has been mass spectrometry, and protein or peptide sequencing. Using this strategy, it has been shown that in addition to Ser^{202}, Ser^{396} and Ser^{404}, PHFτ also is abnormally phosphorylated at Thr^{181}, Thr^{231}, Ser^{235} and Ser^{262} relative to adult postmortem brain-derived CNS τ (Hasegawa *et al.*, 1992). However, it should be noted that Ser^{262} was observed to be phosphorylated only in a fraction of PHFτ. Thus, Ser^{262} may represent a secondary or less constant site of abnormal phosphorylation in PHFτ.

Taken together, these studies indicate that all of the major sites of abnormal phosphorylation in PHFτ are Ser/Pro or Thr/Pro sites. Since there are 17 Ser/Pro or Thr/Pro sites in the largest human brain τ isoform and 14 of these sites are shared by all six CNS τ proteins, it is important now to determine which of these sites are aberrantly phosphorylated in PHFτ (relative to fetal and adult human CNS τ), and which sites interfere with the normal function of τ when they are aberrantly phosphorylated. Given the proximity of Ser^{396} and Ser^{404} to the MT-binding domain just carboxy terminal to the last MT binding repeat, as well as the inverse relationship between the extent to which τ is phosphorylated and the ability of τ to bind MTs, it is not surprising that the aberrant phosphorylation of Ser^{396} in PHFτ relative to normal adult human CNS τ has been implicated in the loss of the ability of PHFτ to bind to MTs (Bramblett *et al.*, 1993). Nonetheless, this loss of function also could result from the abnormal or excessive phosphorylation of other Ser or Thr residues in τ.

Given the important role that abnormal phosphorylation may play in converting normal adult brain τ into PHFτ thereby altering the normal functions of CNS τ, it was germane to compare the phosphorylation state of PHFτ with both adult and fetal CNS τ proteins. When such studies were carried out using post-mortem brain samples as sources of τ proteins, it became apparent that the phosphorylation state of PHFτ partially recapitulates the phosphorylation state of fetal CNS τ (Kanemura *et al.*, 1992; Bramblett *et al.*, 1993; Goedert *et al.*, 1993; Lee *et al.*, 1993). For example, Ser^{202} and Ser^{396} are phosphorylated in PHFτ and in the smallest τ isoform when it is expressed in the fetal CNS, but not in any of the six τ isoforms isolated from post-mortem samples of the normal adult CNS (Bramblett *et al.*, 1993; Goedert *et al.*, 1993). Despite the fact that fetal human CNS τ is phosphorylated at Ser^{202} and Ser^{396} like PHFτ, fetal human τ is capable of binding to MTs (albeit less well than adult CNS τ proteins), while PHFτ completely loses the ability to bind MTs (Bramblett *et al.*, 1993). However, this loss of function is reversible, since dephosphorylation of PHFτ restores the ability of PHFτ to bind MTs. Similar observations have been made in comparative studies of PHFτ and fetal rat τ (Yoshida and Ihara, 1993). In addition, the normal programmed death of large numbers of neurons in the developing CNS occurs without any of the neurofibrillary lesions that are found in the AD brain. These conclusions are based on studies of several different regions of the developing human CNS including regions of the adult brain that are unaffected (e.g. spinal cord; Tohyama *et al.*, 1991), mildly affected (e.g. cerebellum; Yachnis *et al.*, 1993) or severely affected (e.g. hippocampus; Arnold and Trojanowski, 1994) by the accumulation of amyloid and neurofibrillary AD lesions.

The studies summarized above suggest that the accumulation of neurofibrillary lesions in AD may be caused by the aberrant reactivation of fetal protein kinases and the inactivation of fetal protein phosphatases in the AD brain. The absence of any AD lesions in the fetal CNS suggests that these kinases and phosphatases, which are normally present and activated in the fetal nervous system, determine the phosphorylation state of fetal τ but do not generate derivatized forms of τ that are fully equivalent to PHFτ in the AD brain. Accordingly, it is timely now to undertake studies of how the phosphorylation state of fetal τ is regulated in the normal fetal human brain, since such information could help elucidate how adult CNS τ is converted into PHFτ in the AD brain.

In future studies conducted with similar analytical strategies, it should be possible to identify all of the abnormal phosphorylation sites that distinguish PHF τ from normal τ in the fetal and adult CNS. In fact, mass spectrometry and sequencing of fetal and adult rat CNS τ were performed recently; and, despite considerable heterogeneity in the results, these studies showed that fetal τ harboured ten phosphorylated Ser or Thr residues compared to five Ser or Thr phospho-residues in adult brain τ (Watanabe *et al.*, 1993). Notably, almost all of these residues were followed by Pro residues, suggesting that they are phosphorylated by proline-directed kinases. The accomplishment of a similar type of study of human adult and fetal CNS τ is important because it will facilitate efforts to identify the kinases and phosphatases that are involved in the generation of PHFτ in the AD brain (Figure 2). This goal is all the more important to attain now in view of emerging data from studies of human brain biopsy-derived CNS τ, which indicate that freshly isolated adult human τ (i.e. τ proteins isolated with little or no interval between tissue excision and analysis) may be phosphorylated at almost all of the same sites as PHFτ albeit to a far lesser extent at these sites than PHFτ (Garver *et al.*, 1994; Matsuo *et al.*, 1994). Among other things, these data suggest that some of the differences previously noted between normal adult CNS τ and PHFτ that were obtained from studies of post-mortem brain samples may reflect the more rapid dephosphorylation of normal τ versus PHFτ during similar intervals between the death of the patient and the extraction of each of these species of τ protein. Whether or not this is caused by the reduced levels or activity of specific phosphatases in the AD brain remains to be determined.

3.4 Kinases and phosphatases potentially involved in the conversion of normal CNS τ into PHFτ in Alzheimer's disease

Indeed, several kinases have been shown to phosphorylate normal τ *in vitro* at some of the same sites that are phosphorylated in PHFτ (Goedert, 1993; Trojanowski *et al.*, 1993a; Trojanowski and Lee, 1994b). Some of the candidate proline-directed protein kinases that have been implicated in the phosphorylation of PHFτ in this manner include: (1) mitogen-activated protein (MAP) kinases, which also are known as extracellular signal-regulated kinases (ERKs) (Drewes *et al.*, 1992; Goedert *et al.*, 1992a; Ledesma *et al.*, 1992; Trojanowski *et al.*, 1993c; Hyman *et al.*, 1994; Trojanowski and Lee, 1994a); (2) glycogen synthase kinase-3 (Hanger *et al.*, 1992; Mandelkow *et al.*, 1992; Ishiguro *et al.*, 1993); (3) a proline-directed protein kinase (PDPK) (Vulliet *et al.*, 1992); and (4) the cyclin-dependent kinases CDK2 and CDK5 (Baumann *et al.*, 1993; Paudel *et al.*, 1993). In addition, several other kinases have been shown to phosphorylate normal τ *in vitro*, but not at sites that have been documented to be phosphorylated in PHFτ, fetal τ or adult τ (Hasegawa *et al.*, 1992; Watanabe *et al.*, 1993). These include such kinases as calcium/calmodulin-dependent protein kinase or CaMK II (Steiner *et al.*, 1990; Wood *et al.*, 1993) and cyclic AMP-dependent protein kinase (PKA) (Robertson *et al.*, 1993; Scott *et al.*, 1993). Thus, the role of these kinases in regulating the phosphorylation state of τ *in vivo* remains to be clarified. Finally, recent studies suggest that the protein tyrosine kinase *fyn* may regulate the activity of candidate proline-directed kinases involved in the abnormal phosphorylation of CNS τ, and thus may be involved in abnormal phosphorylation cascades in the AD brain (Shirazi and Wood, 1993).

Presently there is even less information on the protein phosphatases that are potentially involved in the conversion of normal CNS τ into aberrantly or excessively phosphorylated PHFτ in the AD brain. However, the preliminary studies conducted to date have suggested that protein phosphatase 2A1 (PP2A1) and protein phosphatase 2B (PP2B or calcineurin) may be implicated in the conversion of normal τ into PHFτ (Goedert *et al.*, 1992a; Drewes *et al.*, 1993; Gong *et al.*, 1993, 1994; Harris *et al.*, 1993; Billingsley *et al.*, 1994 Matsuo *et al.*, 1994). Nonetheless, given the complexity of the biology of Ser/Thr protein phosphatases and their regulatory subunits in different tissues and cell types (reviewed in Mumby and Walter, 1993), and the relatively scant amount of information that exists on the protein phosphatases in the CNS (reviewed in Sim, 1991), it is clear that much more needs to be learned about these enzymes in the brain before their role in AD can be assessed. Indeed, it is highly likely that several different kinases and phosphatases will be shown to be involved simultaneously or sequentially in the progressive conversion of normal τ into PHFτ. Further, normal interactions between a kinase or phosphatase and τ as a substrate could be perturbed by alterations in the axonal transport of τ or of one or more of these enzymes, and this might explain the formation of different types of PHFτ lesions (i.e. tangles versus dystrophic neurites) at different sites (i.e. perikarya versus processes) within neurons. If these speculations prove to be correct, then it may be necessary to elucidate the contributions and inter-relationships of all of these enzymes in the generation of PHFτ before it is possible to understand fully the role of neurofibrillary pathology in the premature and excessive loss of neurons in AD, and the dementia that is so characteristic of this disorder.

3.5 Functional implications of neurofibrillary lesions in Alzheimer's disease

Although current understanding of the detailed pathobiology of AD and the significance of neurofibrillary lesions in this disease is incomplete at this time, the available information suggests that the conversion of normal τ into PHFτ might have deleterious effects on neurons during the progression of AD. For example, it is well known that the accumulation of PHFτ in AD cortex correlates with the abundance of NFTs as well as with diminished levels of normal MT binding competent τ in the CNS (Bramblett *et al.*, 1992, 1993; Khatoon *et al.*, 1992). Further, since abnormal or excessively phosphorylated PHF τ is unable to bind to MTs (Bramblett *et al.*, 1993; Yoshida and Ihara, 1993), we have proposed that the conversion of normal τ into PHFτ could lower the levels of MT binding τ, destabilize MTs, disrupt axonal transport and lead to the 'dying back' of axons in AD (Lee, 1991; Lee and Trojanowski, 1992; Trojanowski *et al.*, 1993a, b; Trojanowski and Lee, 1994a, b). Additionally, disruption of the MT network in neurons could alter the compartmentalization, or targeting and translation of mRNA in different neuronal domains (Bassell *et al.*, 1994). Thus, MTs, CNS τ or PHFτ could be therapeutic targets for the treatment of AD. Specifically, drugs that cross the blood–brain barrier and stabilize axonal MTs, enhance the ability of residual, normal CNS τ to bind and stabilize MTs, or reverse the phosphorylation state of PHFτ, such that it is more like normal τ, might be effective independently or in combination as therapeutic agents to retard the progression of AD.

These speculations are consistent with current views of the regulation of tubulin assembly, the functions of MTs, and the role of τ proteins in promoting the assembly and stability of MTs (Cassimeris, 1993; Drechsel *et al.*, 1992; Goedert *et al.*, 1993; Ludin and Matus, 1993; Matus, 1994). Hence, as a result of the hypothetical scenario outlined here and in Figure 2, the progressive conversion of normal CNS τ into PHFτ in the AD brain could lead to the degeneration or loss of cortico-cortical connections and incremental impairments of synaptic transmission followed by the emergence of cognitive impairments. Furthermore, the accumulation of abundant PHFs in neuronal perikarya and processes could exacerbate the sequence of events outlined above by physically obstructing orthograde and retrograde intraneuronal transport. Whether or not the aberrant phosphorylation of CNS τ is essential for the formation of PHFs and the neurofibrillary lesions in which these abnormal filaments accumulate remains to be determined because the studies of Crowther and co-workers (1994) suggest that recombinant full-length τ can form PHF-like structures when it is in a nonphosphorylated state.

As an extension of these considerations and hypotheses about events leading to the formation of PHFs in AD, we have proposed (Shin *et al.*, 1993; Trojanowski *et al.*, 1993a, b; Trojanowski and Lee, 1994b) that the release of PHFτ from dying neurons or dystrophic neurites could result in interactions between PHFτ and the soluble Aβ normally secreted by neurons or other CNS cells (Seubert *et al.*, 1992; Shoji *et al.*, 1992; Wertkin *et al.*, 1993). Indeed, some experimental data appear to support the notion that the release of PHFτ into the extracellular space could contribute to the formation of SPs (Shin *et al.*, 1993, 1994). Briefly, injections of human PHFτ into the rodent brain with and without aluminium salts induced codeposits of Aβ at the injection site. Similar observations were not obtained following injections of other peptides and polypeptides including several different proteins that are frequently found in NFTs and SPs. For example, cerebral injections of normal adult and fetal human τ, various τ peptides, dephosphorylated PHFτ, high molecular weight neurofilament proteins, α1-antichymotrypsin or ApoE isoforms with or without aluminium salts did not produce codeposits of Aβ. On the other hand, it is interesting that some of the proteins associated with NFTs and SPs in the AD brain such as ubiquitin, ApoE and α1-antichymotrypsin were codeposited with PHFτ at the injection sites in the rodent brain (Shin *et al.*, 1993, 1994). Thus, the chronic but progressive degeneration of neurons or their processes in the AD brain may release PHFτ, other proteins and other materials into the extracellular space, which then are able to recruit or immobilize soluble Aβ into deposits. This same neuronal debris also may play a role in inducing aggregated Aβ to undergo fibrillogenesis and the formation of neuritic amyloid plaques.

Although amyloid fibrils are dominant structural elements in SPs, it is well known that SPs are complex amyloid deposits that are invariably associated with PHFτ, and other proteinaceous or cellular components (Arai *et al.*, 1990, 1992; Masliah *et al.*, 1992; Schmidt *et al.*, 1991, 1994; Trojanowski *et al.*, 1993a). Further, senile plaques appear to disrupt the surrounding neuropil, and they induce gliosis and a microglial response. In contrast, so-called diffuse plaques are comprised almost exclusively of Aβ, and they do not disrupt synaptic profiles or elicit a gliotic response in the cerebellum (Li *et al.*, 1994), hypothalamus (Standaert *et al.*, 1991) and striatum (Gearing *et al.*, 1993), where diffuse plaques occur almost to the exclusion of SPs and neurofibrillary lesions. However, we recently showed by immunohistochemistry and confocal microscopy that the PHF1 antibody labelled a denser meshwork of neuropil threads in AD neocortex than other antibodies to normal τ or PHFτ, and this meshwork of neuropil threads was so extensive in the AD neocortex that nearly all amyloid deposits were permeated by PHF1-positive dystrophic neurites including diffuse plaques (Schmidt *et al.*, 1994). In marked contrast to AD neocortex, the neocortex of normal age-matched controls was nearly devoid of PHF1-positive neuropil threads and it exhibited very few plaques with associated neurofibrillary lesions. This is consistent with earlier suggestions (Arai *et al.*, 1990, 1992) that plaques formed by the convergence of amyloid and neurofibrillary lesions may be more AD specific than other types of amyloid plaques. Further, these studies imply that mechanisms leading to the formation of neuritic amyloid plaques could involve interactions between PHFτ and Aβ, and that these interactions are more likely to occur in the AD neocortex where PHFτ-rich neuropil threads are abundant and widespread abnormalities.

Accordingly, we speculate that interactions between secreted, soluble Aβ and other molecules (including PHFτ and other SP components such as proteoglycans) in the extracellular space could serve as a nidus for the induction of Aβ fibril formation and the generation of complex neuritic plaques. Whether or not the aggregation and fibrillogenesis of Aβ is reversible *in vivo* is uncertain, but there is *in vitro* evidence that the deposition of soluble Aβ in plaques may be reversible (Maggio *et al.*, 1992). Figure 2 depicts a hypothetical scenario summarizing how such events might occur in the AD brain. While the details of this hypothetical scheme remain to be elucidated, it appears increasingly likely that the conversion of soluble Aβ into insoluble amyloid fibrils in the extracellular space leading to the formation of neuritic amyloid plaques is a multistep process that may involve a number of different cofactors (Shin *et al.*, 1993, 1994; Snow and Malouf, 1993; Snow *et al.*, 1994a, 1994b; Trojanowski and Lee, 1994b).

3.6 Concluding remarks

The progressive disruption of several different components of the neuronal cytoskeleton is a common feature of chronic dementias of the elderly, and alterations of the phosphorylation state and degradation of CNS τ in AD is one of the most extensively studied examples of this kind of pathological process. While the disruption of the neuronal cytoskeleton and the incorporation of abnormally phosphorylated CNS τ proteins into pathological inclusions provide markers for AD, these events also could compromise the function and viability of neurons thereby contributing to the emergence of dementia in AD. Thus, we anticipate that advances in understanding the mechanisms whereby neurofibrillary lesions are generated from normal adult τ will provide further insights into how these lesions contribute to neuronal dysfunction and degeneration. Such insights will set the stage for developing strategies for the therapy of AD.

Acknowledgements

We thank our colleagues in the Departments of Pathology and Laboratory Medicine, Neurology, Psychiatry, and the Penn Alzheimer Center for their many contributions to the work reviewed here, which was supported by grants from the NIH and the Dana Foundation.

References

Arai, H., Lee, V.M.-Y., Otvos, L., Greenberg, B., Lowery, D.E., Sharma, S.K., Schmidt, M.L. & Trojanowski, J.Q. (1990) *Proc. Natl Acad. Sci. USA* **87**, 2249–2253.

Arai, H., Schmidt, M.L., Lee, V.M.-Y., Hurtig, H.I., Greenberg, B.D., Adler, C. & Trojanowski, J.Q. (1992) *Neurology* **42**, 1315–1322.

Arnold, S.E. & Trojanowski, J.Q. (1994) *J. Comp. Neurol.* (submitted).

Arriagada, P.A., Growdon, J.H., Hedley-White, E.T. & Hyman, B.T. (1992) *Neurology* **42**, 631–639.

Ashall, F. & Goate, A.M. (1993) *Trends Biochem. Sci.* **19**, 42–46.

Bassell, G.J., Singer, R.H. & Kosik, K. (1994) *Neuron* **12**, 571–582.

Bauman, K., Mandelkow, E.-M., Biernat, J., Piwnica-Worms, H. & Mandelkow, E. (1993) *FEBS Lett.* **336**, 417–424.

Biernat, J., Mandelkow, E.-M., Schroter, C., Lichtenberg-Kraag, B., Steiner, B., Berling, B., Meyer, H., Mercken, M., Vandermeeren, A., Goedert, M., and Mandelkow, E. (1992). The switch of tau protein to an Alzheimer-like state includes the phosphorylation of two-serine-proline motifs upstream of the microtubule binding region. *EMBO J.* **11**, 1593–1597.

Billingsley, M.L., Ellis, C., Kincaid, R.L., Martin, J., Schmidt, M.L., Lee, V.M.-Y. & Trojanowski, J.Q. (1994) *Exper. Neurol.* **126**, 178–184.

Braak, H., Braak, E., Grundke-Iqbal, I. & Iqbal, K. (1986) *Neurosci. Lett.* **66**, 212–222.

Bramblett, G.T., Trojanowski, J.Q. and Lee, V.M.-Y. (1992) *Lab. Invest.* **66**, 212:222.

Bramblett, G.T., Goedert, M., Jakes, R., Merrick, S.E., Trojanowski, J.Q. & Lee, V.M.-Y. (1993) *Neuron* **10**, 1089–1099.

Brion, J.P., Passarier, H., Nunez, J. & Flament-Durand, J. (1985) *Arch. Biol.* **95**, 229–235.

Brion, J.P., Hanger, D.P., Couck, A.M. & Anderton, B. (1991) *Biochem. J.* **279**, 831–836.

Cassimeris, L. (1993) *Cell Motil. Cytoskel.* **26**, 275–281.
Crowther, R.A., Olesen, O.F., Jakes, R. & Goedert, M. (1992) *FEBS Lett.* **309**, 199–202.
Crowther, R.A., Olesen, O.F., Smith, M.J., Jakes, R. & Goedert, M. (1994) *FEBS Lett.* **337**, 135–138.
Couchie, D., Mavilia, C., Georgieff, I.S., Liem, R.K.H., Shelanski, M.L. & Nunez, J. (1992) *Proc. Natl Acad. Sci. USA* **89**, 4378–4381.
Delacourte, A. & Defossez, A. (1986) *J. Neurol. Sci.* **76**, 173–186.
Dickson, D.W., Crystal, H.A., Mattiace, L.M., Masur, D.M., Blau, A.D., Davies, P., Yen, S.-H. & Aronson, M. (1991) *Neurobiol. Aging* **13**, 179–189.
Drechsel, D.N., Hyman, A.A., Cobb, M.H. & Kirschner, M. (1992) *Mol. Biol. Cell* **3**, 1147–1154.
Drewes, G., Lichtenberg-Kraag, B., Doering, F., Mandelkow, E.M., Biernat, J., Goris, J., Doree, M. & Mandelkow, E. (1992) *EMBO J.* **6**, 2131–2138.
Drewes, G., Mandelkow, E.-M., Bauman, K., Goris, J., Merlevede, W. & Mandelkow, E. (1993) *FEBS Lett.* **336**, 425–432.
Garver, T.D., Lehman, R.A.W., Lee, V.M.-Y., Trojanowski, J.Q. & Billingsley, M.L. (1994) *J. Neurochem.* **63**, 2279–2287.
Gearing, M., Wilson, R.W., Unger, E.R., Shelton, E.R. Chan, H.Y., Masters, C.L., Beyreuther, K. & Mirra, S.S. (1993) *J. Neuropath. Exper. Neurol.* **52**, 22–30.
Glenner, G. & Wong, C.W. (1984) *Biochem. Biophys. Res. Commun.* **120**, 885–890.
Goedert, M. (1993) *Trends Neurosci.* **16**, 460–465.
Goedert, M., Wischik, C.M., Crowther, R.A., Walker, J.E. & Klug, A. (1988) *Proc. Natl Acad. Sci. USA* **85**, 4051–4055.
Goedert, M., Spillantini, M.G., Jakes, R., Rutherford, D. & Crowther, R.A. (1989a) *Neuron* **3**, 519–526.
Goedert, M., Spillantini, M.G., Potier, M.C., Ulrich, J. & Crowther, R.A. (1989b) *EMBO J.* **8**, 393–399.
Goedert, M., Cohen, E.S., Jakes, R. & Cohen, P. (1992a) *FEBS Lett.* **312**, 95–99.
Goedert, M., Spillantini, M.G. & Crowther, R.A. (1992b) *Proc. Natl Acad. Sci. USA* **89**, 1983–1987.
Goedert, M., Spillantini, M.G., Cairns, N.J. & Crowther, R.A. (1992c) *Neuron* **8**, 159–168.
Goedert, M., Jakes, R., Crowther, R.A., Six, J., Luebke, U., Vandermeeren, M., Cras, R., Trojanowski, J.Q. & Lee, V.M.-Y. (1993) *Proc. Natl Acad. Sci. USA* **90**, 5066–5070.
Gong, C.-X., Singh, T.J., Grundke-Iqbal, I. & Iqbal, K. (1993) *J. Neurochem.* **61**, 921–927.
Gong, C.-X., Singh, T.J., Grundke-Iqbal, I. & Iqbal, K. (1994) *J. Neurochem.* **62**, 803–806.
Goode, B.L. & Feinstein, S.C. (1994) *J. Cell Biol.* **124**, 769–782.
Greenberg, S.G., Davies, P., Scheim, J.D. & Binder, L.I. (1992) *J. Biol. Chem.* **267**, 564–569.
Grundke-Iqbal, I., Iqbal, K., Tung, Y.-C., Quinlan, M., Wisniewski, H.M. & Binder, L.I. (1986) *Proc. Natl Acad. Sci. USA* **83**, 4913–4917.
Haas, C., Schlossmacher, M., Hung, A.Y., Vigo-Pelfrey, C., Mellon, A., Ostraszewski, B.L., Lieberburg, I., Koo, E.H., Schenk, D., Teplow, D.B. & Selkoe, D. (1992) *Nature* **359**, 322–325.
Hanger, D.P., Hughes, K., Woodgett, J.R., Brion, J.-P. & Anderton, B.H. (1992) *Neurosci. Lett.* **147**, 58–62.
Hardy, J. (1993) *J. NIH Res.* **5**, 46–49.
Harris, K.A., Oyler, G.A., Doolittle, G.M., Vincent, I., Lehman, R.A.W., Kincaid, R.L. & Billingsley, M.L. (1993) *Ann. Neurol.* **33**, 77–87.
Hasegewa, M., Morishima-Kawashima, M., Takio, K., Suzuki, M., Litani, K. & Ihara, Y. (1992) *J. Biol. Chem.* **267**, 17047–17054.
Hyman, B.T., Elvhage, T.E. & Reiter, J. (1994) *Am. J. Pathol.* **144**, 565–572.
Ihara, Y. (1988) *Brain Res.* **459**, 138–144.
Ishiguro, K., Shiratsuchi, A., Sato, S., Omori, A., Arioka, M., Kobayashi, S., Uchida, T. & Imahori, K. (1993) *FEBS Lett.* **325**, 167–172.
Kanemura, K., Takio, K., Miura, R., Titani, K. & Ihara, I. (1992) *J. Neurochem.* **58**, 1667–1675.
Khachaturian, Z.S. (1985) *Arch. Neurol.* **42**, 1097–1106.

Khatoon, S., Grundke-Iqbal, I. & Iqbal, K. (1992) *J. Neurochem.* **59**, 750–753.
Kondo, J., Honda, T., Mori, H., Hamada, Y., Miura, R., Ogawara, M. & Ihara, Y. (1988) *Neuron* **1**, 827–834.
Kosik, K., Joachim, C.L. & Selkoe, D.J. (1986) *Proc. Natl Acad. Sci. USA* **83**, 4044–4048.
Kosik, K.S., Orecchio, L.D., Binder, L., Trojanowski, J.Q., Lee, V.M.-Y. & Lee, G. (1988) *Neuron* **1**, 817–825.
Ksiezak-Reding, H. & Yen, S.-H. (1991) *Neuron* **6**, 717–728.
Lang, E., Szendai, G.I., Lee, V.M.-Y. & Otvos, L., Jr (1992) *Biochem. Biophys. Res. Commun.* **187**, 783–790.
Ledesma, M.D., Correas, L., Avila, J. & Diaz-Nido, J. (1992) *FEBS Lett.* **308**, 218–224.
Lee, J.H., Goedert, M., Hill, W.D., Lee, V.M.-Y. & Trojanowski, J.Q. (1993) *Exp. Neurol.* **121**, 93–105.
Lee, V.M.-Y. (1991) *J. NIH Res.* **3**, 54–56.
Lee, V.M.-Y. & Trojanowski, J.Q. (1992) *Curr. Opin. Neurobiol.* **52**, 183–191.
Lee, V.M.-Y., Balin, B.J., Otvos, L., Jr & Trojanowski, J.Q. (1991) *Science* **251**, 675–678.
Li, Y.-T., Woodruff-Pak, D. & Trojanowski, J.Q. (1994) *Neurobiol. Aging* **15**, 1–9.
Ludin, B. & Matus, A. (1993) *Hippocampus* **3**, 61–72.
Maggio, J.E., Stimson, E.R., Ghilard, J.R., Allen, C.J., Dahl, C.E., Whitcomb, D.C., Vigna, S.R., Vinters, H.V., Labenski, M.E. & Mantyh, P.W. (1992) *Proc. Natl Acad. Sci. USA* **89**, 5462–5466.
Mandelkow, E.M., Drewes, G., Biernat, J., Gustke, N., Van Lint, J., Vandenhaude, J.L. & Mandelkow, E. (1992) *FEBS Lett.* **394**, 315–327.
Masliah, E., Ellisman, M., Carragher, B., Mallory, M., Young, S., Hansen, L., DeTeresa, R. & Terry, R.D. (1992) *Exp. Neurol.* **51**, 404–414.
Masters, C.L., Simms, G., Weinman, N.A., Multhaup, G., McDonald, B.L., Beyreuther, K. (1985) *Proc. Natl Acad. Sci. USA* **82**, 4245–4249.
Matsuo, E.S., Shin, R.-W., Billingsley, M.L., Van der Vorde, A., Trojanowski, J.Q. & Lee, V.M.-Y. (1994) *Neuron* **13**, 989–1002.
Matus, A. (1994) *Trends Neurosci.* **17**, 19–22.
Mullan, M. & Crawford, F. (1993) *Trends Neurosci.* **16**, 398–403.
Mumby, M.C. & Walter, G. (1993) *Physiol. Rev.* **73**, 673–699.
Otvos, L., Jr, Feiner, L., Lang, E., Szendrei, G.I., Goedert, M. & Lee, V.M.-Y. (1994) *J. Neurosci. Res.* **39**, 669–673.
Paudel, H.K., Lew, J., Ali, Z. & Wang, J.H. (1993) *J. Biol. Chem.* **268**, 23512–23518.
Perry, G., Kawai, M., Tabaton, M., Onorato, M., Mulvill, P., Richey, P., Morandi, A., Connolly, J.A. & Gambetti, P. (1990) *J. Neurosci.* **11**, 1748–1755.
Price, D.L., Borchelt, D.R. & Sisodia, S.S. (1993) *Proc. Natl Acad. Sci. USA* **90**, 6381–6384.
Robertson, J., Loviny, T.L.F., Goedert, M., Jakes, R., Murray, K.J., Anderton, B.H. & Hanger, D.P. (1993) *Dementia* **4**, 256–263.
Saunders, A.M., Strittmatter, W.J., Schmechel, D., St George-Hyslop, P.H., Pericak-Vance, M.A., Joo, S.H., Rosi, B.L., Gusella, J.F., Crapper-MacLachlan, D.R., Alberts, M.J., Hulette, C., Crain, B., Goldgaber, D. & Roses, A.D. (1993) *Neurology* **43**, 1467–1472.
Schmidt, M.L., Lee, V.M.-Y. & Trojanowski, J.Q. (1990) *Am. J. Pathol.* **136**, 1069–1075.
Schmidt, M.L., Lee, V.M.-Y. & Trojanowski, J.Q. (1991) *Lab. Invest.* **64**, 352–357.
Schmidt, M.L., Murray, J. & Trojanowski, J.Q. (1993) *Molec. Chem. Neuropath.* **18**, 299–312.
Schmidt, M.L., DiDario, A.G., Lee, V.M.-T. & Trojanowski, J.Q. (1994) *FEBS Lett.* **344**, 69–73.
Scott, C.W., Spreen, R.C., Herman, J.L., Chow, F.P., Davison, M.D., Young, J. & Caputo, C. (1993) *J. Biol. Chem.* **268**, 1166–1173.
Selkoe, D.J. (1993) *Trends Neurosci.* **16**, 403–409.
Seubert, P., Vigo-Pelfrey, C., Esch, F., Lee, M., Dovey, H., Davis, D., Sinha, S., Schlossmacher, M., Whaley, J., Swindlehurst, C., McCormack, R., Wolfert, R., Selkoe, D., Lieberburg, I. & Schenk, D. (1992) *Nature* **359**, 325–327.
Shin, R.-W., Ogomori, K., Kitamoto, T. & Tateishi, J. (1989) *Am. J. Pathol.* **134**, 1365–1371.
Shin, R.-W., Iwaki, T., Kitamoto, T., Sato, Y. & Tateishi, J. (1992) *Am. J. Pathol.* **140**, 937–945.

Shin, R.-W., Bramblett, G.T., Lee, V.M.-Y. & Trojanowski, J.Q. (1993) *Proc. Natl Acad. Sci. USA* **90**, 6825–6828.
Shin, R.-W., Lee, V.M.-Y. & Trojanowski, J.Q. (1994) *J. Neurosci* **14**, 7221–7233.
Shirazi, S.K. & Wood, J.G. (1993) *NeuroReport* **4**, 435–437.
Shoji, M., Golde, T.E., Ghiso, J., Cheung, T.T., Estus, S., Shaffer, L.M., Cai, X.-D., McKay, D.M., Tintner, R., Frangione, B. & Younkin, S.G. (1992) *Science* **258**, 126–129.
Sim, A.T.R. (1991) *Molec. Neurobiol.* **5**, 229–246.
Snow, A.D. & Malouf, A.T. (1993) *Hippocampus* **3**, 257–268.
Snow, A.D., Sekiguchi, R., Nochlin, D., Fraser, P., Kimata, K., Mizutani, A., Arai, M., Schreier, W.A. & Morgan, D.G. (1994a) *Neuron* **12**, 219–234.
Snow, A.D., Sekiguchi, R., Nochlin, D., Kalaria, R.N. & Kimata, K. (1994b) *Am. J. Pathol.* **144**, 337–347.
Standaert, D.G., Lee, V.M.-Y., Greenberg, B.D., Lowery, D.E. & Trojanowksi, J.Q. (1991) *Am. J. Pathol.* **139**, 681–691.
Steiner, B., Mandelkow, E.M., Biernat, J., Gustke, N., Meyer, H.E., Schmidt, B., Mieskes, G., Soling, H.D., Drechsel, D., Kirschner, M.W., Goedert, M. & Mandelkow, E. (1990) *EMBO J.* **9**, 3539–3544.
Strittmatter, W.J., Weisgraber, K.H., Goedert, M., Saunders, A.M., Huang, D., Corder, E.H., Dong, L.-M., Jakes, R., Alberts, M.J., Gilbert, J.R., Han, S.-H., Hulette, C., Einstein, G., Schmechel, D.E., Percak-Vance, M.A. & Roses, A.D. (1994) *Exp. Neurol.* **125**, 163–171.
Szendrei, G.I., Lee, V.M.-Y. & Otvos, L., Jr (1993) *J. Neurosci. Res.* **34**, 243–249.
Tabaton, M., Cammarata, S., Mancardi, G.L., Cordone, G., Perry, G. & Loeb, C. (1991) *Neurology* **41**, 391–394.
Talamo, B.R., Rudell, R., Kosik, K.S., Lee, V.M.-Y., Neff, S., Adelman, L. & Kauer, J.S. (1989) *Nature* **337**, 736–739.
Tohyama, T., Lee, V.M.-Y., Rorke, L.B. & Trojanowski, J.Q. (1991) *J. Comp. Neurol.* **310**, 285–299.
Trojanowski, J.Q. & Lee, V.M.-Y. (1994a) *Am. J. Pathol.* **144**, 449–453.
Trojanowski, J.Q. & Lee, V.M.-Y. (1994b) *Ann. N.Y. Acad. Sci.* **747**, 92–109.
Trojanowski, J.Q., Newman, P.D., Hill, W.D. & Lee, V.M.-Y. (1991) *J. Comp. Neurol.* **310**, 365–376.
Trojanowski, J.Q., Schmidt, M.L., Shin, R.-W., Bramblett, G.T., Goedert, M. & Lee, V.M.-Y. (1993a) *Clin. Neurosci.* **1**, 184–191.
Trojanowski, J.Q., Schmidt, M.L., Shin, R.-W., Bramblett, G.T., Rao, D. & Lee, V.M.-Y. (1993b). *Brain Pathol.* **3**, 45–54.
Trojanowski, J.Q., Mawal-Dewan, M., Schmidt, M.L., Martin, J. & Lee, V.M.-Y. (1993c) *Brain Res.* **618**, 333–337.
Vulliet, R., Halloran, S.M., Braun, R.K., Smith, A.J. & Lee, G. (1992) *J. Biol. Chem.* **267**, 22570–22574.
Watanabe, A., Hasegawa, M., Suzuki, M., Takio, K., Morishima-Kawashima, M., Titani, K., Arai, T., Kosik, R.S. & Ihara, Y. (1993) *J. Biol. Chem.* **268**, 25712–25717.
Wertkin, A.M., Turner, R.S., Pleasure, S.J., Golde, T.E., Younkin, S.G., Trojanowski, J.Q. & Lee, V.M.-Y. (1993) *Proc. Natl Acad. Sci. USA* **90**, 9513–9517.
Wille, H., Drewes, G., Biernat, J., Mandelkow, E.M. & Mandelkow, E. (1992) *J. Cell Biol.* **118**, 573–584.
Wischik, C.M., Novak, M., Thogersen, H.C., Edwards, P.C., Runswick, M.J., Jakes, R., Walker, J., Milstein, C., Roth, M. & Klug. A. (1988) *Proc. Natl Acad. Sci. USA* **85**, 4506–4510.
Wong, C.W., Quaranta, V. & Glenner, G. (1985) *Proc. Natl Acad. Sci. USA* **82**, 8729–8732.
Wood, J.G., Mirra, S.S., Pollock, N.J. & Binder, L.I. (1986) *Proc. Natl Acad. Sci. USA* **83**, 4040–4043.
Wood, J.G., Lu, Q., Reich, C. & Zinsmeister, P. (1993) *Neurosci. Lett.* **156**, 83–86.
Yachnis, A.T., Rorke, L.B., Lee, V.M.-Y. & Trojanowski, J.Q. (1993) *J. Comp. Neurol.* **334**, 356–369.
Yoshida, H. & Ihara, Y. (1993) *J. Neurochem.* **61**, 1183–1186.

Note added in proof to Chapter 3

Since the submission of this Chapter, 5 recent studies from several different research groups independently reported significant increases in the levels of tau in the CSF of AD patients relative to normal controls (2–5). Although additional research is needed to confirm and extend these preliminary data, the provocative findings on CSF tau reported from 5 different research teams in several different countries are remarkably concordant. Hence, measurements of CSF tau by ELISA or other quantitative methods may provide an objective antemortem diagnostic test for AD. Indeed, measurement of tau levels alone or in combination with other proteins that may be detected in CSF (i.e. Aß or fragments of Aß precursor proteins, kinases, phosphatases, proteases, etc.) in multiple CSF samples from the same patient obtained at periodic intervals spanning all stages of AD may yield powerful new strategies for monitoring the progression of this disease and its response to novel therapeutic agents.

References added in proof

1. Arai, H., Masanori, T., Miura, M., Higuchi, S., Muramatsu, T., Machida, N., Seki, H., Takase, S., Clark, C.M., Lee, V.M.-Y., Trojanowski, J.Q., & Sasaki, H. (1995) *Ann. Neruol.*, In press.
2. Hock, C., Golomboswki, S., Naser, W., & Mueller-Spahn, F. (1995) *Ann. Neurol.* **183**, 43–45.
3. Jensen, M., Basum, H., & Lannfelt, L. (1995) *Neurosci. Lett.* **186**, 189–191.
4. Mori, H., Hosoda, K., Matsubara, E., Nakamoto, T., Furiya, Y., Endoh, R., Usami, M., Shoji, M., Maruyama, S., & Hirai, S. (1995) *Neurosci. Lett.* **186**, 181–183.
5. Vigo-Pelfrey, C., Seubert, P., Barbour, R., Blomquist, C., Lee, M., Lee, D., Coria, F., Chang, L., Miller, B., Lieberburg, I., & Shenk, D. (1995) *Neurol.* **45**, 788–793.

CHAPTER 4

THE GENETICS OF ALZHEIMER'S DISEASE AND MUTATIONS IN THE AMYLOID β-PROTEIN PRECURSOR GENE

Frances Busfield and Alison M. Goate

Department of Psychiatry, Washington University School of Medicine, 4940 Children's Place, St Louis, Missouri 63110, USA

Table of Contents

Summary

After heart disease, cancer and stroke, Alzheimer's disease (AD) is the leading cause of death in the developed world. In the US alone, Alzheimer's disease affects 2–3 million people. These individuals eventually require constant care with a majority being institutionalized. AD therefore constitutes a major health problem in developed societies where the average age of the population is increasing. Much effort is being expended in the search for the cause or causes of AD with the hope that a full understanding of the disease will lead to preventive therapies.

Pathobiology of Alzheimer's Disease
ISBN 0-12-286965-5

4.1 Is Alzheimer's disease a genetic disease?

As early as 1925 it was reported that AD ran in families (Meggendorfer, 1925). Many epidemiological studies have since been carried out to try to identify risk factors for AD (Henderson, 1986). The only factor consistently found is the presence of a family history of AD (Brody, 1982; Heyman *et al.*, 1984; van Duijn *et al.*, 1991a). Other risk factors which have been investigated are head trauma, thyroid disease and aluminium intake. Of these, the only one to show any significant correlation with the disease is head trauma (van Duijn *et al.*, 1991b). The most convincing evidence supporting the importance of a genetic component comes from the existence of a number of large, multiply-affected families in which the disease segregates in a manner consistent with fully penetrant autosomal dominant inheritance (Schottky, 1932; Nee *et al.*, 1983; Foncin *et al.*, 1986). The disease in these families is often referred to as familial Alzheimer's disease (FAD). The majority of cases, however, have no clear family history and are classed as sporadic. Recently two subsets of FAD have been recognized on the basis of age of onset. These are late onset (LOAD), with age of onset over 65 years, and early onset (EOAD), with age of onset under 65 years. Two problems confound the determination of genetic contribution to LOAD. Firstly, family members may have died of other causes before reaching the age of onset of AD and, secondly, the appearance of familial clustering can be a chance occurrence owing to the high frequency of AD in the elderly. In FAD pedigrees the age of onset is fairly consistent between affected members of the same pedigree but differs markedly between pedigrees. It was within the EOAD group that the first advances in understanding the molecular basis of AD were made.

4.2 Clinical and pathological features of Alzheimer's disease

AD neuropathology is characterized by the presence of large numbers of senile plaques and neurofibrillary tangles, with extensive neuronal cell loss, primarily in the temporal cortex and hippocampus. Senile plaques are extracellular deposits consisting of a central core of amyloid fibrils, surrounded by a rim of dystrophic neurites together with reactive microglia and astrocytes. The major proteinaceous component of senile plaques is amyloid β-protein (Aβ), a 39–42 amino-acid fragment derived from the amyloid β-protein precursor (APP) by proteolysis (Glenner *et al.*, 1984). APP is a glycoprotein with a single transmembrane domain. Aβ is derived from part of the transmembrane domain and the adjacent extracellular domain (Figure 1) (Kang *et al.*, 1987). More detail of APP processing can be found in Chapter 7. The APP gene consists of 19 exons covering >400 kb (Rooke *et al.*, 1993) with Aβ encoded by exons 16 and 17. Neurofibrillary tangles consist mainly of an abnormally phosphorylated form of the tau protein in paired helical filaments. Neither of these pathological features is specific to AD. Cerebral deposition of Aβ is observed in several disorders including Down's syndrome (DS), hereditary cerebral haemorrhage with amyloidosis-Dutch

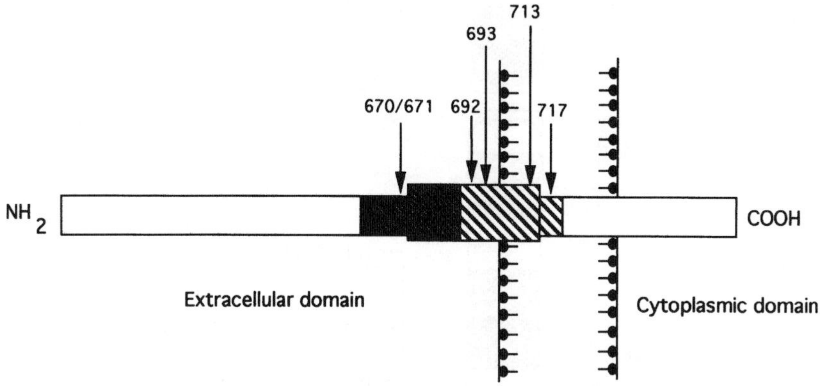

Figure 1 Diagramatic representation of APP showing the position of Aβ and the location of disease associated mutations (APP_{770} numbering).

=Exon 16; =Exon 17; = membrane larger box denotes position of Aβ within APP.

type (HCHWA-D), dementia pugilistica (Roberts *et al.*, 1990) and Guam–Parkinsonism–dementia (Mann and Jones, 1990) as well as during normal ageing. Neurofibrillary tangles have been reported in some instances of Creutzfeldt–Jakob disease (CJD) (Hsiao *et al.*, 1992) and in Guam–Parkinsonism–dementia.

4.3 Other cerebral amyloidoses

AD is not the only dementia involving cerebral amyloidosis to show autosomal dominant inheritance. Prion diseases are a group of transmissible fatal neurodegenerative diseases characterized by the presence of congophilic amyloid plaques. In contrast to AD, these plaques contain the prion protein, mutations in which cause Gerstmann–Scheinker–Straussler syndrome (GSS), CJD or fatal familial insomnia (FFI) (Prusiner, 1994). CJD is represented by both familial and sporadic forms with the familial forms caused by point mutations within the prion protein gene (PrP). The familial forms represent only a small percentage of all cases of CJD; a similar situation exists for AD, where the number of sporadic cases is far greater than the number of familial cases. Amyloid is also deposited in the cerebral angiopathies and, in the case of HCHWA-D, the amyloid deposits consist of Aβ as they do in AD (van Duinen *et al.*, 1990). HCHWA-D also shows an autosomal dominant pattern of inheritance, but patients suffer from stroke and show no signs of dementia. HCHWA-Icelandic type involves amyloid deposition of a cystatin C variant and not Aβ. Aβ deposition is also seen in cases of severe head trauma and in dementia pugilistica caused by repeated blows to the head.

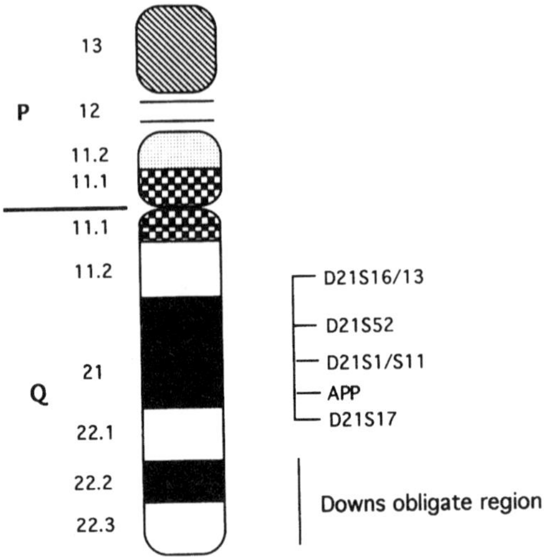

Figure 2 Idiogram of chromosome 21 showing position of relevant markers, APP and the Down's obligate region.

4.4 Down's syndrome and Alzheimer's disease

The first clue suggesting the involvement of chromosome 21 in AD came from the observation that Down's syndrome individuals who have an extra copy of chromosome 21, invariably develop the clinical and pathological features of AD if they live beyond 30 years old (Mann *et al.*, 1986). Studies of post-mortem brains from DS individuals at different ages suggest that Aβ deposition in the form of diffuse plaques is one of the earliest events in the pathogenesis of AD, with neurofibrillary tangles and neuronal death being a secondary occurrence (Mann *et al.*, 1986). With the association of AD neuropathology and trisomy 21, the search for the FAD gene focused on chromosome 21, and especially around the DS obligate region 21q22 (Figure 2).

4.5 Linkage to chromosome 21

The power of linkage analysis had increased as large numbers of DNA markers detecting restriction fragment length polymorphisms (RFLPs) in human DNA were characterized. In 1987, St George-Hyslop *et al.* used a linkage strategy to analyse four large kindreds with histologically proven familial AD (FAD 1, 2, 3 and 4). This kind of strategy had already proved successful in the mapping of genes in other inherited disorders, e.g. Huntington's disease and adult polycistic kidney disease. St George-Hyslop showed evidence of linkage between FAD and two chromosome 21 markers

D21S1/S11 and D21S16. These two markers map to the 21q11.2–21q21 region (Figure 2). D21S16, however, could not be accurately placed and the FAD locus could have been above or below D21S1/S11. Three-point analysis with these loci and AD gave a peak LOD score of 4.25 (a LOD score is the logarithm of the odds ratio, a measure of genetic linkage. A LOD score of 3 is taken to indicate linkage with a 1/1000 chance that the score is a chance occurrence.) Crossover events occurred in both FAD 1 and FAD 4 between the FAD locus and D21S1/S11. The location of an FAD locus in bands 21q11.2–21q21 placed it centromeric to the obligate DS obligate region of 21q22. At the same time as St George-Hyslop reported the location of an FAD locus at 21q11.2–21q21, several groups reported the mapping of the APP gene to the same region of chromosome 21, making APP an excellent candidate gene. A duplication of the chromosomal region around the APP gene was subsequently reported in some sporadic AD cases further strengthening the case for APP (Delabar *et al.*, 1986). However, this result has never been replicated.

4.6 Exclusion of the APP gene

The APP gene appeared to be excluded as the chromosome 21 FAD locus when several studies reported recombination between FAD and the APP locus. The segregaton of an APP polymorphism in two kindreds with EOAD (onset <35 years) and five LOAD kindreds (onset ~65 years), showing apparent autosomal dominant transmission of AD, led Van Broeckhoven *et al.* (1987) to conclude that APP was not closely linked to FAD in these families because recombination events were observed between the APP gene and AD. This demonstrated that APP and AD were genetically distinct unless these crossovers occurred within the APP gene.

A larger study involving 15 families, including seven Volga-German kindreds, was carried out by Schellenberg *et al.* (1988). The Volga Germans are a distinct group of US and Russian families whose German ancestors settled along the Volga river in Russia in the mid-18th century, and in which FAD occurs at a high frequency. FAD did not appear to be caused by mutations in the APP gene even when the data were analysed without the inclusion of the Volga Germans. Indeed, no evidence for linkage to any chromosome 21 markers was observed in these families. This led to the suggestion that FAD is genetically heterogeneous, i.e. that AD can be caused by mutations in more than one gene. It was also suggested that the LOD score of St George-Hyslop may have been a chance occurrence, because only one other report provided evidence supporting a locus on chromosome 21 (Goate *et al.*, 1989).

In 1990, a large collaborative study was undertaken involving the analysis of all available members of 48 pedigrees (St George-Hyslop *et al.*, 1990). They were all genotyped with five markers (D21S16, D21S13, D21S52, D21S1 and D21S11) constituting three independent loci. The data were then analysed using both maximum likelihood and affected pedigree member (APM) methods. APM does not require the specification of mode of transmission of the disease or the use of estimated parameters (e.g. penetrance) eliminating some potential sources of inaccuracy. At D21S1/S11 and D21S16/S13 (0.7 and 0.66% heterozygosity, respectively) maximum likelihood pairwise analysis produced a LOD score of greater than three, the score normally taken to indicate linkage. Not all pedigrees contributed positive scores, a situation which could arise if the disease was heterogeneous. To avoid the confusion of heterogeneity in linkage analysis a predivided sample test requiring families to be grouped according to an identifiable feature of the disease was done using age of onset as the dividing feature. The pedigrees were split into two groups: those with age of onset <65 years and those with an age of onset >65 years. The group with the early age of onset gave a positive LOD score, while those with late onset gave minimal or negative LOD scores. The difference between the two groups was statistically significant. However, even within the EOAD subset giving a positive LOD score, it was not possible to localize the disease gene. Two peak LOD scores were obtained outside the marker map suggesting several possibilities: homogeneity with errors in the data owing to misdiagnosis, nonpaternity or laboratory error, or nonallelic genetic heterogeneity within the EOAD families. A large proportion of the positive LOD score for the EOAD families came from a single large British pedigree (F23), which was to become key to the APP/AD story.

Genetic heterogeneity therefore became a more likely prospect with the probability that the APP gene, which had previously been excluded by data presuming homogeneity, was actually responsible for some of the EOAD cases. With the presence of genetic heterogeneity in EOAD, it became necessary to evaluate each family separately for linkage to the APP gene. Several families were identified showing clear linkage to chromosome 21 with no recombinants between the APP gene and the disease locus, indicating that in these families the APP gene could indeed be the site of a defect.

4.7 Mutations in the APP gene

4.7.1 Exon 17

4.7.1.1 Hereditary cerebral haemorrhage with amyloidosis – Dutch type

Prior to the 1990 collaborative studies that showed FAD to be genetically heterogeneous, a mutation in the APP gene was observed to be linked to another disease – HCHWA-D, which is a rare autosomal dominant disorder causing stroke (Van Broeckhoven *et al.*, 1990). The disease is characterized pathologically by amyloid deposits in cerebral vessel walls and by diffuse plaques in the brain parenchyma. The amyloid protein deposited in HCHWA-D is also Aβ. The two exons encoding Aβ were cloned and sequenced from two patients with HCHWA-D. The sequencing revealed a cytosine to guanine transversion, which results in a single amino-acid change from glutamic acid to glutamine at codon 693 (APP_{770}) of APP (Levy *et al.*, 1990a). The same strategy was used to obtain sequence data from a patient affected with AD and four unaffected individuals, all of whom possessed normal sequence. The point mutation at APP_{693} abolishes an MboII site. Loss of this site from a PCR product and oligospecific hybridization were used to confirm the presence of this mutation. The mutation was not found in any unaffected individuals. It has been postulated that the mutation alters the physical properties of Aβ leading to an increased susceptibility to fibril formation.

HCHWA-I, which occurs in Icelandic families, is histopathologically and clinically similar to HCHWA-D. The brains of HCHWA-I patients demonstrate amyloid angiopathy but, unlike HCHWA-D, the brain parenchyma is not affected and the amyloid fibrils are derived from degradation of a variant of cystatin C, an inhibitor of cysteine proteases (Cohen *et al.*, 1983). The variant cystatin C gene contains a mutation, again resulting in a single amino-acid change. Like cystatin C, one form of APP contains a protease inhibitor domain, although this is a serine not cysteine protease inhibitor. Mutations in these deposited proteins in the cerebral angiopathies pointed again to Aβ and mutations in the APP gene as being a possible cause of AD.

4.7.1.2 Family 23

The conclusions drawn from the 1990 FAD collaborative study and the discovery of the mutation in the APP gene associated with HCHWA-D led investigators to re-examine the possibility of an APP mutation causing FAD. The large British family, F23, that had contributed significantly to the positive LOD score obtained in the 1990 FAD collaborative study was studied more closely. An extensive analysis of the segregation of polymorphic markers spanning up to 55% of the long arm of chromosome 21 was undertaken in this family (Goate *et al.*, 1991). Two-point linkage analysis between all the markers gave a maximum LOD score of 2.9 between AD and an EcoRI polymorphism within the APP gene. Two individuals with recombinant chromosomes provided evidence for the position of the AD locus (Figure 3). The same allele is shared at D21S17 between all affected members and an unaffected individual 15 years over the mean age of onset (in 1991). This excluded the disease locus from being telomeric to this marker. The second individual with a recombinant chromosome was unaffected but only 2 years over the mean age of onset, and (in 1991) shared the same alleles as affected individuals between D21S13 and D21S1/S11 but not at the APP locus. This then was a key individual, who, if he remained healthy, would place the disease gene telomeric of those markers or, if he were to develop AD, would exclude APP as the disease locus (Figure 3). Since this analysis could not exclude APP as the disease locus, the APP gene was analysed further in this family. Because of the location of the mutation in HCHWA-D in Aβ, and the extensive deposition of Aβ in all AD cases, the initial effort was focused on sequencing the two exons of APP that contribute to this peptide, exons 16 and 17 (Figure 1).

Sequencing of exon 17 from affected individuals revealed a C-T transition at base

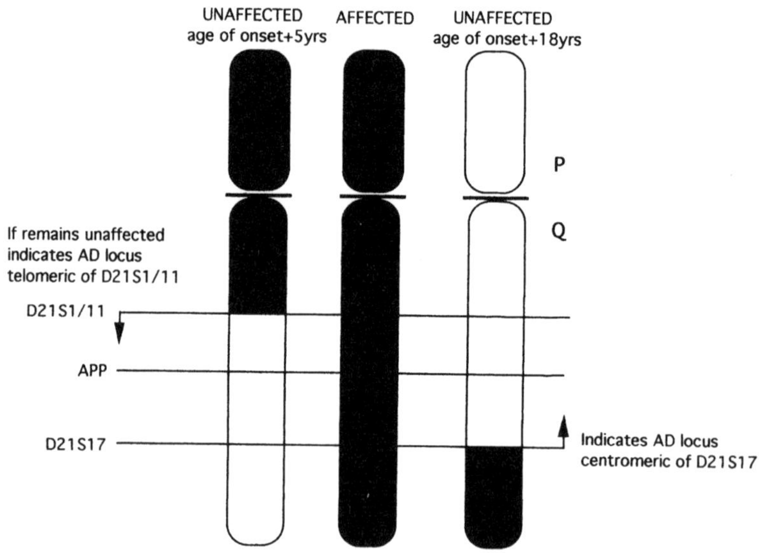

Figure 3 Chromosome 21 comparisons of the key individuals in F23 who provided an indication for the possible location of an FAD locus. The unaffected individuals are 5 and 18 years over the mean age of onset (correct for 1994) and are unaffected to date. Black = portion inherited from chromosome carrying disease locus.

2149 resulting in a valine (val) to isoleucine (ile) change at amino acid 717 (APP_{770}). This base transition was probably the result of deamination of a methylated cytosine situated 5′ to guanine, a relatively common mutation in human DNA. The transition creates a BclI restriction site, which allows the detection of the mutation by digestion of an exon 17 PCR product. Two-point linkage analysis between AD and this polymorphism gave a maximum LOD score of 3.37 and showed that it segregated with the disease. Screening of several hundred unrelated normal individuals failed to reveal anyone with this substitution. The group went on to screen 14 cases from nine families of LOAD, none of whom had this particular mutation. However, two out of 18 cases from 16 EOAD families were found to have the BclI restriction site (Goate *et al.*, 1991). These two affected individuals came from a pedigree collected by the Duke research group (F372), which showed a modest positive LOD score between chromosome 21 markers and AD. The rare EcoRI polymorphism at the 3′ end of the APP gene was genotyped in all affected individuals from F23 and F372. In F23 the disease cosegregates with the rare allele, which is present in only 5% of the population. In contrast, the disease segregates with the common allele in F372. The two pedigrees are therefore unlikely to be related as only 20 kb separates the mutation and the polymorphism. This meant that the mutation had arisen more than once, giving credibility to the idea that it is a pathogenic mutation. The failure to detect the substitution in normal chromosomes also supports this theory.

4.7.1.3 Other APP_{717} mutations

Following the discovery of the first mutation to be linked to AD exon 17 was sequenced in many EOAD families. Another eight families (three Japanese, one British, one Canadian, one Australian and two Italian), multiply affected with AD, have been shown to segregate to have this particular mutation (Naruse *et al.*, 1991; Yoshioka *et al.*, 1991; Fidani *et al.*, 1992; Karlinsky *et al.*, 1992; Sorbi *et al.*, 1993). The sequencing also revealed two additional allelic variants at codon 717 (Chartier Harlin *et al.*, 1991; Murrell *et al.*, 1991), which also segregated with AD. The amino-acid changes produced by these point mutations are valine to glycine or phenylalanine. The discovery of these variants provided more support for the hypothesis that the mutations are pathogenic.

4.7.1.4 Other exon 17 mutations

A family with a mixed AD/HCHWA-D phenotype was described by Hendriks *et al.* (1992). Four individuals have the symptoms of HCHWA-D with onset around 40 years, whereas seven other individuals present with presenile dementia with onset around 49 years. All these affected members were found to have a mutation at codon 692 in exon 17 of the APP gene, with alanine being replaced by glycine. Thus two distinct phenotypes appear to be caused by a single mutation. The pathogenicity of this mutation is confused, however, by the presence of one individual presenting with dementia but not having the mutation. The age of onset at 61 years for this individual was significantly higher than the mean. Also the individual does not share the same allele of the D21S210 marker close to APP, suggesting that the disease in this individual may be a phenocopy. The factors determining the type of disease an individual displays remain to elucidated and could be genetic and/or environmental in nature. A similar situation has been observed in pedigrees carrying the Asp to Asn mutation at codon 178 of the PrP gene. Families with this mutation present with either CJD or FFI, two clinically and pathologically distinct diseases (Goldfarb *et al.*, 1992). The phenotypic expression of this mutation is thought to be determined by the alleles at the codon 129 polymorphism within the PrP gene. The methionine/valine (met/val) 129 polymorphism is common in Caucasian populations and homozygosity for either allele predisposes individuals to the infectious form of prion disease, iatrogenic and sporadic CJD (Owen *et al.*, 1990; Collinge *et al.*, 1991). It also predisposes to earlier ages of onset in inherited forms of prion disease such as GSS (Dlouhy *et al.*, 1992). The phenotype seen with the codon 178 mutation depends upon the allele at codon 129 on the mutant chromosome. Codon 129 met is associated with FFI and codon 129 val with CJD. Not only do the codon 129 alleles affect the disease phenoptype but homozygosity at codon 129 leads to a significantly shorter disease duration in both FFI and CJD (Goldfarb *et al.*, 1992). In CJD homozygosity at codon 129 significantly reduces the age of onset, an effect not seen in FFI. Thus two distinct diseases are caused by a single point mutation at codon 178 but the phenotype is modified by the genotype at a different codon within the same gene. Since all affected people in the APP_{692} family inherit the same mutant APP gene but will inherit one of two normal APP genes from the unaffected parent, heterozygosity versus homozygosity of a polymorphism elsewhere within the APP gene could modify the disease phenotype. Alternatively, polymorphism within a gene unlinked to APP or an environmental factor could also cause this modification of the phenotype.

A mutation at codon 713 involving an amino-acid substitution has been identified in two individuals. The first of these to be described was in a chronic schizophrenic case with cognitive defects (Jones *et al.*, 1992). Further screening of 100 schizophrenics and 100 nondemented controls failed to detect the alanine to valine substitution. Another study of 86 unrelated chronic schizophrenics also failed to detect the 713 mutation (Coon *et al.*, 1993). It seems unlikely therefore that this mutation is a primary cause of chronic schizophrenia. Linkage analysis does not indicate involvement of the APP gene in schizophrenia but, if heterogeneity of schizophrenia is considerable, the power of linkage analysis is reduced. Further evidence of the nonpathogenicity of mutations at this codon came with the identification of this mutation in a single individual with early onset AD (59 years). This individual had a double mutation, one at codon 713 and the other at codon 715. These were both G-A transitions resulting in an alanine to threonine change at 713 and a silent change at 715 (Carter *et al.*, 1992). The double mutation was also present in four siblings (two over 62 years) and an aunt (88 years), all of whom had normal neuropsychological profiles. A total of 123 unaffected individuals were screened for these changes but none were found. The 713 mutation is therefore either nonpathogenic or has incomplete penetrance. It is the only APP missense mutation so far identified which does not segregate directly with the AD or HCHWA-D phenotype.

A silent C-T base change has also been reported at codon 708 (Balbin *et al.*, 1992; Kamino *et al.*, 1992) as well as two point mutations and a six base pair deletion in intronic sequence surrounding exon 17 (Kamino *et al.*, 1992). The 708 base change has been found in normal individuals as well as in two AD patients. The six base pair deletion is in intron 17 and was also found in a 92-year-old nondemented control. The deletion does not segregate with AD and none of these changes appear to be pathogenic.

4.7.2 Exon 16

Two Swedish EOAD families were identified in which the APP gene was shown to be tightly linked to the disease locus using the GT12 (D21S210) marker. Exons 16 and 17 were then sequenced in one affected individual, and a double mutation was identified at codons 670 and 671 (APP_{770}) in exon 16 of the APP gene (Mullan *et al.*, 1992a). The mutations are caused by two base pair transversions of G-T and A-C, which result in lysine and methionine being replaced by asparagine and leucine at codons 670 and 671, respectively. The base changes result in the loss of an MboII site, which was used to identify the presence of the mutations in all of the affected members of both families, a result confirmed in four cases by sequencing. This double mutation was not identified in 50 normal chromosomes, suggesting that the mutation is pathogenic for AD in these families. It has since been shown that the two families originating from the same county in Sweden are in fact related. It is not known if both point mutations are required for pathogenicity. *In vitro* transfection experiments suggest that the codon 670 mutation produces the same change in Aβ production as the double mutation (Citron *et al.*, 1992). The substitution at 671 may therefore be a polymorphic variant present in the normal population. However, since no convenient restriction digest exists to detect this change, no systematic population screening has been undertaken.

4.8 Importance of APP mutations in Alzheimer's disease

The presence of Aβ in AD plaques and the implication from DS that Aβ deposition is a primary event in AD pathology has led to speculation that APP has a central role to play in all types of AD. The discovery of mutations in the APP gene cosegregating with the disease at first seemed to support the importance of Aβ deposition in AD. However, in the 3 years since the discovery of the first mutation it has become clear that the APP mutations account for only a very small percentage of AD. These cases account for only a small proportion of the early-onset subset of FAD (Schellenberg *et al.*, 1991b; Kamino *et al.*, 1992; Tanzi *et al.*, 1992). The APP V717I mutation appears to be of more significance in certain populations with three of the 12 reported families being Japanese and the others European or of European origin. Other APP mutations have only been reported in single families, so no geographic significance can be recognized. Pathogenic APP mutations have not been reported in any LOAD or sporadic cases of AD, and the APP gene has also been excluded as the disease locus in many EOAD pedigrees by genetic linkage analysis (Tanzi *et al.*, 1987; Schellenberg *et al.*, 1988, 1991b; St George-Hyslop *et al.*, 1990; Kamino *et al.*, 1992). Although APP mutations are important because they represent the first known cause of AD, they represent a small percentage of total AD cases and the focus of AD genetics has turned to the search for other genes involved in AD. It may be that the other genes involved play a role in the expression or metabolism of APP, and cause Aβ deposition as a result of their loss or change of function. APP processing could therefore be the final step in a chain of events in which disruption at any point could result in the abnormal deposition of Aβ.

To date no mutations which cosegregate with disease have been found in any of the other 17 exons, introns or the promoter of the APP gene. Several base changes mentioned above have been identified in and around exon 17 (Balbin *et al.*, 1992; Kamino *et al.*, 1992; Zubenko *et al.*, 1992) but these do not segregate with the disease. However, since insufficient samples are available from most AD families for linkage analysis, and systematic screening has only been carried out on exon 16 and 17, mutations in APP may have been under-estimated as a cause of AD.

4.9 Pathogenicity of the APP mutations

It is not yet known with absolute certainty that the APP mutations are pathogenic, although the genetic evidence supports this hypothesis with odds of $1{:}10^{30}$. The mutations are found only in AD cases and individuals at 50% risk for AD. Haplotype analysis shows that all of the APP mutations are independent events, i.e. none of the families are related to one another. The mutations in APP could exert their effect in one of several ways. Since overexpression of APP appears to result in the premature deposition of Aβ, the mutations may alter the expression or degradation rate of APP, leading to accumulation of Aβ. This effect could be exerted at the transcriptional or translational level, or through the stability of mRNA or protein. Another way in which the mutations may exert their effect is through alteration of post-translational modification, alteration in the proteolytic processing of APP, or in the interaction of APP or products of APP with other cellular components, e.g. ApoE. It is also possible that the mutations which are within Aβ could change its physico-chemical properties, making it more likely to assume the β-sheet conformation found in the amyloid plaques. The 670/671 and 717 mutations do not lie within Aβ but flank it, lying very close to the sites of secretase cleavage which release Aβ from APP. This is in contrast to most hereditary systemic amyloidoses in which the mutated residue is within the deposited amyloid (Levy *et al.*, 1990b; Nichols *et al.*, 1990; Benson, 1992; Benson *et al.*, 1993; Pepys *et al.*, 1993). It has therefore been suggested that these mutations may have an effect upon the proteolytic processing of APP. Cell studies have been undertaken to try to elucidate the effect of the mutations upon processing of APP. These studies indicate that the 670/671 and 717 mutations appear to affect different processing pathways. Transfection of human neuroblastoma cell lines with constructs containing the mutations at 670/671 result in an increase of soluble Aβ derivatives, whereas constructs containing 717 mutations have no effect upon soluble Aβ levels (Citron *et al.*, 1992; Cai *et al.*, 1993). Recent *in vitro* work suggests that 717 mutations may exert their effect by increasing the amount of the longer forms of Aβ (Suzuki *et al.*, 1994). Soluble Aβ is a normal product of cell metabolism (Haass *et al.*, 1992). *In vitro* transfection studies have shown that the Swedish mutation leads to elevated levels of soluble Aβ (Citron *et al.*, 1992), possibly because the mutant APP is a better substrate for the β-secretase enzyme which cleaves between the methionine at 671 and the aspartate at 672. Recent preliminary data have shown that the HCHWA-D mutation, which is within Aβ, may alter the secondary structure resulting in premature Aβ deposition (Fabian *et al.*, 1993). There is presently no adequate explanation for the vasculotropic action of APP_{693} whilst APP_{717} and $APP_{670/671}$ cause preferential deposition in the brain parenchyma. Although the mutations in APP associated with AD have not been found in any unaffected individuals, it is not possible to say that they are indeed the cause of the pathological changes. The only sure way of demonstrating the pathogenicity of any of these mutations is by the production of transgenic mice expressing mutant APP, which spontaneously develop neurodegenerative disease, possibly with AD neuropathology. Mice transgenic for PrP102, a variant thought to cause GSS, have been shown to undergo adult-onset spontaneous neurodegeneration, indicating that this mutation is pathogenic (Hsiao *et al.*, 1990). This mutation was introduced by microinjection and overexpression of the mutant protein led to the disease phenotype.

Since AD pathology is seen in DS individuals several attempts have been made to create an overexpression model for AD by microinjection of normal APP cDNAs. Of these only one, expressing the full length 751 amino acid APP isoform, has shown any AD-like pathology (Quon *et al.*, 1991). These mice exhibit diffuse extracellular Aß immunoreactive deposits, similar to those of young DS adults, but no mature AD plaques (Higgins *et al.*, 1994).

The development of YAC transgenic technology enabled the introduction of entire genes spanning several hundred kilobases as transgenes with all the intronic and flanking sequences intact. This removes the need for a heterologous promoter and ensures the expression of all possible isoforms. Two groups have introduced the entire APP gene (650Kb) into the mouse genome using YACs (Lamb *et al.*, 1993; Pearson *et al.*, 1993). Although both groups report the expression of human APP mRNA isoforms and APP protein there have been no reports of AD pathology.

While the overexpression of normal APP led to limited success in the production of AD pathology the production of transgenic mice expressing mutant APP has also been attempted. Games *et al.*, (1995) reported the production of transgenic mice overexpressing a human APP cDNA containing the 717 valine-phenylalanine mutation. The human APP minigene was under the control of a platelet-derived growth factor promoter and all three major splicing variants of human APP mRNA were expressed. These mice exhibited a more classical AD like pathology. This included, extracellular Aß deposits, dystrophic neuritic components, gliosis and loss of synaptic density with regional specificity resembling that of AD. While this represents the creation of an animal model which consistently produces AD-like neuropathology neither the Quon nor the Games' models however exhibit the formation of NFT's another classical pathological feature of AD. Since overexpression of APP may be sufficient to cause AD-type pathology, by analogy with DS, the pathogenicity of the APP mutations has still not been demonstrated unequivocally. Gene targeting, a technique that enables the introduction of specific point mutations into the mouse genome without change in copy number provides the best means to test this, since the mutant protein would be expressed correctly and resulting pathology would be a consequence of the mutation and not inappropriate expression from multiple transgenes.

4.10 Clinical and pathological features associated with APP mutations

There is great variability in the clinical and pathological manifestations of AD. The variability led some investigators to believe that subsets of AD with differing aetiologies may exist, which would be represented by specific clinical and/or pathological manifestations. As early as 1989, before the discovery of the APP mutations and while the controversy over linkage to chromosome 21 was still going on, 24 FAD kindreds were compared with respect to their clinical and neuropathological features (Bird *et al.* by 1989). They concluded that there was considerable phenotypic heterogeneity within AD and supported the grouping into LOAD and EOAD on the basis of age of onset. They found considerable overlap in the neuropathological characteristics of FAD families but suggested that the predominance of one pathological feature, e.g. neurofibrillary tangles (NFT), or the occurrence of additional pathological characteristics such as anterior horn cell loss may be useful distinguishing features. Since the identification of the 717 mutations, however, several studies have been undertaken to look at the association of clinical and pathological features of patients with these mutations (Karlinsky *et al.*, 1992; Lantos *et al.*, 1992; Mann *et al.*, 1992; Mullan *et al.*, 1993). The age of onset for patients with the 717 val–ile mutation is at the higher end of the range of EOAD with a mean age of onset of 52 years (Mullan *et al.*, 1993). Despite differing genetic backgrounds and geographic location, the age of onset is fairly consistent between the families ranging from 49.4 years in F372 to 54.9 years in F23. In comparison, the age of onset for chromosome 14-linked FAD cases is significantly earlier, indicating that age of onset may be linked to disease aetiology. In contrast, the progression of the disease is variable even within a single family, indicating that factors other than the mutation are involved (Mullan *et al.*, 1993). Anecdotal evidence suggests that genotype at the ApoE locus can modify age of onset of AD in the APP mutation families with ε4 alleles decreasing the age of onset and ε2 alleles delaying onset (Hardy *et al.*, 1993; St George-Hyslop *et al.*, 1994). In general, the Japanese families showed a shorter survival period, but this could be as a result of later diagnosis or a difference in nursing practices, and not necessarily as a result of a more rapid rate of progression (Mullan *et al.*, 1993).

The neuropathology in APP_{717} families is generally typical of severe AD with numerous plaques and tangles, neuronal loss, neurofibrillary degeneration and mild angiopathy (Karlinsky *et al.*, 1992; Lantos *et al.*, 1992; Mullan *et al.*, 1993). In one case, cortical and subcortical Lewy bodies were also present but comparison of cytoskeletal pathology between this case and sporadic AD cases revealed no substantial differences (Lantos *et al.*, 1992). Since not all 717 cases have Lewy bodies, this suggests that their presence is nonspecific or that the AD phenotype could be broader than previously suspected. Large variation in Aβ deposition in members of F372 who had similar disease duration also suggests that this feature of the pathology is independent of the mutation. Indeed, a recent study has shown that the amyloid burden in LOAD is proportional to the ApoE-ε4 allele dosage (Schemechel *et al.*, 1993). The relationship between the molecular aetiology and specific neuropathology is not, therefore, a direct one, a situation also observed in some prion dementias (Collinge *et al.*, 1990). Apart from the age of onset therefore no other clinical or neuropathological features could be associated with the mutation. Other cases involving the 717 val–gly and 717 val–phe showed typical AD pathology at autopsy with no apparent association of any feature with a particular mutation (Mann *et al.*, 1992).

4.11 Linkage of AD to other loci

Since the percentage of EOAD which could be accounted for by the APP mutations was very small and APP linkage to LOAD was absent, the search for linkage to other chromosomes continued. In 1992, Schellenberg *et al.* reported linkage to chromosome 14 markers in EOAD families which were not linked to chromosome 21. The families studied had three or more affected subjects in two or more generations, and at least one autopsy documented case. The families were analysed in two separate groups: the Volga Germans and early-onset non-Volga German (ENVG), and these sets combined. All these groupings gave negative LOD scores for APP, markers on 21 centromeric of APP and chromosome 19. A total of 64 markers with at least one on each chromosome were analysed and two markers on chromosome 14 gave weak linkage to the disease in the ENVG group. On analysis with two further markers they found strong positive LOD scores with D14S43 in seven out of nine families. Linkage was not observed in the Volga-German group of families. Analysis for heterogeneity between the two groups was highly significant with no evidence for heterogeneity within the groups. In the same year, three other papers all reported linkage to the same region of chromosome 14 (Mullan *et al.*, 1992b; St George-Hyslop *et al.*, 1992; Van Broeckhoven *et al.*, 1992). The consensus of these papers was that linkage of AD to chromosome 14 existed in EOAD families, and that the flanking markers were D14S53 and D14S52. More recently this region has been narrowed down to 7 centimorgans between D14S289 and D14S61 (Figure 4). The only known genes in the region were excluded so research efforts focussed upon creating new markers to reduce the distance between flanking markers and to clone the entire region into yeast artificial chromosomes (YAC's) (Clark *et al.*, 1995). Recently Sherrington et al (1995) reported the isolation of several cDNA's from this region. Five mutations were reported within one gene (S182) that segregated with the disease in eight early onset FAD pedigrees. Six novel mutations have since been identified segregating with disease in thirteen early onset FAD pedigrees (Clark personal communication). It would appear therefore that mutations within this gene account for a greater percentage of early onset FAD cases than mutations within the APP gene. The gene has been localised to a 75Kb region between D14S77 and D14S71 (Figure 4) and codes for a 467 amino acid protein. Hydrophobicity plots of the protein predict six to eight transmembrane domains. Absence of a signal peptide suggests that the N-teminus of the protein is not on the external surface of the membrane. Data base searches have revealed expressed sequence tags (ESTs) with high homology to S182 indicating the presence of a gene family. These ESTs map to chromosome 1 and may provide the location of yet another AD locus.

Some of the familial LOAD cases showed linkage to chromosome 19 (Pericak-Vance *et al.*, 1991) and the gene for Apo E has been implicated (Figure 4). ApoE is one of many proteins which is found deposited in senile plaques. Genetic association studies have demonstrated association between the ApoE-ε4 allele and AD in familial and sporadic cases (Corder *et al.*, 1993; Van Duijn *et al.*, 1994). The risk of developing AD is dosage dependent: individuals carrying two ε4 alleles have a higher

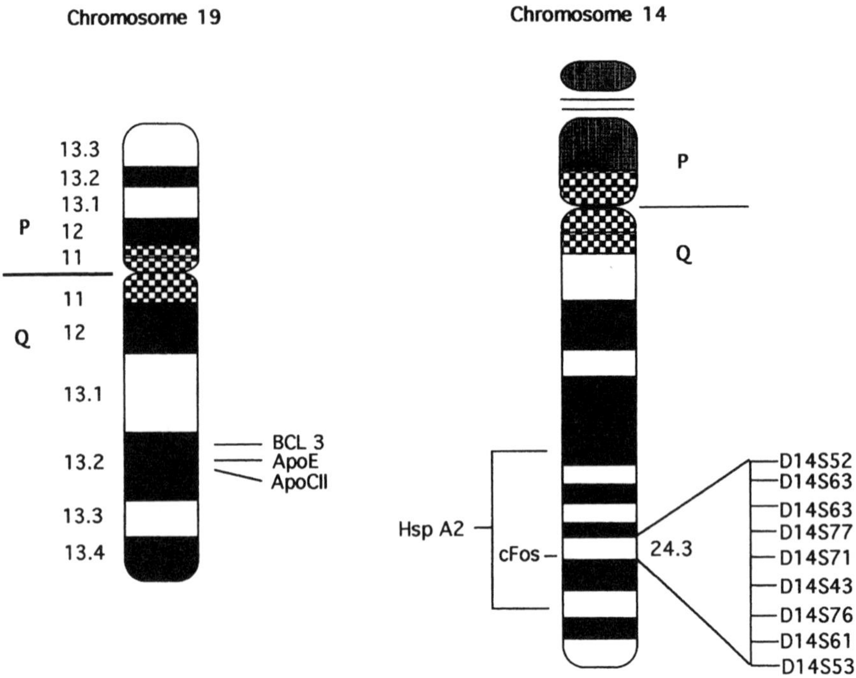

Figure 4 Idiograms for chromosomes 14 and 19 showing the positions of relevant markers and genes.

risk and earlier onset of disease than those with one ε4 allele who have a higher risk and earlier onset of disease than those with no ε4 alleles. A second allele at the same locus (ApoE-ε2) appears to be protective against AD (Corder *et al.*, 1994; Talbot *et al.*, 1994), supporting the hypothesis that the association between AD and ApoE has biological significance. In contrast to the two EOAD loci which cause AD, the ApoE-ε4 allele is probably not sufficient to cause disease. It merely predisposes the individual to AD. The precise mechanism is not yet known but AD cases show an ApoE-ε4 dose-dependent burden of Aβ in the brain suggesting that ApoE-ε4 may increase Aβ deposition or decrease clearance of Aβ deposits. *In vitro* studies have shown that ApoE binds to Aβ (Strittmatter *et al.*, 1993). However, *in vivo* ApoE is usually complexed with lipid and it is not yet known whether these lipoprotein complexes bind to Aβ. ApoE-ε4 has recently been shown to be associated with an increased risk of developing another cerebral amyloidosis, CJD (Amouyel *et al.*, 1994). This observation suggests that ApoE-ε4 may have a more general role in enabling amyloidogenic proteins to take up a β-sheet conformation since there is no similarity between Aβ and PrP at the amino-acid sequence level. AD and ApoE is discussed in detail in Chapter 13.

4.12 Conclusions

Over the past 5 years there have been many advances in the field of AD research. The first genetic cause of AD, discovered in 1991, was heralded as a great breakthrough. In subsequent years it has become clear that these mutations account for a minority of AD cases. Genetic heterogeneity has created a much more complicated picture than was at first imagined with at least three loci causing EOAD, on chromosomes 21, 14 and 19. At least one other locus must exist because the Volga German families show no linkage to any of the known loci. Finally, association studies have revealed that ε4 alleles at the ApoE locus predispose to both familial and sporadic AD. Since there are many cases of AD carrying no ApoE-ε4 alleles, there must be other genetic or environmental factors that predispose to AD.

References

Amouyel, P., Vidal, O., Laonay, J.M., Laplanche, J.L., for the French Research Group on Epidemiology of Human Spongiform Encephalopathies. (1994) *The Lancet* **344**, 1315–1318.

Balbin, M., Abrahamson, M., Gustafson, L., Nilsson, K., Brun, A. & Grubb, A. (1992) *Hum. Genet.* **89**, 580.

Benson, M.D. (1992) *J. Intern. Med.* **232**, 525–527.

Benson, M.D., Liepnieks, J., Uemichi, T., Wheeler, G. & Correa, R. (1993) *Nature Genet.* **3**, 252–255.

Bird, T.D., Sumi, S.M., Nemens, E.J., Nochlin, D., Schellenberg, G., Lampe, T.H., Sadovnick, A., Chui, H., Miner, G.W. & Tinklenberg, J. (1989) *Ann. Neurol.* **25**, 12–25.

Brody, J. (1982) *Am. J. Epidemiol.* **115**, 155–162.

Cai, X.-D., Golde, T.E. & Younkin, S.G. (1993) *Science* **259**, 514–516.

Carter, D.A., Desmarais, E., Bellis, M., Campion, D., Clerget-Darpoux, F., Brice, A., Agid, Y., Jaillard-Serradt, A. & Mallet, J. (1992) *Nature Genet.* **2**, 255–256.

Chartier Harlin, M.C., Crawford, F., Houlden, H., Warren, A., Hughes, D., Fidani, L., Goate, A., Rossor, M., Roques, P., Hardy, J. & Mullan, M. (1991) *Nature* **353**, 844–846.

Citron, M., Oltersdorf, T., Haass, C., McConlogue, L., Hung, A.Y., Seubert, P., Vigo-Pelfry, C., Lieberburg, I. & Selkoe, D.J. (1992) *Nature* **360**, 672–674.

Clark, R.F., Cruts, M., Korenblat, K., He, C., Talbot, C., Van Broeckhoven, C. & Goate A.M. (1995) *Human Molecular Genetics* **4**, 1347–1354.

Cohen, D.H., Feiner, H., Jensson, O. & Frangione, B. (1983) *J. Exp. Med.* **158**, 623.

Collinge, J., Owen, F., Poulter, M., Leach, M., Crow, T.J., Rossor, M.N., Hardy, J., Mullan, M.J., Janota, I. & Lantos, P.L. (1990) *Lancet* **336**, 7–9.

Collinge, J., Palmer, M.S. & Dryden, A.J. (1991) *Lancet* **337**, 1441–1442.

Coon, H., Hoff, M., Holik, J., Delisi, L., Crowe, T., Freedman, R., Shields, G., Boccio, A., Lerman, M., Gershon, E., Gejman, P., Leppert, M. & Byerley, W. (1993) *Am. J. Med. Gen. (Neuropsych. Genet.)* **48**, 6–39.

Corder, E., Saunders, A., Strittmatter, W., Schmechel, D., Gaskell, P., Small, G., Roses, A., Haines, J. & Pericak-Vance, M. (1993) *Science* **261**, 921–923.

Delabar, J.M., Lamour, Y., Gegonne, A., Davous, P., Roudier, M., Nicole, A., Ceballos, I., Amouyel, P., Stehelin, D. & Sinet, P.M. (1986) *Ann. Genet.* **29**, 226–228.

Dlouhy, S., Hsiao, K., Farlow, M., Foroudi, T., Conneally, P., Johnson, P., Prusiner, S., Hodes, M. & Ghetti, B. (1992) *Nature Genet.* **1**, 64–67.

Fabian, H., Szendrei, G., Mantsch, H. & Otvos, L., Jr (1993) *Biochem. Biophys. Res. Commun.* **191**, 232–239.
Fidani, L., Rooke, K., Chartier Harlin, M.C., Hughes, D., Tanzi, R., Mullan, M., Roques, P., Rossor, M., Hardy, J. & Goate, A.M. (1992) *Hum. Mol. Genet.* **1**, 165–168.
Foncin, J.F., Salmon, D. & Bruni, A.C. (1986) In Brilley, M., Kato, A. and Weber, M. (eds) *New Concepts in Alzheimer's Disease*, pp. 242–256. London, Macmillan.
Glenner, G.G., Wong, C.W., Quaranta, V. & Eanes, E.D. (1984). *Appl. Pathol.* **2**, 357–369.
Goate, A.M., Haynes, A.R., Owen, M.J., Farrall, M., James, L.A., Lai, L.Y., Mullan, M.J., Roques, P., Rossor, M.N., Williamson, R. & Hardy, J. (1989) *Lancet* **1**, 352–355.
Goate, A., Chartier Harlin, M.C., Mullan, M., Brown, J., Crawford, F., Fidani, L., Giuffra, L., Haynes, A., Irving, N., James, L. Mant, R., Newton, P., Rooke, K., Roques, P., Talbot, C., Pericak-Vance, M., Roses, A., Williamson, R., Rosser, M., Owen, M. & Hardy, J. (1991) *Nature* **349**, 704–706.
Goldfarb, L., Petersen, R., Tabaton, M., Brown, P., LeBlanc, A., Montagna, P., Cortelli, P., Julien, J., Vital, C., Pendelbury, W., Haltia, M., Wills, P., Hauw, J., McKeever, P., Monari, L., Schrank, B., Swergold, G., Autilio-Gambetti, L., Gajdusek, K.D., Lugaresi, E. & Gambetti, P. (1992) *Science* **258**, 806–808.
Haass, C., Schlossmacher, M., Hung, A., Vigo-Pelfrey, C., Mellon, A., Ostaszewski, B., Lieberburg, I., Koo, E., Schenk, D., Teplow, D. & Selkoe, D. (1992) *Nature* **359**, 322–325.
Hardy, J., Houlden, H., Collinge, J., Kennedy, A., Newman, S., Rossor, M., Lannfelt, L., Lilius, L., Winblad, B., Crook, R. & Duff, K. (1993) *Lancet* **342**, 737–738.
Henderson, A.S. (1986) *Br. Med. Bull.* **42**, 3–10.
Hendriks, L., van Duijn, C., Cras, P., Cruts, M., Van Hul, W., van Harskamp, F., Warren, A., McGinnis, M., Antonorakis, S. & Van Broeckhoven, C. (1992) *Nature Genet.* **1**, 218–221.
Heyman, A., Wilkinson, W.E., Stafford, J.A., Helms, M.J., Sigmon, A.H. & Weinberg, T. (1984) *Ann. Neurol.* **15**, 335–341.
Hsiao, K., Scott, M., Foster, D., Groth, D.F., DeArmond, S.J. & Prusiner, S.B. (1990) *Science* **250**, 1587–1590.
Hsiao, K., Dlouhy, S., Farlow, M., Cass, C., Da Costa, M., Conneally, P., Hodes, M., Ghetti, B. & Prusiner, S. (1992) *Nature Genet.* **1**, 68–71.
Jones, C., Morris, S., Yates, C., Moffoot, A., Sharpe, C., Brock, D. & St Clair, D. (1992) *Nature Genet.* **1**, 306–309.
Kamino, K., Orr, H.T., Payami, H., Wijsman, E.M., Alonso, E., Pulst, S.M., Anderson, L., O'Dahl, S., Nemens, E., White, J.A., Sadovnick, A.D., Ball, M.J., Kaye, J., Warren, A., McInnis, M., Antonarakis, S.E., Korenberg, J.R., Sharma, V., Kukull, W., Larson, E., Heston, L.L., Martin, G.M., Bird, T.D. & Schellenberg, G.D. (1992) *Am. J. Hum. Genet.* **51**, 998–1014.
Kang, J., Lemaire, H.G., Unterbeck, A., Salbaum, J.M., Masters, C.L., Grzeschik, K.H., Multhaup, G., Beyreuther, K. & Muller Hill, B. (1987). *Nature* **325**, 733–736.
Karlinsky, H., Vaula, G., Haines, J.L., Ridgley, J., Bergeron, C., Mortilla, M., Tupler, R.G., Percy, M.E., Robitaille, Y., Noldy, N.E., Yip, T.C.K., Tanzi, R.E., Gusella, J.F., Becker, R., Berg, J.M., Crapper-McLachlan, D.R. & St George-Hyslop, P.H. (1992) *Neurology* **42**, 1445–1453.
Lantos, P.L., Luthert, P.J., Hanger, D., Anderton, B.H., Mullan, M. & Rossor, M. (1992) *Neurosci. Lett.* **137**, 221–224.
Levy, E., Carman, M.D., Fernandez Madrid, I.J., Power, M.D., Lieberburg, I., van Duinen, S.G., Bots, G.T., Luyendijk, W. & Frangione, B. (1990a) *Science* **248**, 1124–1126.
Levy, E., Haltia, M., Fernandez-Madrid, I.J., Koivunen, O., Ghiso, J., Prelli, F. & Frangione, B. (1990b) *J. Exp. Med.* **172**, 1865–1867.
Mann, D.M.A. & Jones, D. (1990) *Neurosci. Lett.* **109**, 68–75.
Mann, D.M., Yates, P.O., Marcyniuk, B. & Ravindra, C.R. (1986) *Neuropathol. Appl. Neurobiol.* **12**, 447–457.

Mann, D.M.A., Jones, D., Snowden, J.S., Neary, D. & Hardy, J. (1992) *Neurosci. Lett.* **137**, 225–228.
Meggendorfer, F. (1925) *Zentralbl. Neurol. Psychiat.* **40**, 359.
Mullan, M., Crawford, F., Axelman, K., Houlden, H., Lilius, L., Winblad, B. & Lannfelt, L. (1992a) *Nature Genet.* **1**, 345–347.
Mullan, M., Houlden, H., Windelspecht, M., Fidani, L., Lombardi, C., Diaz, P., Rossor, M., Crook, R., Hardy, J., Duff, K. & Crawford, F. (1992b) *Nature Genet.* **2**, 340–342.
Mullan, M., Tsuji, S., Miki, T., Katsuya, T., Naruse, S., Kaneko, K., Shimizu, T., Kojima, T., Nakano, I., Ogihara, T., Miyatake, T., Ovenstone, I., Crawford, F., Goate, A., Hardy, J., Roques, P., Roberts, G., Luthert, P., Lantos, P., Clark, C., Gaskell, P., Crain, B. & Roses, A. (1993) *Neurobiol. Aging.* **14**, 407–419.
Murrell, J., Farlow, M., Ghetti, B. & Benson, M. (1991) *Science* **254**, 97–99.
Naruse, S., Igarashi, S., Kobayashi, H., Aoki, K., Inuzuka, T., Kaneko, K., Shimizu, T., Iihara, K., Kojima, T., Miyatake, T. & Tsuji, S. (1991) *Lancet* **337**, 978–979.
Nee, L.E., Polinsky, R.J., Eldridge, R., Weingartner, H., Smallberg, S. & Ebert, M. (1983) *Arch. Neurol.* **40**, 203–208.
Nichols, W.C., Gregg, R.E., Brewer, H.B.J. & Benson, M.D. (1990) *Genomics* **8**, 318–323.
Owen, F., Poulter, M., Collinge, J. & Crow, T. (1990) *Am. J. Hum. Genet.* **46**, 1215–1216.
Pepys, M.B., Hawkins, P.N., Booth, D.R., Vigushin, D.M., Tennent, G.A., Soutar, A.K., Totty, N., Nguyen, O., Blake, C.C.F., Terry, C.J., Feest, T.G., Zalin, A.M. & Hsuan, J.J. (1993) *Nature* **362**, 553–557.
Pericak-Vance, M.A., Bebout, J.L., Gaskell, P.C., Jr, Yamaoka, L.H., Hung, W., Alberts, M.J., Walker, A.P., Bartlett, R.J., Haynes, C.A., Walsh, K.A., Earl, N.L., Heyman, A., Clark, C.M. & Roses, A.D. (1991) *Am. J. Hum. Genet.* **48**, 1034–1050.
Prusiner, S. & Hsiao, K. (1994) *Neurology* **35**, 385–395.
Roberts, G.W., Allsop, D. & Bruton, C. (1990) *J. Neurol. Neurosurg. Psychiat.* **53**, 373–378.
Rooke, K., Talbot, C., James, L., Anand, R., Hardy, J. & Goate, A. (1993) *Mammal. Genome* **4**, 662–669.
Schellenberg, G.D., Bird, T.D., Wijsman, E.M., Moore, D.K., Boehnke, M., Bryant, E.M., Lampe, T.H., Nochlin, D., Sumi, S.M., Deeb, S.S., Beyreuther, K. & Martin, G.M. (1988) *Science* **241**, 1507–1510.
Schellenberg, G.D., Anderson, L., O'Dahl, S., Wijsman, E., Sadovnick, A.D., Ball, M.J., Larson, E.B., Kukull, W.A., Martin, G.M., Roses, A.D. & Bird, T.D. (1991a) *Am. J. Hum. Genet.* **49**, 511–517.
Schellenberg, G.D., Pericak Vance, M.A., Wijsman, E.M., Moore, D.K., Gaskell, P.C.J., Yamaoka, L.A., Bebout, J.L., Anderson, L., Welsh, K.A., Clark, C.M. *et al.* (1991b) *Am. J. Hum. Genet.* **48**, 563–583.
Schellenberg, G.D., Bird, T.D., Wijsman, E.M., Orr, H.T., Anderson, L., Nemens, E., White, J.A., Bonnycastle, L., Weber, J.L., Alonso, M.E., Potter, H., Heston, L.L. & Martin, G.M. (1992) *Science* **258**, 668–671.
Schmechel, D., Saunders, A., Strittmatter, W., Crain, B., Hulette, C., Joo, S., Pericak-Vance, M., Goldgaber, D., Roses, A.D. (1993) *Proc. Natl. Acad. Sci. USA* **90**, 9649–9653.
Schottky, J. (1932) *Z. Neurol. Psychol.* **140**, 333–397.
Sorbi, S., Nacmias, B., Forleo, P., Piacentini, S. & Amaducci, L. (1993) *Nature Genet.* **4**, 10.
St George-Hyslop, P.H., Tanzi, R.E., Polinsky, R.J., Haines, J.L., Nee, L., Watkins, P.C., Myers, R.H., Feldman, R.G., Pollen, D., Drachman, D., Growdon, J., Bruni, A., Foncin, J.-F., Salmon, D., Frommelt, P., Amaducci, L., Sorbi, S., Piacantini, S., Stewart, G.D., Hobbes, W.J., Conneally, P.M. & Gusella, J.F. (1987) *Science* **235**, 885–890.
St George-Hyslop, P.H., Haines, J.L., Farrer, L.A., Polinsky, R., Van Broeckhoven, C., Goate, A., McLachlan, D.R., Orr, H., Bruni, A.C., Sorbi, S., Rainero, I., Foncin, J.-F., Pollen, D., Cantu, J.-M., Tupler, R., Voskresenskaya, N., Mayeux, R., Growdon, J., Fried, V.A., Myers, R.H., Nee, L., Backhovens, H., Martin, J.-J., Rossor, M., Owen, M.J., Mullan, M., Percy,

M.E., Karlinsky, K., Rich, S., Heston, L., Montesi, M., Mortilla, M., Nacmias, N. & Gusella, J.F. (1990) *Nature* **347**, 194–197.

St George-Hyslop, P., Haines, J., Rogaev, E., Mortilla, M., Vaula, G., Pericak-Vance, M., Foncin, J.-F., Montesi, M., Bruni, A., Sorbi, S., Rainero, I., Pinessi, L., Pollen, D., Polinsky, R., Nee, L., Kennedy, J., Macciardi, F., Rogaeva, E., Liang, Y., Alexandrova, N., Lukiw, W., Schlumpf, K., Tanzi, R., Tsuda, T., Farrer, L., Cantu, J.-M., Duara, R., Amaducci, L., Bergamini, L., Gusella, J., Roses, A. & Crapper-McLachlan, D. (1992) *Nature Genet.* **2**, 330–334.

St George-Hyslop, P., Crapper McLachlan, D., Tuda, T., Rogaev, E., Karlinsky, H., Lippa, C.F. & Pollen, D. (1994) *Science* **263**, 441–580.

Strittmatter, W.J., Saunders, A.M., Schmechel, D., Pericak-Vance, M., Enghild, J., Salvesen, G.S. & Roses, A.D. (1993) *Proc. Natl. Acad. Sci. USA* **90**, 1977–1981.

Tanzi, R.E., St George Hyslop, P.H., Haines, J.L., Polinsky, R.J., Nee, L., Foncin, J.F., Neve, R.L., McClatchey, A.I., Conneally, P.M. & Gusella, J.F. (1987) *Nature* **329**, 156–157.

Tanzi, R., Vaula, G., Romano, D., Mortilla, M., Huang, T., Tupler, R., Wasco, W., Hyman, B., Haines, J., Jenkins, B., Kalaitsidaki, M., Warren, A., McGinnis, M., Antonorakis, S., Karlinsky, H., Percy, M., Connor, L., Growdon, J., Crapper-Mclachlan, D., Gusella, J. & St George-Hyslop, P. (1992) *Am. J. Hum. Genet.* **51**, 273–282.

Van Broeckhoven, C., Genthe, A.M., Vandenberghe, A., Horsthemke, B., Backhovens, H., Raeymaaker, P., Van Hul, W., Wehnert, A., Gheuen, J., Cras, P., Bruyland, M., Martin, J.-J., Salbaum, M., Multhaup, G., Masters, C.L., Beyreuther, K., Gurling, H.M.D., Mullan, M.I., Holland, A., Barton, A., Irving, N., Williamson, R., Richards, S.J. & Hardy, J.A. (1987) *Nature* **329**, 153–155.

Van Broeckhoven, C., Haan, J., Bakker, E., Hardy, J.A., Van Hul, W., Wehnert, A., Vegter Van der Vlis, M. & Roos, R.A. (1990) *Science* **248**, 1120–1122.

Van Broeckhoven, C., Backhovens, H., Cruts, M., De Winter, G., Bruyland, M., Cras, J.-J. (1992) *Nature Genet.* **2**, 335–339.

van Duijn, C.M., Clayton, D., Chandra, V., Fratiglioni, L., Graves, A.B., Heyman, A., Jorm, A.F., Kokmen, E., Kondo, K., Mortimer, J.A., Rocca, W.A., Shalat, S.L., Soininen, H. & Hofman, A. (1991a) *Int. J. Epidemiol.* **20** (Suppl. 2), S13–S20.

van Duijn, C.M., Stijnen, T. & Hofman, A. (1991b) *Int. J. Epidemiol.* **20** (Suppl. 2), S4–S12.

van Duijn, C., de Knijff, P., Wehnert, A., Havekes, L., Hofman, A. & Van Broekhoven, C. (1994) *Nature Genet.* **7**, 74–77.

van Duinen, S.G., Castano, E., Prelli, F., Bots, G.T., Luyendijk, W. & Franglione, B. (1987) *Proc. Natl. Acad. Sci. USA* **84**, 5991–5994.

Yoshioka, K., Miki, T., Katsuya, T., Ogihara, T. & Sakaki, Y. (1991) *Biochem. Biophys. Res. Commun.* **178**, 1141–1146.

Zubenko, G.S., Farr, J., Stiffler, J.S., Hughes, H.B. & Kaplan, B.B. (1992) *J. Neuropath. Exp. Neurol.* **51**, 459–463.

CHAPTER 5

TRANSCRIPTIONAL CONTROL OF AMYLOID β-PROTEIN PRECURSOR EXPRESSION

Wolfgang W. Quitschke and Dmitry Goldgaber

Department of Psychiatry and Behavioral Science, State University of New York at Stony Brook, Stony Brook, New York 11794–8101, USA

Table of Contents

Introduction

The human amyloid β-protein precursor (APP) gene is located on chromosome 21 (Goldgaber *et al.*, 1987; Kang *et al.*, 1987; Robakis *et al.*, 1987; Tanzi *et al.*, 1987) and the mouse homologue on chromosome 16 (Lovett *et al.*, 1987). The APP gene has a complex transcriptional unit with three polyadenylation signals (Kang *et al.*, 1987). It contains at least 19 exons distributed over approximately 400 kilobases (Lemaire *et al.*, 1989; Yoshikai *et al.*, 1990; Rooke *et al.*, 1993). Multiple transcripts derived from the APP gene by alternative splicing have been identified (Kang *et al.*, 1987; Kitaguchi *et al.*, 1988; Ponte *et al.*, 1988; Tanzi *et al.*, 1988; DeSauvage and Octave, 1989; Golde *et al.*, 1990; Jacobsen *et al.*, 1991; König *et al.*, 1992). These mRNA isoforms are transcribed from the same promoter, but differentially expressed in cells and tissues. The predominant isoforms in most cells are APP_{695}, APP_{751} and APP_{770}, which all contain the amyloid β-protein (Aβ) domain located on exons 16 and 17 (Kang *et al.*, 1987; Kitaguchi *et al.*, 1988; Ponte *et al.*, 1988; Tanaka *et al.*, 1988; Tanzi *et al.*, 1988; Golde *et al.*, 1990).

Pathobiology of Alzheimer's Disease
ISBN 0-12-286965-5

5.1 Cell and tissue distribution of APP mRNA

APP mRNA was detected in all major tissues and the expression pattern is similar between species (Selkoe *et al.*, 1987; Manning *et al.*, 1988; Yamada *et al.*, 1989). The highest levels of APP or its transcripts are generally observed in brain and kidney, and the lowest, barely detectable, levels in liver (Manning *et al.*, 1988; Neve *et al.*, 1988; Tanzi *et al.*, 1988). It is also expressed at varying levels in numerous cell lines (Quitschke and Goldgaber, 1992), as well as in primary human myoblasts and myotubes (Zimmermann *et al.*, 1988). APP mRNA is also found in circulating platelets (Gardella *et al.*, 1990).

In brain, APP is primarily expressed in neurons and endothelial cells (Goedert, 1987; Bahmanyar *et al.*, 1987; Bendotti *et al.*, 1988; Card *et al.*, 1988; Manning *et al.*, 1988; Shivers *et al.*, 1988; Yamada *et al.*, 1989). However, APP has also been observed in some glial cells (Schmechel *et al.*, 1988) and also at increased levels in reactive astrocytes following neuronal damage (Siman *et al.*, 1989). APP_{695} is the primary isoform in brain, whereas APP_{751} and APP_{770} are predominantly expressed in peripheral tissues (Neve *et al.*, 1988; Mita *et al.*, 1989; Tanaka *et al.*, 1989; Golde *et al.*, 1990; Kang and Müller-Hill, 1990).

5.2 APP expression in Alzheimer's disease

The analysis of APP expression is more complex in Alzheimer's disease because of the reliance on post-mortem material for evaluation, which in particular presents problems with mRNA stability. Furthermore, the pathological condition is frequently well advanced with the consequence that neuronal damage has already occurred. The level of APP transcripts or the level of APP itself may therefore be secondarily affected by the pathological condition. However, a number of diverse results have emerged. For example, studies have shown that there is an increase in the level of APP transcripts in specific brain regions in Alzheimer's disease (Cohen *et al.*, 1988; Higgins *et al.*, 1988). This may be caused by a specific increase in the level of transcripts containing the Kunitz-type protease inhibitor in hippocampal pyramidal neurons (Johnson *et al.*, 1990) and other brain regions (Tanaka *et al.*, 1989). In contrast, the increase in the level of APP transcripts in nucleus basalis and locus ceruleus neurons was found to be attributable to the APP transcript lacking the Kunitz domain (Palmert *et al.*, 1988).

5.3 APP expression in Down's syndrome

The most striking instance in which overexpression of APP may serve as a contributing factor in amyloid formation is observable in Down's syndrome. Down's syndrome is characterized by a trisomy of chromosome 21, the same chromosome on which the APP gene is located, and afflicted individuals gradually develop Aβ plaques at a comparatively early age (Mann *et al.*, 1990). The level of APP mRNA was found to be 4–5-fold higher in patients with Down's syndrome than in control individuals (Neve *et al.*, 1988). For the purpose of this discussion, it is assumed that the higher level of APP message is attributable, at least in part, to a higher level of gene expression. It is not clear whether the extra copy of the APP gene in Down's syndrome accounts for the entire increase in expression, since the difference in mRNA levels is higher than would be predicted on the basis of gene dosage alone. However, similar results have also been observed in aneuploid mice with trisomy 16 (the chromosome carrying the mouse APP gene), suggesting that three copies of the gene definitely enhance APP expression (Bendotti *et al.*, 1988; O'Hara *et al.*, 1989). In addition, transplanted mouse hippocampal tissue with trisomy 16 displays evidence of immunoreactivity to Aβ (Richards *et al.*, 1991). These observations suggest that the overexpression of APP, attributable to an extra copy of the gene, causes the signs of early Alzheimer-type pathology in this model system. Overexpression of APP is also likely to be related to the early development of Aβ plaques in individuals with Down's syndrome, who develop neuropathological changes similar to those observed in Alzheimer's disease decades earlier than the general population (Coyle *et al.*, 1988; Rumble *et al.*, 1989; Mann *et al.*, 1990).

5.4 APP expression in response to activating factors

It was observed that interleukin-1 (IL-1) causes a twofold increase in APP mRNA levels in human umbilical vein endothelial cells (HUVEC) (Goldgaber *et al.*, 1989). In a different study, IL-1 was found to increase APP mRNA expression by 50% in human, and in murine brain-derived endothelial cells (Forloni *et al.*, 1992). Neuronal cells responded with a 175% increase in APP transcript levels whereas glial cells appeared to be unaffected by IL-1 (Forloni *et al.*, 1992). Treatment of astrocytes with transforming growth factor-beta 1 resulted in a sixfold increase in total APP mRNA levels whereas exposure to IL-1 had little effect on these cells (Gray and Patel, 1993a).

Interleukin-6 (IL-6) has been implicated in having a role in Alzheimer pathogenesis, possibly by stimulating the synthesis of the protease inhibitor α2-macroglobulin (Ganter *et al.*, 1991). However, there are conflicting results with regard to the effect of IL-6 on neuronal APP expression (Altstiel and Sperber, 1991; Bauer *et al.*, 1993).

Nerve growth factor has been reported to increase the level of APP transcripts in developing hamster brain (Mobley *et al.*, 1988). However, no response to NGF of APP mRNA levels was observed in rat brain or neuronal cultures (Forloni *et al.*, 1993), although it stimulates the release of APP from PC-12 cells (Refolo *et al.*, 1989) and it induces differential splicing of the APP transcripts (Smith *et al.*, 1991).

Basic fibroblast growth factor (bFGF) produces a 5–10-fold increase in APP mRNA levels in C6 glioma cells but causes only a slight increase in PC-12 cells (Quon *et al.*, 1990). A twofold increase in APP mRNA was observed in HUVEC cultures with this factor (Goldgaber *et al.*, 1989). Similarly, bFGF treatment of cultured fibroblasts results in a 4.5-fold increase in APP mRNA and epidermal growth factor causes a twofold increase (Gray and Patel, 1993b).

Heat shock induces a specific 1.6-fold increase in APP mRNA in human lymphoblastoid cells (Abe *et al.*, 1991) and, in response to stimulation with phytohaemagglutinin, synthesis of APP is induced in human peripheral blood leucocytes (Mönning *et al.*, 1990). Furthermore, evidence has been presented that APP expression in culture is also dependent on cell density and the developmental stage of the cells. For example, there is a dramatic increase in APP mRNA levels in confluent HUVEC cultures as compared to cells grown at lower densities (Goldgaber *et al.*, 1989). In a related observation, senescent primary fibroblasts displayed greatly elevated levels of APP mRNA compared to early passage cells. Elevated APP mRNA levels could also be observed in early passage cells that had been growth arrested by serum deprivation (Adler *et al.*, 1991).

These observations suggest that APP gene expression is subject to regulation both in a cell- and tissue-specific manner and in response to external activating factors. Evidence that the APP promoter region *per se* contributes prominently to the regulation of APP gene expression was obtained in transgenic mice. A 4.5 kb DNA fragment from the 5′ end of the human APP gene was placed 5′ to the reporter gene *LacZ*. The expression pattern from the APP promoter was found to closely resemble the distribution pattern of the endogenous APP mRNA (Wirak *et al.*, 1991). This suggests that the APP promoter contains sequence elements which are targets for regulatory transcription factors that mediate cell type specific expression.

5.5 Sequence elements of potential regulatory significance in the APP promoter

The human, mouse and rat APP promoter regions have been cloned and sequenced (Salbaum *et al.*, 1988; La Fauci *et al.*, 1989, Izumi *et al.*, 1992; Chernak, 1993). The sequence of the human (Salbaum *et al.*, 1988) and mouse (Izumi *et al.*, 1992) APP promoter region from position –350 upstream from the transcriptional start site to position +100, is presented in Figure 1. The published sequence of the corresponding rat APP promoter was adapted from Chernak (1993). The APP promoter is devoid of apparent CCAAT and TATA boxes. The main transcriptional start site was mapped by primer extension and nuclease S1 protection in the human promoter and it is indicated at position +1 (Salbaum *et al.*, 1988; LaFauci *et al.*, 1989) (Figure 1). Additional

Figure 1 Sequence of the human (Salbaum *et al.*, 1988; La Fauci *et al.*, 1989), mouse (Izumi *et al.*, 1992) and rat (Chernak, 1993) APP promoter as adapted from the respective references. The numbering system in the rat promoter was here changed to coincide with the transcriptional start site (+1) as determined for the human promoter (Salbaum *et al.*, 1988). Selected putative recognition sequences for regulatory factors are boxed and have been adapted directly

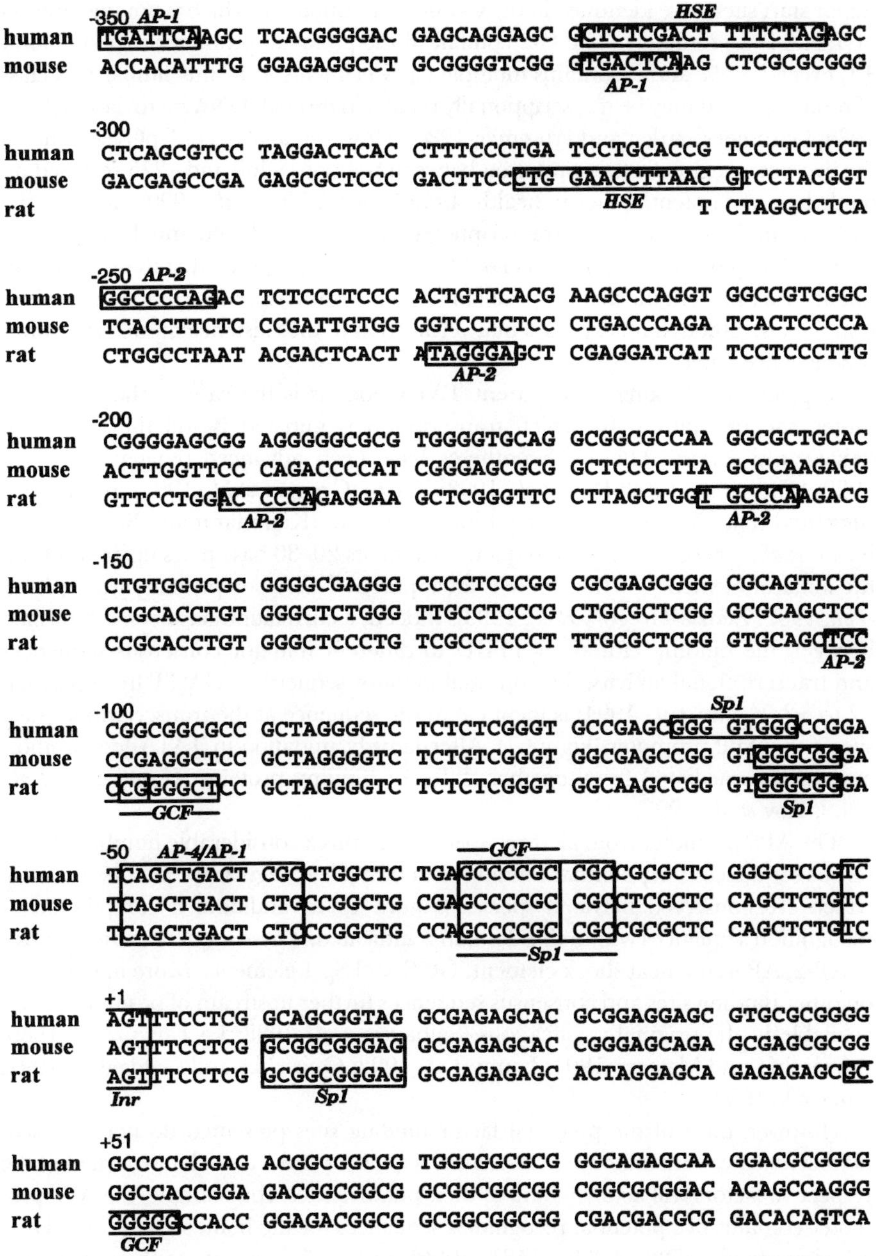

from the respective species as they appear in the original references (Salbaum *et al.*, 1988; La Fauci *et al.*, 1989; Izumi *et al.*, 1992; Quitschke and Goldgaber, 1992; Chernak, 1993). The initiator element between positions –2 and +3 in the vicinity of the main transcriptional start site (+1) is indicated (Inr).

minor start sites were identified in the vicinity of position +1, which is not uncommon in TATA-less promoters. The GC content of the proximal promoter region (–96 to +1) exceeds 75% and it contains multiple CpG elements. It is thus similar to a class of promoters that may be transcriptionally regulated through DNA methylation (Bird, 1986; Gardiner-Garden and Frommer, 1987). However, analysis of CpG elements in the promoter region further upstream between positions –460 and –275 revealed no methylation of potential sites in healthy brain tissue (Milici *et al.*, 1990). Some of the elements in the vicinity of the transcriptional start site may be potentially capable of forming hairpin structures (La Fauci *et al.*, 1989). It was proposed that these structures might either prevent the inappropriate initiation of transcription or function as initiation signals, although no experimental verification of the existence of such structures has been presented.

In promoters lacking an apparent TATA box, as is the case in the APP promoter, the mechanism by which transcription is initiated is still the subject of debate and several different hypotheses have been advanced (Kageyama *et al.*, 1989; Wiley *et al.*, 1992; Roy *et al.*, 1993; Zenzie-Gregory *et al.*, 1993). It has been suggested that TFIID binds to the initiator element (Kaufmann and Smale, 1994; Kraus *et al.*, 1994), as well as to sequence elements 20–30 base pairs upstream from the initiator element whether there is an apparent TATA box present or not (Wiley *et al.*, 1992; Zenzie-Gregory *et al.*, 1993). Indeed, a correlation has been established between the binding affinity of TFIID to different initiator consensus sequences and transcriptional activity. The optimal initiator sequence TCAGT from position –2 to +3 (Kraus *et al.*, 1994) is identical to the sequence at the transcriptional start site of the APP promoter (Figure 1). Alternatively, initiation in TATA-less promoters may be mediated by altogether different binding proteins (Kageyama *et al.*, 1989; Roy *et al.*, 1993).

The APP promoter from all three species contains a considerable number of consensus sequences for potential binding sites of nuclear regulatory factors, many of which are conserved between species (Figure 1). In addition, a compilation of recognition sequences within this domain is indicated. These sequences include AP-1, AP-2, AP-4, the heat shock element, GCF, and Sp 1 elements. More information on other binding sites and consensus sequences further upstream of position –350 is available in the original publications (Salbaum *et al.*, 1988; La Fauci *et al.*, 1989; Beyreuther and Masters, 1991; Izumi *et al.*, 1992; Quitschke and Goldgaber, 1992; Chernak, 1993).

However, most of the potential factor binding sites presented do not represent perfect matches with established consensus sequences and it is not known whether the respective factors will actually bind to their putative sites. For example, the APP promoter contains five potential recognition sequences for the homeobox protein Hox-1.3 at positions –2598, –2278, –1471, –1436, and –913. However, of these sites, only the sequences at positions –1471 and –2598 were shown actually to bind the corresponding transcription factor at an affinity resembling the binding affinity to the optimal consensus sequence (Goldgaber *et al.*, 1991). However, the functional significance of this binding activity is not known.

5.6 Analysis of APP promoter regulation

The primary method used to analyse APP promoter regulation has been to transiently transfect selected 5′ deletions or other modified promoter constructs into various cell lines. The constructs are cloned 5′ to a reporter gene, and the gene product is quantified either enzymatically or immunochemically. To account for experimental variations, the reporter gene is usually cotransfected with a control plasmid that expresses a different gene. The following sections briefly review the results from studies on APP promoter expression that have been performed to date. As will be apparent, the results and interpretations from the different studies show a certain degree of variability.

5.6.1 Analysis of the effect of IL-1 on human APP promoter activity

An increase in the level of IL-1 has been reported in the brains of individuals with Alzheimer's disease and Down's syndrome (Griffin *et al.*, 1989). In addition, IL-1, phorbol 12-myristate 13-acetate, and heparin-binding growth factor I (HBGF-I) were found to increase the level of APP transcript in HUVEC cells (Goldgaber *et al.*, 1989). Of these agents, interleukin-1 displayed the most rapid response with a more than twofold induction after 6 h of exposure. Based on these observations, the effect of interleukin-1 on the expression from the APP promoter was analysed. Promoter deletions were cloned 5′ to the human growth hormone gene and transiently transfected into the mouse neuroblastoma cell line AB1 (Goldgaber *et al.*, 1989; Donnelly *et al.*, 1990). It was found that interleukin-1 stimulated APP promoter expression about twofold. Similar results were obtained in stably transfected AB1 cells (Donnelly *et al.*, 1990).

The sequence element responsible for this effect was traced to a promoter region between positions −485 and −305. This promoter region contains both an AP-1 element at position −350 and a heat shock element at position −317 (Figure 1). It was proposed that interleukin-1 acts on the AP-1 site via a protein kinase C-mediated pathway possibly involving cfos/cjun, since protein kinase inhibitors abolish the response to interleukin-1 (Goldgaber *et al.*, 1989).

5.6.2 The effect of 5′ deletions on human APP promoter activity in HeLa and PC-12 cells

In the human APP promoter, 5′ deletions between positions −751 and −10 were analysed by transient transfection into HeLa and PC-12 cells (Lahiri and Robakis, 1991). The promoter constructs were placed 5′ to the reporter gene chloramphenicol acetyl transferase (CAT). The constructs were cotransfected with a plasmid containing the β-galactosidase gene transcribed from the RSV promoter. The activity from this plasmid was used for normalization of experimental data.

The results indicated a somewhat variable effect on the activity between promoter deletions in PC-12 and HeLa cells. However, in both cell lines the activity of promoter

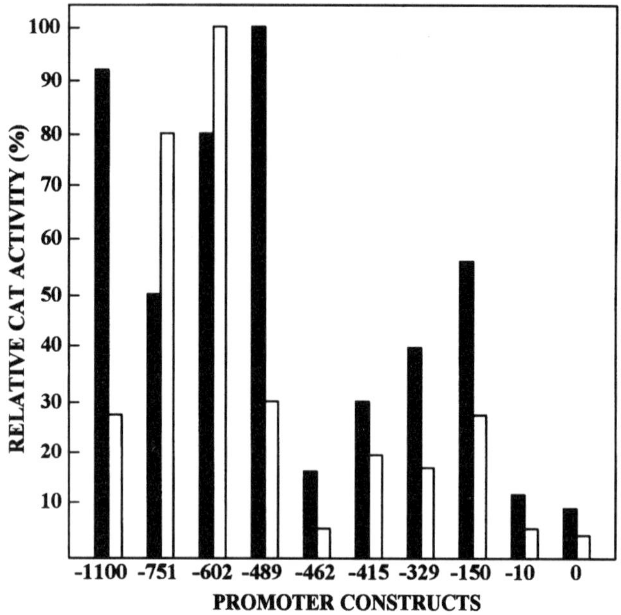

Figure 2 Relative CAT activities generated by transient transfections of human APP promoter deletions into HeLa (shaded bars) and PC-12 cells (white bars). The promoter construct with the highest CAT activity within each cell line was assigned the value 100%. The promoter deletions are indicated by their position upstream from the transcriptional start site (+1). The negative control plasmid (0) does not contain a promoter element. The data were adapted from Lahiri and Robakis (1991).

construct (–751) was the same as the activity of promoter construct (–489) (Figure 2). A further deletion to position –462 resulted in a sharp decline in activity. Activity was gradually, albeit partially, recovered through deletions at positions –415, –329 and –150. An additional deletion at position –10 caused another precipitous decline in promoter activity (Figure 2). These results were interpreted as indicating the existence of two regulatory blocks. One of these blocks is located between positions –600 and –460, which acts as a positive regulator; hence its removal results in a decline in promoter activity. The other element(s) is located between positions –450 and –150, and acts as a negative regulator. Deletions across this domain gradually restore some of the promoter activity.

5.6.3 Analysis of mouse APP promoter activity in mouse neuroblastoma and L cells

A similar deletion analysis was performed on the mouse APP promoter (Izumi *et al.*, 1992). The APP promoter constructs were placed 5′ to the reporter gene CAT. However, in this study the cotransfected control plasmid contained the firefly

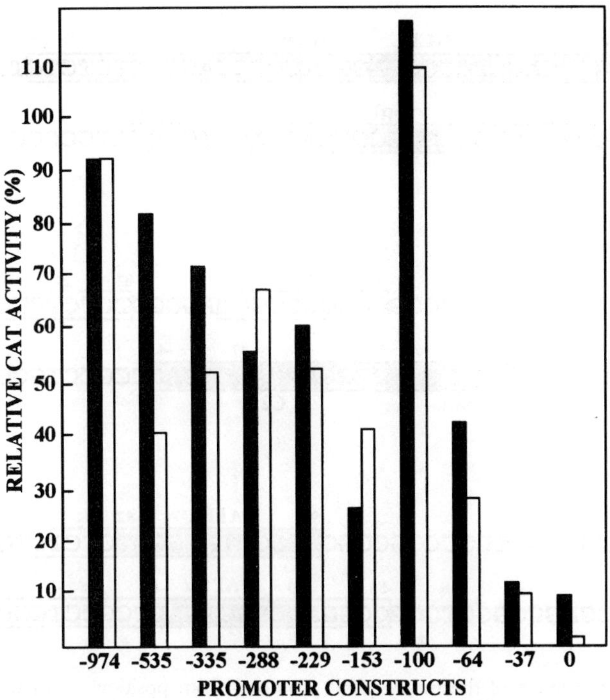

Figure 3 Relative CAT activities of mouse APP promoter deletions analysed by transient transfection into mouse neuroblastoma NB41A3 cells (shaded bars) and L cells (white bars). The activities of the APP promoter deletions were expressed as values relative to a positive control plasmid in which the CAT gene was transcribed from the SV40 promoter (100%). The negative control plasmid (0) is without promoter.

luciferase gene, which was transcribed from the RSV promoter. The activities of APP promoter deletions from position −974 to position −37 were determined by transient transfection in mouse neuroblastoma and L cells. In both cell lines gradual deletions from position −974 to position −153 caused a decrease in promoter activity from 92% (relative to an SV-40 promoter-containing plasmid) to 26% in the neuroblastoma cells and to 41% in L cells (Figure 3). Deleting the promoter to position −100 increased promoter activity to 117% and 109% in neuroblastoma cells and L cells, respectively. Further deletions to position −64 and −37 caused a stepwise decline to near background activity (Figure 3). These results were interpreted to show the existence of a negative regulatory element between positions −153 and −100 and a positive regulatory element between positions −100 and −37.

In order to identify nuclear factors that bind to this domain, a DNase I protection assay was performed with nuclear extracts from neuroblastoma cells and a promoter fragment extending from positions −229 to +1. This revealed the presence of three protected domains indicating the binding of nuclear factors. The protected sequence

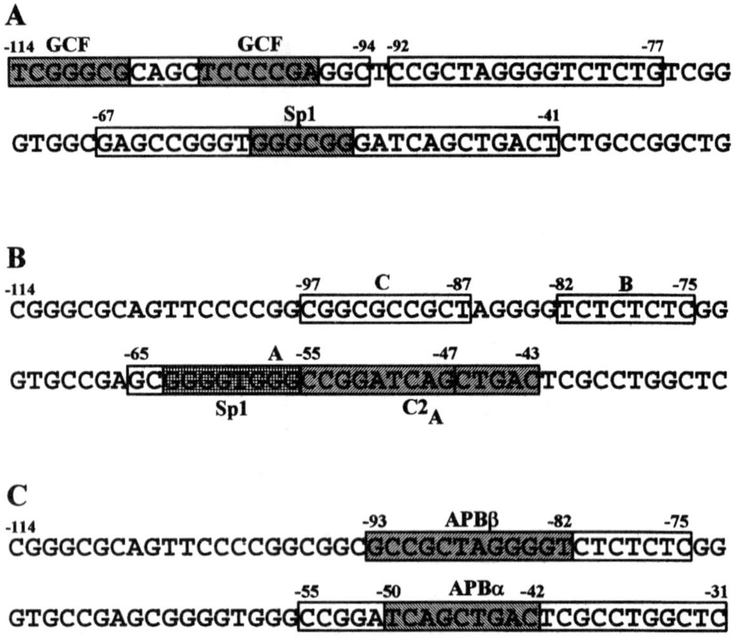

Figure 4 (A) Sequence of the mouse APP promoter from positions –114 to –31. Boxed sequence elements from positions –114 to –94, –92 to –77, and –67 to –41 indicate DNase I protected domains. Within the boxed sequences, shaded domains indicate potential binding sites for regulatory factors GCF and Sp 1. Adapted from Izumi *et al.* (1992). (B) The same sequence as in (A) of the human APP promoter. Boxed sequences show DNase I protected domains C (positions –97 to –87), B (–82 to –75), and A (–65 to –47). The region around domain A is further subdivided into the Sp 1 recognition sequence (shaded) and the DNase I protected domain for $C2_A$ alone, which extends from positions –55 to –43. Adapted from results described by Pollwein *et al.* (1992) and Pollwein (1993). (C) Sequence of the human APP promoter. Boxed sequences show the DNase I protected domain of APBα from positions –55 to –31 and the sequence delineated by methylation interference of APBβ from positions –93 to –82. Shaded areas indicate the core recognition sequences of the respective nuclear binding factors. Adapted from Quitschke and Goldgaber (1992) and Quitschke (1994).

elements were located from positions –114 to –94, –92 to –77, and –67 to –41 (Figure 4A). The binding domain between position –114 and –94 coincides with the negative regulatory element postulated from the deletion analysis. An oligonucleotide containing this element was placed in the control plasmid. It was shown that this element shows some degree of reduction in promoter activity when placed either 5′ to the SV40 promoter, or 3′ to the CAT gene and SV40 enhancer of the control plasmid.

The sequence within this element contains some similarities to the negative regulatory element GCF (consensus sequence GCGGGGC) (Kageyama and Pastan, 1989). Mobility shift gel electrophoresis was performed with oligonucleotides containing this domain and nuclear extracts from mouse liver, kidney and brain. The results demonstrated that the nuclear factor that binds to this domain was present in nuclear extract from liver and kidney, but not from brain. It was also shown by mobility shift electrophoresis that Sp 1 binds to the sequence element between positions –67 and –41 (Figure 4A).

5.6.4 Analysis of human APP promoter activity in HeLa cells

In this study, the largest APP promoter fragment extended to position –488 (Pollwein *et al.*, 1992). This fragment and subsequent deletions were cloned 5′ to the luciferase gene, which was here used as a reporter gene. Plasmid pCH110, which contains the β-galactosidase gene activated by the SV40 promoter, was used as a control plasmid. The promoter constructs were analysed by transient transfection in HeLa cells and the deletion at position –380 was assigned the relative activity of 100%.

A deletion at position –200 resulted in a 50% decline in activity. Promoter activity then remained unchanged at this level to position –94. A further decline to near background levels was observed with deletions at positions –40 and –25 (Figure 5A).

The main decline in promoter activity here was observed between positions –94 and –40 suggesting the presence of an activating domain in this region. To determine if this region can activate a heterologous promoter, the fragment from positions –94 to –35 was cloned 5′ to the TATA box of the herpes simplex virus *tk* gene. This construct showed a 40-fold activation over the TATA box alone. However, the activating effect was only observable if the –94 to –35 fragment was placed close to the transcriptional start site suggesting positional constraints. DNase footprint analysis of the proximal APP promoter region revealed three protected domains: A (–47 to –65), B (–75 to –82), and C (–87 to –97) (Figure 4B). When these domains were placed 5′ to a heterologous TATA box, it was found that domain C is the primary contributor to transcriptional activity in that construct (Figure 5B). In addition, domain A alone increases the level of expression significantly above the level observed with the TATA box alone. Adding domain B to domain A resulted in the same level of activity as domain A alone (Figure 5B). The proximal domain A was interpreted as being a 19 base-pair-long GC-rich sequence, whereas domain C is a palindrome of the sequence GCGGCGCCGC (Pollwein *et al.*, 1992).

In an extension of this study, the proximal domain A was subjected to further analysis (Pollwein, 1993). It was found by mobility shift electrophoresis, DNase foot printing and methylation interference that two factors, $C1_A$ and $C2_A$, bind to this domain. One of these factors, $C1_A$, specifically recognizes the sequence GGGGTGGG. Competition and mutation experiments indicated that the factor $C1_A$ was identical to Sp 1. The other factor, $C2_A$, was found to bind slightly downstream of Sp 1 (Figure 4B). The identity of this factor has not yet been established but the binding site overlaps with the recognition sequence for transcription factor AP-4. However, competition experiments called into question whether the two factors are identical.

Point mutations that either abolished binding to the Sp 1 site, the $C2_A$ site, or both were introduced into the APP promoter fragment extending to position –94. The wild-type sequence and the mutations were analysed by transient transfection in

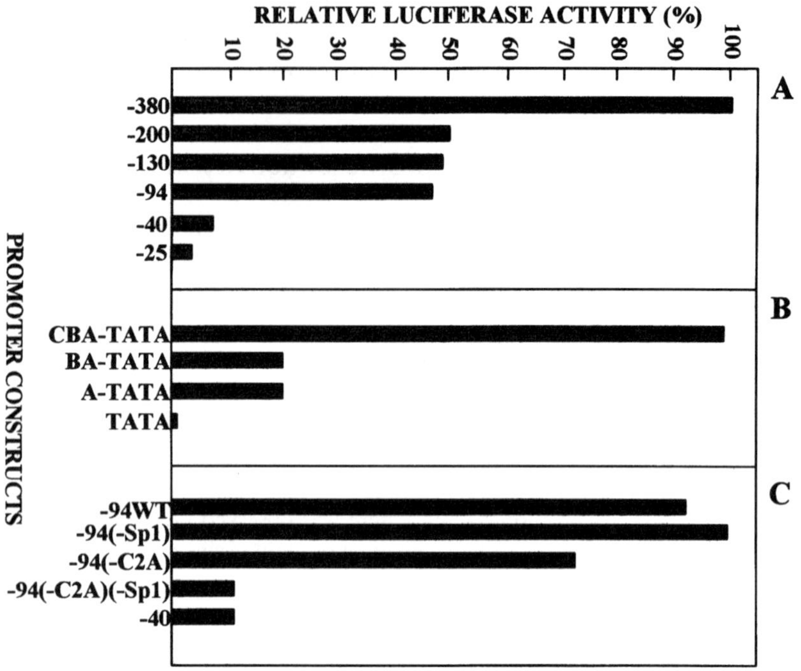

Figure 5 Relative luciferase activities of human promoter constructs transcribing the luciferase gene analysed in HeLa cells. The promoter construct with the highest activity was assigned the value 100% in each series of experiments. (A) Deletion analysis of the APP promoter with the position of deletions indicated. (B) Analysis of APP promoter region containing combinations of DNAse I protected domains C, B, or A placed 5′ to a heterologous promoter fragment containing a TATA box. Adapted from Pollwein *et al.* (1992). (C) Analysis of the human APP promoter region extending to position –94. –94WT represents the unmodified wild-type sequence, –94(–Sp 1) contains mutations that eliminate Sp 1 binding, and –94(–$C2_A$) contains mutations that eliminate $C2_A$ binding. In construct –94(–$C2_A$)(–Sp 1), both binding sites have been eliminated by substitution mutation. A deletion at position –40 is shown for comparison. These results were adapted in modified form from Pollwein (1993).

HeLa cells. The results showed that the wild-type fragment ending at position –94 had an eightfold higher activity than the fragment ending at position –40. Mutations that disrupted either the Sp 1 binding site or the $C2_A$ binding site alone had little effect on promoter activity. However, disrupting both binding sites caused the activity to decline to the level of the deletion at position –40 (Figure 5C).

5.6.5 Comparative analysis of the human APP promoter activity in cultured cell lines

The largest promoter fragment in this study extended to position –2832. Additional 5′ deletions were introduced at positions –1359, –488, –303, –204, –96 and –49

(Quitschke and Goldgaber, 1992). The deletion constructs were introduced into the polycloning site of the specially designed expression vector pCAT2bGAL, which was used for transfection studies. This plasmid contains both the transcriptional unit of the bacterial β-galactosidase gene transcribed from the chicken β-actin promoter and the reporter gene CAT transcribed from APP promoter constructs in a colinear arrangement. The β-actin promoter is devoid of enhancer elements, and it mediates high levels of constitutive expression via a CCAAT and a TATA domain in most cell lines and tissues (Seiler-Tuyns *et al.*, 1984; Quitschke *et al.*, 1989). It is therefore ideally suited as a control for experimental variation. Having both the reporter gene and the control gene with the control promoter on the same plasmid guarantees that the copy number of the two genes is the same in transfected cells.

The APP promoter constructs were placed 5′ to the reporter gene CAT and analysed by transient transfection in PC-12, H4 (human glioma), C6 (rat glioma), 10T1/2 (mouse mesodermal stem cell), and C2C12 (mouse skeletal muscle myoblast) cells. When normalized to the activity of the β-actin promoter, the highest level of expression from the APP promoter was observed in PC-12 cells (Figure 6A) and the lowest level in C2C12 cells. Furthermore, the APP promoter could be deleted from position −2832 to position −94 without showing any significant effect on promoter activity in all cell lines. However, deleting the promoter to position −49 causes at least at 7–10-fold decline in promoter activity in all cell lines (Figure 6A). This suggested that an activator domain located between positions −49 and −96 was removed. Mobility shift electrophoresis, DNase foot printing, and methylation interference with a fragment from positions −96 to +50 revealed a factor binding domain that extended approximately from positions −55 to −31. This binding site will here be referred to as APBα (Figure 4C). Transverse mutations revealed the core recognition sequence of nine base pairs TCAGCTGAC. Block mutations of three base pairs at a time within this domain completely eliminate factor binding. Mutating the three base pairs to the 5′ side of the core (GGA) causes a significant decline in binding activity. The nuclear factor binding to this domain was detected in nuclear extracts from all cell lines examined (Quitschke and Goldgaber, 1992).

However, no effect on promoter activity was observed when these block mutations that eliminated binding to APBα were introduced into the promoter region by site-directed mutagenesis and analysed by transient transfection in HeLa and PC-12 cells (Figure 6B) (Quitschke, 1994). From this observation it was concluded that the contribution of domain APBα to APP promoter activity was marginal, at least within the context of the region extending to position −94. In order to find a second site that is more important for promoter activity, a series of internal deletions of ten base pairs each from positions −31 to −80 and a 5′ deletion at position −77 were analysed to determine their effect on promoter activity. It was found that the 5′ deletion at position −77 was the only construct that caused a significant decline in promoter activity (Figure 6B). This suggested the presence of an upstream activator domain, which here will be designated APBβ. Using methylation interference and mobility shift electrophoresis with wild-type and mutated oligonucleotides across this domain, a 12 base-pair core recognition sequence was identified. Transverse block mutations of six

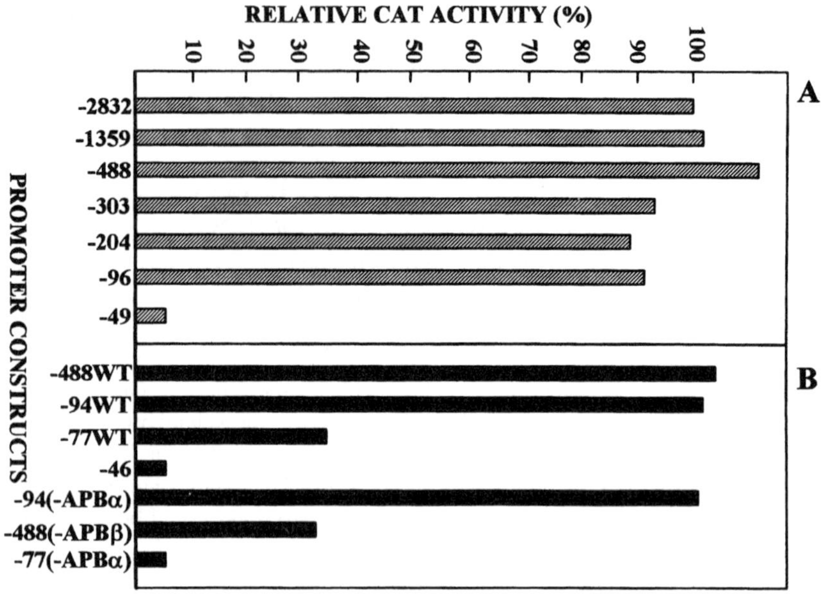

Figure 6 Relative CAT activities of human APP promoter constructs analysed by transient transfection. (A) Deletion analysis of the APP promoter from positions –2832 to –49. The deletions are defined by the positions of the respective restriction sites. The actual position of the deletion may be 1–4 base pairs shorter due to subsequent blunt-ending with T4 DNA polymerase. The activity of the longest promoter construct extending to position –2832 was assigned the value 100%. The presented results are derived from transient transfection of PC-12 cells. However, essentially the same results were obtained in other cell lines as discussed in the text. Results were adapted in modified form from Quitschke and Goldgaber (1992). (B) Analysis of APP promoter constructs containing substitution mutations that eliminate either binding to APBα or APBβ. Construct –94(–APBα) contains only binding site APBβ, –488(–APBβ) contains only binding site APBα. Construct –77(–APBα) contains neither binding site. For comparison, the respective activities of the wild-type deletions are shown from positions –488 to –46. Adapted in modified form from Quitschke (1994).

base pairs each eliminate sequence-specific binding to the domain GCCGCTAGGGGT (Figure 4C). Block mutations 5′ to this sequence have no effect on binding activity, whereas a block mutation on the 3′ side results in a small reduction in binding activity. When the bindng and nonbinding mutations across domain APBβ were analysed by transient transfection, the results showed that the effect on APP promoter activity correlated with *in vitro* binding activity. Specifically, nonbinding mutations showed the same promoter activity as a deletion at position –77, whereas mutations that do not affect binding *in vitro* show the same promoter activity as the deletion at position –94 (Figure 6B). These results suggested that the nuclear factor that binds to APBβ is the primary contributor to APP promoter activity.

Although there is a significant decline in APP promoter activity as a result of removing APBβ, the activity from the deletion at position −77 was still found to be about fourfold higher than the deletion at position −46. This suggests that a small but consistent residual activity is observed in the absence of APBβ. To examine if this residual activity is attributable to binding domain APBα, the block mutations across this domain were analysed by transient transfection within the context of the deletion at position −77, which lacks the upstream APBβ element. In this case, the activity of the constructs containing the block mutations across APBα indeed correlates with the *in vitro* binding activity. The block mutations across the core recognition sequence, which eliminate binding activity, display a promoter activity similar to the deletion at position −46. In contrast, a block mutation which has no effect on binding activity has the same promoter activity as the deletion at position −77. This suggests that APBα may contribute to the activity of the APP promoter. However, in this assay system the contribution of APBα is measurable only in the absence of APBβ.

In summary, 94 base pairs 5′ to the transcriptional start site are sufficient for full cell-type-specific promoter activity. Two nuclear factors bind to this region in a sequence-specific manner (Figure 4C). One of the two factors, here designated as APBα, binds to a recognition sequence located between positions −30 and −58 upstream from the main transcriptional start site (+1). The other factor, APBβ, recognizes a sequence between positions −70 and −94. In quantitative terms, APBβ contributes approximately 70–90% and APBα contributes approximately 10–30% to the transcriptional activity from the APP promoter in all cell lines that were analysed.

5.7 Conclusion

The studies described in the previous sections all address the issue as to which sequence elements in the APP promoter are relevant in regulating gene expression. However, the results and their interpretation are to some degree variable. In four of the studies the APP promoter was analysed by transfecting stepwise deletion constructs into various cell lines under standard culture conditions. The single point of agreement here is that a region between positions −100 and −40 contains one or several activator elements essential for high levels of expression from the APP promoter.

The role of sequence elements upstream from position −100 is subject to different interpretations. Analysis of promoter deletions in HeLa and PC-12 cells (Lahiri and Robakis, 1991) showed a great deal of variation in activity between different deletions upstream from position −150 (Figure 2). In addition, upstream from position −462, identical deletions displayed considerable differences in activity between HeLa and PC-12 cells. No such fluctuations were observed in the other two studies analysing the human APP promoter (Pollwein *et al.*, 1992; Quitschke and Goldgaber, 1992). Pollwein *et al.* (1992) reported a 50% decline in a deletion at position −200 as compared to a deletion at positions −380 (Figure 5A). In contrast, Quitschke and Goldgaber (1992) observed no significant change in activity in promoter constructs ranging from position −2832 to position −96 (Figure 6A). Analysis of the mouse APP promoter yielded yet another set of different results. Here a gradual decline in promoter activity was observed between deletions from position –947 to position –153. This was followed by a 2–4-fold increase in activity in a deletion at position –100 (Figure 3). It was argued that the increase in activity between positions –153 and –100 was caused by the removal of a negative regulatory element, possibly the factor GCF, which binds to the consensus sequence GCGGGGC. However, the consensus sequence in the mouse promoter does not match this sequence perfectly. A similar sequence is also present in the same position in the human promoter and it appears to have no effect on promoter activity (Lahiri and Robakis, 1991; Pollwein *et al.*, 1992; Quitschke and Goldgaber, 1992).

These perplexingly diverse results on the contribution of sequence elements upstream from position –100 to APP promoter regulation require some consideration. All studies agree that constructs carrying only sequence elements between positions –46 and –10 are largely inactive. The difference in activity between these most proximal deletions and deletions upstream of position –100 are in the range of 6–20-fold and consequently quite dramatic. It is therefore conceivable that the effect of removing upstream sequences is considerably more subtle. Therefore, measurements of promoter activities from deletions upstream of position –100 may to a greater extent be subject to experimental variations such as transfection efficiency, which can be quite substantial. Differences in activities in the range of 50% between such deletions could then be difficult to interpret even when normalized to a cotransfected control plasmid. Alternatively, regulatory factors may act on the upstream APP promoter by mechanisms that are not reproducible between different investigators. These factors may be contributed by media components such as serum, subtle differences in the cell lines used, or by other unidentified growth conditions.

Although there is agreement on the importance of the proximal domain downstream of position –100 for APP expression, there are some different interpretations as to which sequences confer this effect. In one study on the human APP promoter (Quitschke and Goldgaber, 1992; Quitschke, 1994), the two nuclear factor binding sites APBα and APBβ were characterized (Figure 4C). Of these, APBβ is essential for high levels of expression in HeLa and PC-12 cells (Quitschke, 1994). The potential core recognition sequence for APBβ extends from positions –93 to –82. Furthermore, methylation interference and mobility shift assays suggest that sequence elements downstream of the core may also have a limited influence on binding activity. This binding domain has yet to be characterized by DNase foot printing. However, based on the minimal length of oligonucleotides required to support binding, it is likely that the sequence occupied by the binding factor extends at least ten base pairs on either side of the core domain (unpublished observation). In view of this, binding domain APBβ partially overlaps with two adjacent binding domains observed by DNase foot printing in the mouse APP promoter (Izumi *et al.*, 1992) (Figure 4C). One of these was suggested to contain negative regulatory elements. It also overlaps with a binding domain C (Figure 4B) described by Pollwein *et al.* (1992), an element that was found to be essential for high levels of transcription when fused to a heterologous TATA box (Figure 5B). Although the interpretations of the recognition sequence vary somewhat, it is possible that binding domain C and APBβ are the same binding site. It is also possible that at least one of the protected domains in the mouse APP promoter corresponds to binding domain C and APBβ (Figure 4).

The other binding site, APBα (Quitschke and Goldgaber, 1992), was mapped by DNAse I foot printing to extend from positions –55 to –31. This site overlaps with binding domain A from positions –65 to –47 (Pollwein *et al.*, 1992), and a proximal site in the mouse promoter from positions –67 to –41, which was shown to bind Sp 1 (Izumi *et al.*, 1992). Binding domain A was shown to contain the binding sites for Sp 1 and a second factor designated $C2_A$. Indeed, DNase foot printing of $C2_A$ revealed a protected domain from positions –55 to –43, which more closely matches the domain established for APBα. These results indicate that the binding sites for factors $C2_A$ and APBα may be identical (Figure 4).

It was demonstrated that, if both the binding sites for $C2_A$ and Sp 1 were eliminated by substitution mutagenesis from the promoter region extending to position –94, promoter activity was essentially abolished even in the presence of domain C. However, eliminating either binding site alone showed little effect on promoter activity (Pollwein, 1993) (Figure 5C). In a different study (Quitschke, 1994), it was shown that, in promoter constructs without APBβ, promoter activity was greatly reduced. However, a residual activity was observed and found to be attributable to APBα (Figure 6B). Hence the Sp 1 site alone could not compensate for the loss of both binding sites APBβ and APBα. The two studies are not necessarily in disagreement since the function of the Sp 1 and APBα sites in the presence of APBβ was not specifically addressed.

In summary, expression from the APP promoter requires at least 94 base pairs upstream from the transcriptional start site. Three factor binding sites were identified in this proximal region that have some role in promoter activity. These are an upstream element APBβ (Quitschke, 1994) or binding domain C (Pollwein *et al.*, 1992), Sp 1 (Izumi *et al.*, 1992; Pollwein, 1993), and $C2_A$ (Pollwein, 1993) or APBα (Quitschke, 1994) (Figure 4). Elements located upstream of position –94 have been proposed to modulate APP promoter activity. These include positive and negative elements in both human and mouse promoters (Lahiri and Robakis, 1991; Izumi *et al.*, 1992). One region between positions –485 and –305 was identified that responds to induction by interleukin-1 (Goldgaber *et al.*, 1989; Donnelly *et al.*, 1990).

Addendum

Since the writing of this chapter additional reports have been published with particular relevance to transcriptional regulation of APP expression. Hoffman and Chernak (1994) have identified two sequence elements in the rat APP promoter that affect promoter activity. Deletion of one of these elements, located between positions –260 and –248 upstream from the *translational* start site, reduces promoter activity by 85%.

Deletion of a second element located between positions −223 and −192 reduces activity by 30%. Both the location of these elements and their effect on promoter activity are remarkably similar to the APBβ and APBα domains described by Quitschke (1994) (see section 6.5).

Trejo *et al.* (1994) have reported the activation of APP expression by protein kinase C with phorbol 12-myristate 13-acetate in a human glioma cell line. It was found that the activation was likely to be mediated by the binding of c-jun homodimers to the upstream AP-1 binding site at position −350 rather than the downstream binding site at position −45 (see Figure 1). These results extend the observations made by Goldgaber *et al.* (1989) and Donnelly *et al.* (1990) (see section 6.1).

Vostrov *et al.* (1995) have presented evidence based on thermostability, relative binding affinity, electrophoretic mobility and antibody recognition that the cellular factors that bind to APBα (see section 6.5) are heterodimers, of USF43 and USF44.

References

Abe, K., St George-Hyslop, P.H., Tanzi, R.E. & Kogure, K. (1991) *Neurosci. Lett.* **125**, 169–171.

Adler, M.J., Coronel, C., Shelton, E., Seegmiller, J.E. & Dewji, N.N. (1991) *Proc. Natl Acad. Sci. USA* **88**, 16–20.

Altstiel, L.D. & Sperber, K. (1991) *Progr. Neuro-Psychopharmac. Biol. Psych.* **15**, 481–495.

Bahmanyar, S., Higgins, G.A., Goldgaber, D., Lewis, D.A., Morrison, J.H., Wilson, M.C., Shankar, S.K. & Gajdusek, D.C. (1987) *Science* **237**, 77–80.

Bauer, J., Strauss, S., Schreiter-Gasser, U., Ganter, U., Schlegel, P., Witt, I., Yolk, B. & Berger, M. (1993) *FEBS Lett.* **285**, 111–114.

Bendotti, C., Forloni, G.L., Morgan, R.A., O'Hara, B.F., Oster-Granite, M.L., Reeves, R.H., Gearhart, J.C. & Coyle J.T. (1988) *Proc. Natl Acad. Sci. USA* **85**, 3628–3632.

Beyreuther, K. & Masters, C.L. (1991) *Brain Res. Rev.* **16**, 86–88.

Bird, A.P. (1986) *Nature* **321**, 209–213.

Card, J.P., Meade, R.P. & Davis, L.G. (1988) *Neuron* **1**, 835–846.

Chernak, J.M. (1993) *Gene* **133**, 255–260.

Cohen, M.L., Golde, T.E., Usiak, M.F., Younkin, L.H. & Younkin, S.G. (1988) *Proc. Natl Acad. Sci. USA* **85**, 1227–1231.

Coyle, J.T., Oster-Granite, M.L., Reeves, R.G., Gearhart, J.D. (1988) *TINS* **11**, 390–394.

DeSauvage, F. & Octave, J. (1989) *Science* **245**, 651–653.

Donnelly, R.J., Friedhoff, A.J., Beer, B., Blume, A.J. & Vitek, M.J. (1990) *Cell. Mol. Neurobiol.* **10**, 485–495.

Forloni, G., Demicheli, F., Giorgi, S., Bendotti, C. & Angeretti, N. (1992) *Brain Res. Mol. Brain Res.* **16**, 128–134.

Forloni, G., Del Bo, R., Angeretti, N., Smiroldo, S., Gabellini, N. & Vantini, G. (1993) *Brain Res.* **620**, 292–296.

Ganter, U., Strauss, S., Jonas, U., Weidemann, A., Bayreuther, K., Volk, B., Berger, M. & Bauer, J. (1991) *FEBS Lett.* **282**, 127–131.

Gardella, J.E., Ghiso, J., Gorgone, G.H., Marratta, D., Kaplan, A.P., Frangione, B. & Gorevic, P.D. (1990) *Biochem. Biophys. Res. Commun.* **173**, 1292–1298.

Gardiner-Garden, M. & Frommer, M. (1987) *J. Mol. Biol.* **196**, 261–282.

Goedert, M. (1987) *EMBO J.* **6**, 3627–3632.

Golde, T.E., Estus, S., Usiak, M., Younkin, L.H. & Younkin, S.G. (1990) *Neuron* **4**, 253–267.

Goldgaber, D., Lerman, M.I., McBride, O.W., Saffiotti, U. & Gajdusek, D.C. (1987) *Science* **235**, 877–880.
Goldgaber, D., Harris, H.W., Hla, T., Maciag, T., Donnelly, R.J., Jacobsen, J.S., Vitek, M.P. & Gajdusek, D.C. (1989) *Proc. Natl Acad. Sci. USA* **86**, 7606–7610.
Goldgaber, D., Schmechel, D.E. & Odenwald, W.F. (1991) *Brain Res. Rev.* **16**, 89–92.
Gray, C.W. & Patel, A.J. (1993a) *Mol. Brain Res.* **19**, 251–256.
Gray, C.W. & Patel, A.J. (1993b) *Neuroreport* **4**, 811–814.
Griffin, W.S.T., Stanley, L.C., Jing, C., White, L., Macleod, V., Perrot, L.J., White, C.L., III & Araoz, C. (1989) *Proc. Natl Acad. Sci. USA* **86**, 7611–7615.
Higgins, G.A., Lewis, D.A., Bahmanyar, S., Goldgaber, D., Gajdusek, D.C., Young, W.G., Morrison, J.H. & Wilson, M.C. (1988) *Proc. Natl Acad. Sci. USA* **85**, 1297–1301.
Hoffman, P.W. & Chernak, J.M. (1994) *Biochem. Biophys. Res. Commun.* **201**, 610–617.
Izumi, R., Yamada, T., Yoshikai, S.-i., Sasaki, H., Hattori, M. & Sakaki, Y. (1992) *Gene* **112**, 189–195.
Jacobsen, J.S., Muenkel, H.A., Blume, A.J. & Vitek, M.P. (1991) *Neurobiol. Aging* **12**, 575–583.
Johnson, S.A., McNeill, T., Cordell, B. & Finch, C.E. (1990) *Science* **248**, 854–856.
Kageyama, R. & Pastan, I. (1989) *Cell* **59**, 815–825.
Kageyama, R., Merlino, G.T. & Paston, I. (1989) *J. Biol. Chem.* **264**, 15508–15514.
Kang, J. & Müller-Hill, B. (1990) *Biochem. Biophys. Res. Commun.* **166**, 1192–1200.
Kang, J., Lemaire, H.-G., Unterbeck, A., Salbaum, M.J., Masters, C.L., Grzeschik, K.H., Multhaup, G., Beyreuther, K. & Müller-Hill, B. (1987) *Nature* **325**, 733–736.
Kaufmann, J. & Smale, S. (1994) *J. Cell. Biochem.* Suppl. 18C, p. 20, Abstract L121.
Kitaguchi, N., Takahashi, Y., Tokushima, Y., Shiojiri, S. & Ito, H. (1988) *Nature* **331**, 530–532.
König, G., Mönning, U., Czeck, C., Prior, R., Baniti, R., Schreiter-Gasser, U., Bauer, J., Masters, C.L. & Beyreuther, K. (1992) *J. Biol. Chem.* **267**, 10804–10809.
Kraus, R.J., Zink, N.M., Murray, E.E., Wiley, S.R., Loritz, K.J. & Mertz, J.E. (1994) *J. Cell. Biochem.* Suppl. 18C, p. 20, Abstract L123.
La Fauci, G., Lahiri, D.K., Salton, S.R.J. & Robakis, N.K. (1989) *Biophys. Res. Commun.* **159**, 297–304.
Lahiri, D.K. & Robakis, N.K. (1991) *Mol. Brain Res.* **9**, 253–257.
Lemaire, H.G., Salbaum, J.M., Multhaup, G., Kang, J., Bayney, R.M., Unterbeck, A., Beyreuther, K. & Müller-Hill, B. (1989) *Nucl. Acids Res.* **17**, 517–522.
Lovett, M., Goldgaber, D., Ashley, P., Cox, D.R., Gajdusek, D.C. & Epstein, C.J. (1987) *Biochem. Biophys. Res. Commun.* **144**, 1069–1075.
Mann, D.M., Jones, D., Prinja, D. & Purkiss, M.S. (1990) *Acta Neuropathol.* **80**, 318–327.
Manning, R.W., Reid, C.M., Lampe, R.A. & Davis, L.G. (1988) *Mol. Brain Res.* **3**, 293–298.
Milici, A., Salbaum, J.M. & Beyreuther, K. (1990) *Biochem. Biophys. Res. Commun.* **169**, 46–50.
Mita, S., Schon, E.A. & Herbert, J. (1989) *Am. J. Path.* **134**,1253–1261.
Mobley, W.C., Neve, R.L., Prusiner, S.B. & McKinley, M.P. (1988) *Proc. Natl Acad. Sci. USA* **85**, 9811–9815.
Mönning, U., König, G., Prior, R., Mechler, H., Schreiter-Gasser, U., Masters, C.L. & Beyreuther, K. (1990) *FEBS Lett.* **277**, 261–266.
Neve, R.L., Finch, E.A. & Dawes, L.R. (1988) *Neuron* **1**, 669–677.
O'Hara, B.F., Fisher, S., Oster-Granite, M.L., Gearhart, J.D. & Reeves, R.H. (1989) *Dev. Brain Res.* **49**, 300–304.
Palmert, M.R., Golde, T.E., Cohen, M.L., Kovacs, D.M., Tanzi, R.E., Gusella, J.F., Usiak, M.F., Younkin, L.H. & Younkin, S.G. (1988) *Science* **241**, 1080–1084.
Pollwein, P. (1993) *Biochem. Biophys. Res. Commun.* **190**, 637–647.
Pollwein, P., Masters, C.L. & Beyreuther, K. (1992) *Nucl. Acids Res.* **20**, 63–68.
Ponte, P., Gonzalez-DeWhitt, P., Schilling, J., Miller, J., Hsu, D., Greenberg, B., Davis, K., Wallace, W., Lieberburg, I. & Fuller, F. (1988) *Nature* **331**, 525–527.
Quitschke, W.W. (1994) *J. Biol. Chem.*, **269**, 21229–21233.
Quitschke, W.W. & Goldgaber, D. (1992) *J. Biol. Chem.* **267**, 17362–17368.

Quitschke, W.W., Lin, Z.-Y., DePonti-Zilli, L. & Paterson, B.M. (1989) *J. Biol. Chem.* **264**, 9539–9546.
Quon, D., Catalano, R. & Cordell, B. (1990) *Biochem. Biophys. Res. Commun.* **167**, 96–102.
Refolo, L.M., Salton, S.R.J., Anderson, J.P., Mehta, P. & Robakis, N.K. (1989) *Biochem. Biophys. Res. Commun.* **164**, 664–670.
Richards, S.-J., Waters, J.J., Beyreuther, K., Masters, C.L., Wischik, C.M., Sparkman, D.R., White, C.L., III, Abraham, C.R. & Dunnet, S.B. (1991) *EMBO J.* **10**, 297–303.
Robakis, N.K., Wisniewski, H.M., Jenkins, E.C., Devine-Gage, E.A., Houck, G.E., Yao, X.L., Ramakrishna, N., Wolfe, G., Silverman, W.P. & Brown, W.T. (1987) *Lancet* **8529**, 384–385.
Rooke, K., Talbot, C., James, L. Hardy, J.A. & Goate, A.M. (1993) *Mammalian Genome* **4**, 662–669.
Roy, A.L., Malik, S., Meisterernst & Roeder, R.G. (1993) *Nature* **365**, 355–359.
Rumble, B., Retallack, R., Hilbich, C., Simms, G., Multhaup, G., Martins, R., Hockey, A., Montgomery, P., Beyreuther, K. & Masters, C.L. (1989) *N. Engl. J. Med.* **320**, 1446–1452.
Salbaum, J.M., Weidemann, A., Lemaire, H.-G., Masters, C.L. & Beyreuther, K. (1988) *EMBO J.* **7**, 2807–2813.
Schmechel, D.E., Goldgaber, D., Burkhart, D.S., Gilbert, J.R., Gajdusek, D.C. & Roses, A.D. (1988) *Alzheimer Dis. Assoc. Disord.* **2**, 96–111.
Selkoe, D.J., Bell, D.S., Podlisny, M.B., Price, D.L. & Cork, L.C. (1987) *Science* **235**, 873–77
Seiler-Tuyns, A., Eldridge, J.D. & Paterson, B.M. (1984) *Proc. Natl Acad. Sci. USA* **81**, 2980–2984.
Shivers, B.D., Hilbich, C., Multhaup, G., Salbaum, M., Beyreuther, K. & Seeburg, P.H. (1988) *EMBO J.* **7**, 1365–1370.
Siman, R., Card, J.P., Nelson, R.B., Davis, L.G. (1989) *Neuron* **3**, 275–285.
Smith, C.J., Wion, D. & Brachet, P. (1991) *Mol. Brain Res.* **10**, 351–354.
Tanaka, S., Nakamura, S., Ueda, K., Kameyama, M., Shiojiri, S., Takahashi, Y., Kitaguchi, N. & Ito, H. (1988) *Biochem. Biophys. Res. Commun.* **157**, 472–479.
Tanaka, S., Shiojiri, S., Takahashi, Y., Kitaguchi, N., Ito, H., Kameyama, M., Kimura, J., Nakamura, S. & Ueda, K. (1989) *Biochem. Biophys. Res. Commun.* **165**, 1406–1414.
Tanzi, R.E., Gusella, J.F., Watkins, P.C., Bruns, G.A.P., St George-Hyslop, P., Van Keuren, M.L., Patterson, D., Pagan, S., Kurnit, D.M. & Neve, R.L. (1987) *Science* **235**, 880–884.
Tanzi, R.E., McClatchey, A.I., Lamperti, E.D., Villa-Komaroff, L., Gusella, J.F. & Neve, R.L. (1988) *Nature* **331**, 528–530.
Trejo, J., Massamiri, T., Deng, T., Dewji, N.N., Bayney, R.M. & Brown, J.H. (1994) *J. Biol. Chem.* **269**, 21682–21690.
Wiley, R.S., Kraus, R.J. & Mertz, J.E. (1992) *Proc. Natl Acad. Sci. USA* **89**, 5814–5818.
Wirak, D.O., Bayney, R., Kundel, C.A., Lee, A., Scangos, G.A., Trapp, B.D. & Unterbeck, A.J. (1991) *EMBO J.* **10**, 289–296.
Vostrov, A.A., Quitschke, W.W., Vidal, F., Schwarzman, A. & Goldgaber, D. (1995) *Nucleic Acids Res.* **23**, In Press.
Yamada, T., Sasaki, H., Dohura, K., Goto, I. & Sakaki, Y. (1989) *Biochem. Biophys Res. Commun.* **158**, 906–912.
Yoshikai, S.-i., Sasaki, H., Doh-ura, K., Furuya, H. & Sakaki, Y. (1990) *Gene* **87**, 257–263.
Zenzie-Gregory, B., Khachi, A., Garraway & Smale, S.T. (1993) *Mol. Cell. Biol.* **13**, 3841–3849.
Zimmermann, K., Herget, T., Salbaum, J.M., Schubert, W., Hilbich, C., Cramer, M., Masters, C.M., Multhaup, G., Kang, J., Lemaire, H.-G., Beyreuther, K. & Starzinski-Powitz, A. (1988) *EMBO J.* **7**, 367–372.

CHAPTER 6

POST-TRANSLATIONAL CONTROL OF THE AMYLOID β-PROTEIN PRECURSOR PROCESSING

Joseph D. Buxbaum

Laboratory of Molecular and Cellular Neuroscience, The Rockefeller University, New York, New York 10021, USA

Table of Contents

6.1 Introduction

6.1.1 APP and Aβ in Alzheimer's disease

A hallmark of Alzheimer's disease (AD) is the build-up of an amyloid β-protein (Aβ) (Glenner and Wong, 1984; Masters *et al.*, 1985) in the brain parenchyma and in the cerebrovasculature (Tomlinson and Corsellis, 1984). Aβ is derived from a large transmembrane precursor, the amyloid β-protein precursor (APP) (Goldgaber *et al.*, 1987; Kang *et al.*, 1987; Robakis *et al.*, 1987; Tanzi *et al.*, 1987, 1988; Kitaguchi *et al.*, 1988; Ponte *et al.*, 1988). For a variety of reasons, many researchers believe that the build-up of Aβ in the brain causes the synaptic loss and associated dementia that occurs in AD. These reasons include the observation that one of the several mutations (hereafter referred to as the Swedish mutation; Mullan *et al.*, 1992) in APP that cosegregate with AD is associated with an abnormally high production of Aβ (Citron *et al.*, 1992; Cai *et al.*, 1993). It therefore seems plausible to argue that increased production of Aβ might underlie the symptoms of AD in individuals bearing this mutation. More recently it has been shown that an allele of apolipoprotein E (ApoE4) is associated with

Pathobiology of Alzheimer's Disease
ISBN 0-12-286965-5

forms of AD (Corder *et al.*, 1993; Strittmatter *et al.*, 1993). The different alleles of ApoE differ in their ability to induce the aggregation and precipitation of Aβ *in vitro* (La Du *et al.*, 1995). In the case of individuals with ApoE4 it is possible that there is an associated increase in Aβ deposition (Schmechel *et al.*, 1993), which again might underlie the symptoms of AD. Most recently, it was shown that three APP mutations secrete increased levels of longer, more aggregatable, Aβ peptide (Suzuki *et al.*, 1994). Thus, there is evidence to suggest that both increased Aβ production and decreased Aβ clearance may contribute to AD. From these findings it is a small jump to argue that decreasing Aβ formation and/or increasing Aβ clearance might slow the progression of AD.

For some researchers this argument is sufficiently compelling to justify the study of the physiological regulation of APP processing and Aβ production. Regulation of APP processing might also give us insights into the function of APP and of its various products. For example, it might suggest that the secreted form of APP (APP_S) has an important role in the body's response to injury; the ability to regulate the secretion of APP_S would then have important ramifications for its function.

6.1.2 Methods for studying the regulation of APP processing

To date there have been several methods used to study the regulation of APP processing. One of the most frequently used is pulse-chase analysis (Weidemann *et al.*, 1989). In this method, cells are first incubated with a radioactive amino acid for a brief period and then incubated in regular medium containing an excess of unlabelled amino acids. During this latter incubation (the chase), test compounds may be added to study their effects on APP processing. The advantage of this method is that by studying only the labelled APP molecules, synthesized before the addition of the test substance, one can dissociate post-translational effects of the test compounds from any other effects. This is particularly important because, as will be detailed below, many of the compounds that affect APP processing also affect APP expression. Other methods designed to distinguish post-translational effects of test compounds from other effects include experiments carried out with a short time course and the inclusion of protein synthesis inhibitors in such experiments.

Another consideration regarding the methods used to assay the regulation of APP processing is the choice of antibodies used to detect the APP. It is advantageous to have antibodies against the amino- and carboxyl-terminal of APP as well as against Aβ in order to simplify analysis of the effects of various test compounds on APP processing. It is also important to use antibodies that can select for APP without cross-reacting with other members of the APP family (APLPs), particularly if one is interested in Aβ production (which is derived only from APP). It is obvious from the homology among the various APP homologues that there could be cross-reactivity between the various antibodies, and this has in fact been observed in practice (e.g. Hunt *et al.*, 1994). A further complicating factor in the choice of antibodies is the fact that, in different cell types and in cells expressing different APP mutations, the cleavage of APP can be different; if the antibody used discriminates between the different products, the analysis of the results may be complex. For example, CHO cells and cells expressing the

Swedish mutation produce significantly more APP_S cleaved at the amino terminal of Aβ domain (β-cleavage) than other cell types. Analysing APP_S from these cells using an antibody (like 6E10; Kim *et al.*, 1990) which cross-reacts with the amino terminal of the Aβ region, would only recognize the normally cleaved (α-cleaved) product.

6.2 Protein kinase C and APP

6.2.1 Protein kinase C regulates APP processing

Soon after the discovery of APP it was noted that the cytoplasmic domain of this molecule could be an efficient subtrate for diverse protein kinases (Gandy *et al.*, 1988). Phosphorylation had been shown to regulate the proteolysis of other transmembrane proteins, thus raising the possibility that generation of Aβ might be regulated by protein phosphorylation. Since those early studies on APP peptides, phosphorylation of APP has been demonstrated with intact APP both *in vitro* and *in vivo. In vitro* phosphorylation has been shown using purified protein kinase C and intact membrane-associated APP (Suzuki *et al.*, 1992). *In vivo* phosphorylation at the Thr^{654}, Ser^{655} and Thr^{668} (the numbering system follows Kang *et al.*, 1987) has been unequivocally demonstrated in both living cells and rat brain using phosphorylation-state specific antibodies (M. Oishi, unpublished observation). The potential significance of these phosphorylation sites is underscored by the fact that they are conserved to varying degrees amongst the different members of the APP family (see Figure 1). Thus, the phosphorylation site at Thr^{668} is conserved across all known APP homologues, while the phosphorylation site at Ser^{655} is conserved in all homologues except APL-1 from *C. elegans*; the site at Thr^{654} seems restricted to vertebrate APP.

At about the same time as the first demonstration that APP may be a phosphoprotein it was shown that APP undergoes cleavage and secretion (Weidemann *et al.*, 1989). This immediately raised the question as to whether this secretion was regulated by protein phosphorylation. To study this question, specific antibodies against the various parts of the APP molecule were used. The first experiments directed at studying the regulation of APP processing made use of a carboxyl-terminal antibody, which precipitated both full-length APP and the carboxyl terminal of APP, which remains behind after the cleavage and secretion of APP_S (we now know that this antibody also cross-reacts with other members of the APP family, see Hunt *et al.*, 1994). It was observed that, when cells were treated with either phorbol dibutyrate (PBt_2), an agent which activates protein kinase C, and/or okadaic acid, an agent which inhibits protein phosphatases 1 and 2A, the levels of mature (fully glycosylated and sulphated) APP diminished significantly, while the levels of carboxyl-terminal derivatives of APP increased (Buxbaum *et al.*, 1990). Because APP_S production from full-length APP involves the loss of full-length APP and the generation of the carboxyl-terminal fragment, it was tempting to speculate that these results could be explained by arguing that

Consensus:	KKKQY.**T**I..HG.VEVD...**T**......PEERHL.KMQ.NGYE**NPTY**KFFEQMQ.
Mammalian APP:	KKKQY**TS**IH-HGVVEVDAAV**T**------PEERHLSKMQQNGYE**NPTY**KFFEQMQN
Xenopus APP:	KKKQY**TT**IH-HGVVEVDAAV**T**------PEERHLTKMQQNGYE**NPTY**KFFEQMQN
Human APLP2/APPH:	RKRQYG**T**IS-HGIVEVDPML**T**------PEERHLNKMQNHGYE**NPTY**KYLEQMQI
Mouse APLP:	KKKPYG**T**IS-HGVVEVDPML**T**------LEEQQLRELQRHGYE**NPTY**RFLEE-RP
C. elegans APL-1:	RRRA-----MRGFIEVD-VY**T**------PEERHVAGMQVNGYE**NPTY**SFFDS-KA
Drosophila APPL:	KWRTSR**S**PHAQGFIEVDQNV**T**THHPIVREEKIVPNMQINGYE**NPTY**KYFEV-KE

Figure 1 Potential phosphorylation sites in the cytoplasmic domain of APP-related proteins. The optimally aligned cytoplasmic domains of the known APP-related proteins are shown. Phosphorylation sites at Thr^{654}, Ser^{655} and Ser^{668} are indicated (numbering follows Kang *et al.*, 1987), as is the highly conserved NPTY sequence involved in targeting APP within the cell. The sequences were taken from the following sources: mammalian APP, Kang *et al.* (1987); *Xenopus* APP, Okado and Okamoto (1992); APLP2/APPH, Sprecher *et al.* (1993), Wasco *et al.* (1993); APLP, Wasco *et al.* (1992); *C. elegans* APL-1, Daigle and Li (1993); *Drosophila* APPL, Rosen *et al.* (1989).

APP_S production was regulated by protein phosphorylation. When cells were incubated with H-7, an inhibitor of several protein kinases including protein kinase C, an increased recovery of cell-associated mature APP was observed, suggesting that protein phosphorylation played a role in basal APP processing in naive (untreated) cells.

Maturation (glycosylation and sulphation) of APP, which is accompanied by a shift in apparent molecular weight in cultured cells, was not affected by phorbol esters, but was affected by okadaic acid. Furthermore, phorbol esters and okadaic acid, especially in combination, caused an increased recovery of an unusually large carboxyl-terminal fragment of APP in PC12 cells. The significance of these latter two effects of protein phosphorylation on APP metabolism is still unknown.

Subsequent to the first study of phosphorylation on APP processing, the effects of protein phosphorylation on APP_S production were studied by several groups. Using antibodies against the amino terminal of APP (which again later were shown to cross-react with other members of the APP family) and studying the APP_S released into the extracellular space, it could be convincingly shown that phorbol esters and/or okadaic acid stimulate the production of APP_S (Caporaso *et al.*, 1992). That this was true for APP *per se* awaited studies carried out in transfected cells (where APP was the predominant representative of the APP gene family; e.g. see Gillespie *et al.*, 1992) and/or studies making use of APP-specific antibodies (e.g. see Buxbaum *et al.*, 1993a). Most recently, H-7 was also demonstrated to inhibit APP_S production (Gabuzda *et al.*, 1993), again suggesting that basal APP processing in naive cells is under the control of protein phosphorylation.

The regulation of APP processing and secretion by phorbol esters and okadaic acid was observed in many cell types, and appears to be a ubiquitous means of regulating APP cleavage and secretion. These include various cell types that might model the cells important in producing Aβ in the brain, such as human endothelial, neuroblastoma and glioma cells (see Buxbaum *et al.*, 1992, 1994).

After it was shown that Aβ was produced normally by cultured cells, the role of protein kinase C, and protein phosphatases 1 and 2B in the regulation of APP processing was extended to include a role in regulating the production of Aβ as well as p3. p3 is an Aβ fragment which is assumed to be derived after normal (α-) cleavage of APP (Haass *et al.*, 1993). Several laboratories have recently shown that activation of protein kinase C and/or inhibition of protein phosphatases 1 and 2A lead to dramatically decreased production of Aβ with increased production of p3 (Buxbaum *et al.*, 1993a; Gabuzda *et al.*, 1993; Hung *et al.*, 1993). If p3 is in fact derived after α-cleavage of APP, then it is not surprising that increasing α-cleaved APP_S production is associated with increased p3 production. The decrease in Aβ production may be in part due to the amounts of full-length APP being limiting: increasing α-cleavage would then decrease the amounts of APP available for β cleavage (see Buxbaum, *et al.*, 1993a, and below). Stimulation of APP_S formation by protein kinase C decreases the levels of carboxyl-terminal APP fragments containing full-length Aβ (Fukushima *et al.*, 1993), consistent with the hypothesis that Aβ is derived from such fragments. As before, the effects of protein kinase C activation on Aβ production have been observed in a variety of cell types, including primary human astrocytes (Gabuzda *et al.*, 1993).

H-7 could apparently cause decreased p3 production (accompanied by decreased APP_S formation) and increased Aβ production in naive cells (Gabuzda *et al.*, 1993). Cells expressing any of several mutations in APP which cosegregate with AD, including the Swedish mutation, still respond to protein kinase C activation with decreased Aβ production (Buxbaum *et al.*, 1993a; Hung *et al.*, 1993), suggesting that regulating the production of Aβ as a therapeutic approach may be possible even in individuals with such mutations.

In summary, activators of protein kinase C stimulate APP_S and p3 formation with a concomitant decrease in Aβ production and in the levels of cell-associated full-length APP. An inhibitor of protein kinases has the opposite effect, while an inhibitor of protein phosphatases acts in a manner similar to activators of protein kinase C. Thus, increasing net phosphorylation in the cell is associated with increased production of APP_S and decreased production of Aβ while decreasing net phosphorylation is associated with decreased APP_S production and increased production of Aβ.

Okadaic acid inhibits both protein phosphatases 1 and 2B. By the use of the phosphatase inhibitor calyculin A, it was recently shown that protein phosphatase 1 is the likely regulator of APP processing (da Cruz e Silva *et al.*, 1995). Phorbol esters can activate several different protein kinase C isozymes: there is currently evidence suggesting that protein kinase C-α may be an example of an isozyme of protein kinase C which can regulate APP processing (Slack *et al.*, 1993).

6.2.2 The cytoplasmic domain of APP does not mediate the effects of protein kinase C

The mechanism(s) by which protein kinase C activation and protein phosphatases 1 and/or 2B inhibition regulated APP processing were assumed to involve the phosphorylation of APP in the cytoplasmic domain by protein kinase C or other protein kinases. To test this assumption, APP molecules mutated in the cytoplasmic domain were studied for their response to protein phosphorylation. Point mutations of potential phosphorylation sites in the cytoplasmic domain had no effect on the phorbol ester regulation of APP_S and Aβ formation (da Cruz e Silva *et al.*, 1993; Hung and Selkoe, 1994). Even deletion of the entire cytoplasmic domain of APP did not affect the phorbol ester-induced secretion of APP_S. Thus, the well-characterized phosphorylation sites in the APP cytoplasmic domain apparently are not necessary in the regulation of APP processing by protein kinase C. A large portion of the extracellular domain (between residues of 78 and 590 of APP_{695}), which includes the major site of phosphorylation of APP *in vivo*, is also without an obvious role in the regulation of APP processing by protein kinase C (da Cruz e Silva *et al.*, 1993; Hung and Selkoe, 1994). Furthermore, there is no major change in the phosporylation of APP caused by kinase C activation (Gabuzda *et al.*, 1993; Hung and Selkoe, 1994), although using phosphorylation-state specific antibodies, changes have been shown to occur (M. Oishi, unpublished observation). These results indicate that protein kinase C regulates APP processing by the phosphorylation of some component of the processing pathway other than the amyloid precursor protein. This conclusion in turn raises two questions: (1) what is the role of the highly conserved potential phosphorylation sites in the cytoplasmic domain; and (2) what is the substrate for protein kinase C which is responsible for its effects on APP processing?

A role for phosphorylation has recently been proposed (Suzuki *et al.*, 1994) for what appears to be the most highly conserved phosphorylation site in the cytoplasmic domain of APP, i.e. Thr^{668}. It has been shown that cell-cycle-dependent changes in cdc2 kinase correlate with changes in the migration of certain APP species in PAGE, the levels of APP_S and APP carboxyl-terminal fragment, and the phosphorylation of full-length APP and APP carboxyl-terminal fragment. A possible causal relationship between changes in phosphorylation of APP and changes in its processing is supported by the observation that okadaic acid treatment leads to activation of cdc2 kinase, and increased levels of immature APP and APP carboxyl-terminal fragment; these same changes in APP processing also occur at the G_2/M phase of the cell cycle where cdc2 is active.

The mechanism by which protein kinase C regulates APP processing may be analogous to the way this enzyme regulates the processing of other transmembrane proteins, such as pro-TGFα, and the CSF-1 and TNF receptors, where phosphorylation of the transmembrane protein is apparently also not necessary. Studies with permeabilized and/or broken cells are being carried out to determine whether activation of a protease and/or modulation of a trafficking protein are the means by which protein kinase C regulates APP processing.

The effects of okadaic acid on APP processing have not yet been examined to the same degree as those of phorbol esters. While it may be assumed that the effects of okadaic acid parallel those of phorbol esters, it has been shown that okadaic acid treatment, in contrast to phorbol ester treatment, is associated with dramatically increased phosphorylation of full-length APP and APP carboxyl-terminal fragment *in vivo* (Suzuki *et al.*, 1994).

6.3 Regulation of APP processing by intracellular messengers

6.3.1 Regulation of APP processing by phospholipase C-linked first messengers

With the demonstration of a role for protein kinase C in the regulation of APP processing, it was predicted that various first messengers, which activate the phospholipase C/protein kinase C cascade, would also be capable of regulating APP processing. This prediction was confirmed in several subsequent studies. Cholinergic agonists were shown to regulate APP_S formation by several groups: the effects could be mediated by muscarinic receptors, particularly muscarinic receptors known to be coupled to the phospholipase C/protein kinase C cascade (Buxbaum *et al.*, 1992; Nitsch *et al.*, 1992). Acetylcholine is altered in Alzheimer's disease brain, making these studies par-

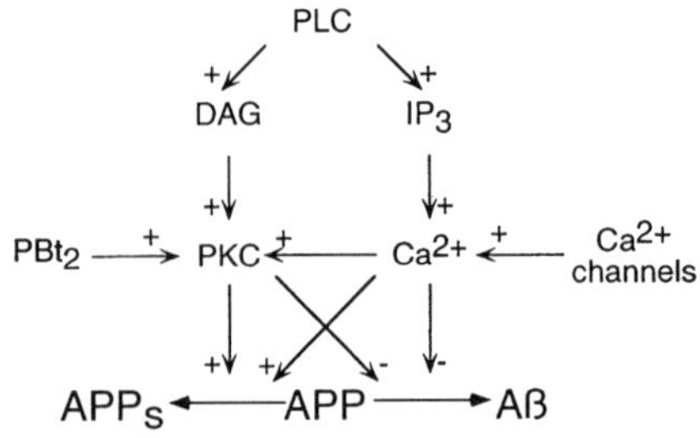

Figure 2 Scheme for the redundant regulation of APP processing by phospholipase C activation. PLC, phospholipase C; IP_3, inositol tris-phosphate; DAG, diacylglycerol; Ca^{2+} cytoplasmic calcium; PBt_2, phorbol dibutyrate; PKC, protein kinase C. Adapted from Buxbaum *et al.* (1993).

ticularly interesting. Interleukin-1 is also altered in Alzheimer's disease brain and cerebrospinal fluid (Griffin *et al.*, 1989; Cacabelos *et al.*, 1991) and can activate the phospholipase C/protein kinase C cascade. In these early studies interleukin-1 was also shown to be able to regulate APP_S formation (Buxbaum *et al.*, 1992; also Buxbaum *et al.*, 1994).

With the observation that Aβ is normally produced by cells, the effects of activation of the phospholipase C/protein kinase C cascade on Aβ formation were studied. Direct activation of this cascade by mastoparan and mastoparan X increased the formation of APP_S while decreasing the formation of Aβ (Buxbaum *et al.*, 1993a). Similarly, muscarinic agonists could decrease Aβ production in cells overexpressing the M1 or M3 muscarinic receptors (Hung *et al.*, 1993; Buxbaum *et al.*, 1994). In human neuroglioma cells cholinergic and muscarinic agonists, as well as interleukin-1, could again regulate Aβ production to varying degrees (Buxbaum *et al.*, 1994). The surprise came when the protein kinase C in cells overexpressing muscarinic receptors was down-regulated. In such cells it was predicted that cholinergic agonists should have little effect, however, no difference was observed in cells lacking phorbol ester-stimulated protein kinase C when compared to control cells, when examined for muscarinic agonist regulation of APP processing (Buxbaum *et al.*, 1994). This was interpreted as suggesting that the effects of the muscarinic agonists could be mediated by either the phospholipase C/protein kinase C cascade or by the phospholipase C/calcium cascade (see Figure 2).

In summary, various protein kinase C/phospholipase C linker first messengers have been shown to regulate APP processing. These include acetylcholine, other cholinergic agonists and interleukin-1, as well as bradykinin, thrombin and ATP. A recent report raises the possibility that the effects of phospholipase C-linked first messengers on APP processing may involve the activation of phospholipase A_2 (Emmerling *et al.*, 1993). While the activation of protein kinase C by various phospholipase C-linked first messengers may be sufficient to mediate the effects of these first messengers on APP processing, it may not be necessary as increased cellular calcium in response to these first messengers may have similar effects as activation of protein kinase C. Finally, comparing the relative efficiencies of different first messengers for their ability to regulate APP_S and Aβ production indicates that, for a given cell, compounds that are better able to stimulate APP_S production are generally better able to inhibit Aβ formation (see Table 2 in Buxbaum *et al.*, 1994), consistent with APP being rate-limiting in the formation of these metabolites (Buxbaum *et al.*, 1993a).

6.3.2 Hormonal and growth factor regulation of APP processing and/or transcription

Several physiological compounds, including cytokines, growth hormones and oestrogen, have been studied for their effects on APP levels over several days. As mentioned previously, it is not always easy to distinguish between transcriptional and post-translational effects of the test compounds in these sorts of studies. For example, incubating PC12 cells with NGF for 2 or more days increases the amount of APP_S recovered in conditioned medium (Refolo *et al.*, 1989; Schubert *et al.*, 1989; Fukuyama, 1993). This may be caused by both effects of NGF on APP transcription (Mobley *et al.*, 1988; Smith 1993; Ohyagi, 1993) as well as on APP processing [as judged by the relative changes in holoAPP and APP_S following shorter NGF treatment (Refolo *et al.*, 1989; Schubert *et al.*, 1989; but see Fukuyama *et al.*, 1993)]. Prolonged EGF treatment of cells can also lead to increased levels of APP_S (Refolo *et al.*, 1989) as can prolonged TNFα, TFGβ and PDGF treatment (Vaughan and Cunningham, 1993). Several of these growth factors have been shown to regulate APP transcription (Gray and Patel, 1993a,b). Oestrogen treatment of cultured cells also increases APP_S levels (Jaffe, *et al.*, 1994).

6.4 Regulation of APP processing by calcium

6.4.1 A role for calcium in the regulation of APP processing in cultured cells

A potential role for intracellular calcium in the regulation of APP processing was first demonstrated using the calcium ionophore A23187. Treating B-104 neuroblastoma cells or differentiated PC12 cells with A23187 led to increased production of APP_S (Loffler and Huber, 1993). It was recently shown that A23187 treatment increased the production of Aß in 293 cells (Querforth and Selkoe, 1994). These data suggest that voltage or ligand-gated calcium channels could regulate APP processing. Calcium released from intracellular stores has also been implicated in the regulation of APP processing. Treating various cells with thapsigargin or cyclopiazonic acid, compounds which inhibit the endoplasmic reticulum calcium ATPase, leading to the release of calcium into the cytoplasm, caused increased APP_S formation (Buxbaum *et al.*, 1994).

Significantly, the effects of these compounds on APP_S formation were still observed in cells which had been treated for 24 h with phorbol esters to down-regulate protein kinase C, suggesting that the effects of these compounds are independent of protein kinase C. Under most conditions, increased APP_S formation induced with thapsigargin and cyclopiazonic acid was accompanied by decreased Aβ formation. However, in the presence of low concentrations of thapsigargin, increases in Aβ, or an Aβ-like peptide, were observed. Similarly, caffeine increased Aβ production in 293 cells (Querforth and Selkoe, 1994).

Evidence for a physiological role of calcium in the regulation of APP_S and Aβ formation was mentioned above. In cells transfected with the M3 receptor and treated for 24 h with phorbol esters to down-regulate protein kinase C, carbachol was able to stimulate APP_S formation and inhibit Aβ production. This indicates that the phospholipase C/calcium cascade is able to regulate APP processing in a protein kinase C-independent manner (see Buxbaum *et al.*, 1994 and Figure 2). Similar results were observed with cells treated for 24 h with phorbol esters to down-regulate protein kinase C and subsequently treated with ATP.

In summary, there is now compelling evidence that calcium, derived from either extracellular or intracellular sources, can regulate APP processing. The mechanism by which calcium exerts its effects on APP processing is as yet unknown. The role of the cytoplasmic phosphorylation sites of APP in calcium-regulated APP processing is under investigation; preliminary studies indicate that they may not be necessary. It is interesting to consider the analogies between pro-TGFα and APP: both proteins can undergo secretory processing, which is stimulated by protein kinase C or calcium (see Pandiella and Massague, 1991) and for neither protein is phosphorylation of the holoprotein apparently required to mediate the effects of the stimulatory agent(s).

6.4.2 A role for calcium in the regulation of APP processing in platelets

Calcium can apparently also regulate APP processing and secretion in platelets. Incubating platelets with either thrombin, calcium ionophore or collagen stimulates the release of APP or APP fragments from the cell (Bush *et al.*, 1990; Gardella *et al.*, 1990; Smith *et al.*, 1990, Van Nostrand *et al.*, 1990; Schlossmacher *et al.*, 1992). The release of APP from platelets involves cleavage of full-length APP, probably within the Aβ domain, although some evidence for the release of full-length APP upon platelet activation has been reported by some researchers (Bush *et al.*, 1990; Gardella *et al.*, 1990), but not others (Schlossmacher *et al.*, 1992). The effects of thrombin and calcium ionophores on platelets probably involve a mechanism which is different from that which mediates their effects on neuronal and cultured cells because, for platelets, the APP is localized to α-granules where it is released upon degranulation. Interestingly, hyperacidification of platelets from patients with severe Alzheimer's disease in response to thrombin has recently been reported (Davies *et al.*, 1993); this abnormality may cause abnormal granule, and hence APP, secretion.

6.5 Regulation of APP processing *in vivo*

6.5.1 Studying APP processing *in vivo*

Several studies have examined the processing of APP in both the peripheral and central nervous system. Such studies, when carried out on mice and rats, have the limitation that it has been impossible (at the time of writing) to distinguish APP from APLPs. With cells in culture, one can overexpress APP or use APP-specific antibodies to identify APP. For *in vivo* studies, one must use either: transgenic animals (which have only recently become available, e.g. Lamb *et al.*, 1993; Pearson and Choi, 1993); APP-specific antibodies, which cross-react with mouse or rat APP (the two widely used APP-specific antibodies, P2-1 and 6E10, preferentially recognize human, and not mouse or rat, APP; e.g. Buxbaum *et al.*, 1993b); or use other species in which APP shares sufficient sequence homology with human APP to cross-react well with existing APP-specific antibodies (for example, guinea pig APP will cross-react with the APP-specific antibody 6E10). For these reasons, unless otherwise indicated, all of the *in vivo* results should be taken as studying APP *and/or* APLP processing.

6.5.2 Normal processing of APP in the brain

Studies on the transport of APP *in vivo* indicate that APP is transported as a full-length molecule via fast axonal transport, after which it can be metabolized (Koo *et al.*, 1990; Morin *et al.*, 1993; Sisodia *et al.*, 1993). The first indication of regulation of processing of APP in the central nervous system was made when APP processing was studied in brain slices (Nitsch *et al.*, 1993). In these experiments, it was shown that electrical stimulation caused the release of APP_S into the superfusate. This effect could be inhibited by tetrodotoxin.

It is of interest to determine whether the various means of regulating APP processing identified in cultured cells also are important in the brain. Several groups are pursuing this question in brain slices as well as in living animals. As an example of the latter approach, neurotoxin treatment associated with persistent protein kinase C activation is correlated with increased APP_S formation in rat brains, which is apparently caused, at least in part, by changes in APP processing (F. Cattabeni, unpublished observations). A related discovery demonstrated that lithium treatment, which alters neurotransmitter function, alters APP_S levels in humans (Clarke *et al.*, 1993a, b).

6.5.3 Changes in APP processing in ageing

The levels of various APP-derived proteins are altered in the brain and cerebrospinal fluid (CSF) in ageing and in AD (Palmert *et al.*, 1990; Nordstedt *et al.*, 1991; Van Nostrand *et al.*, 1992). These changes might reflect alterations in APP processing, as evidenced by differential changes with ageing in the levels of holoAPP and in carboxy-terminal fragments of APP in brain (Nordstedt *et al.*, 1991), and differential changes with ageing in the levels of two different soluble APP derivatives in CSF (Palmert *et al.*, 1990). Changes in APP processing associated with ageing may be more pronounced in AD and may correlate with the degree of dementia (Palmert *et al.*, 1990).

6.5.4 Regulation of APP processing by injury

Many groups have noted changes in APP levels and/or distribution following brain injury (Siman *et al.*, 1989; Roberts *et al.*, 1990; Abe *et al.*, 1991; Kawarabayashi *et al.*, 1991; Otsuka *et al.*, 1991; Iverfeldt *et al.*, 1993; Kalaria *et al.*, 1993; Wallace, *et al.*, 1993; Griffin, *et al.*, 1994). These changes reflect, in part, changes in APP expression and/or changes in the constellation of cells found at the injury site. However, there is some evidence for changes in APP processing following brain injury (Iverfeldt *et al.*, 1993). By studying APP processing in areas of nerve terminal degeneration, far from the lesion site, it could be shown that changes in APP processing are caused by neuronal degeneration affecting APP processing, rather than altered APP processing affecting neuronal degeneration. The importance of these observations is underscored when one realizes that head injury is a risk factor for AD (e.g. Mortimer *et al.*, 1991; van Duijn *et al.*, 1992; Mayeux *et al.*, 1993); other possible risk factors for AD, such as epilepsy and critical coronary artery disease (McAreavey *et al.*, 1992; Sparks *et al.*, 1993), are also associated with changes in APP expression and/or processing (Sparks *et al.*, 1993; Sheng *et al.*, 1994a).

One possible cause for the changes in APP levels and/or distribution following brain injury is that injury-related changes in interleukin-1 levels alter APP expression (see Goldgaber *et al.*, 1989; Donnelly *et al.*, 1990) and/or processing (Buxbaum *et al.*, 1992, 1994), or alter the constellation of cells surrounding the site of injury (e.g. Giulian *et al.*, 1988). Interleukin-1 levels are altered in AD (Griffin *et al.*, 1989; Cacabelos *et al.*, 1991), and the levels of interleukin-1 correlate with the degree of dementia (Cacabelos *et al.*, 1991). Correlation between increased interleukin-1 levels and increased APP levels have been observed in other neuropathological conditions, including HIV associated neurodegeneration, temporal lobe epilepsy and head injury (Griffin *et al.*, 1994; Sheng *et al.*, 1994a; Stanley *et al.*, 1994). To study the effects of interleukin-1 on APP *in vivo*, interleukin-1 has been injected into the brains of experimental animals (Sheng *et al.*, 1994b). APP levels, as measured by Western blotting, were elevated following interleukin-1 injection; studies aimed at detecting changes in APP processing under these conditions are currently under way.

6.6 Summary

APP processing appears to be under complex regulation. This regulation is apparently important under both normal and pathological conditions. Of direct clinical interest is the observation that Aβ formation can be regulated by various means. This raises the possibility that altered APP processing may cause an increase in Aβ formation in

AD, and suggests that it may be possible to regulate the production of Aβ as a therapeutic approach in AD. As an example of the utility of the latter approach, consider a patient carrying the Swedish APP mutation. If it is true that the cause of AD in such a patient is a result of increased Aβ production, then decreasing Aβ production should delay the onset of the disease. Even in individuals where increased Aβ formation is not the cause of AD but some other cause, such as the presence of ApoE4, which causes Aβ accumulation and hence synaptic loss, decreasing Aβ formation may be beneficial. It is of course a very long way from *in vitro* experiments to therapy. The current emphasis on studying APP processing and its regulation *in vivo* represents the next step towards this goal. The observation that lithium treatment alters APP_S levels in CSF of patients supports regulating APP processing as a therapeutic approach in AD.

References

Abe, K., Tanzi, R.E. & Kogure, K. (1991) *Neurosci. Lett.* **125**, 172–174.

Bush, A.I., Martins, R.N., Rumble, B., Moir, R., Fuller, S., Milward, E., Currie, J., Ames, D., Weidemann, A., Fischer, P., Multhaup, G., Beyreuther, K. & Masters, C.L. (1990) *J. Biol. Chem.* **265**, 15977–15983.

Buxbaum, J.D., Gandy, S.E., Cicchetti, P., Ehrlich, M.E., Czernik, A.J., Fracasso, R.P., Ramabhadran, T.V., Unterbeck, A.J. & Greengard, P. (1990) *Proc. Natl Acad. Sci. USA* **87**, 6003–6006.

Buxbaum, J.D., Oishi, M., Chen, H.I., Pinkas-Kramarski, R., Jaffe, E.A., Gandy, S.E. & Greengard, P. (1992) *Proc. Natl Acad. Sci. USA* **89**, 10075–10078.

Buxbaum, J.D., Koo, E.H. & Greengard, P. (1993a) *Proc. Natl Acad. Sci. USA* **90**, 9195–9198.

Buxbaum, J.D., Christensen, J.L., Ruefli, A.A., Greengard, P. & Loring, J.F. (1993b) *Biochem. Biophys. Res. Commun.* **197**, 639–645.

Buxbaum, J.D., Ruefli, A.A., Parker, C.A., Cypess, A.M. & Greengard, P. (1994) *Proc. Natl Acad. Sci. USA* **91**, 4489–4493.

Cacabelos, R., Barquero, M., Garcia, P., Alvarez, X.A. & Varela de Seijas, E. (1991) *Meth. Find. Exp. Clin. Pharmacol.* **13**, 455–458.

Cai, X.D., Golde, T.E. & Younkin, S.G. (1993) *Science* **259**, 514–516.

Caporaso, G.L., Gandy, S.E., Buxbaum, J.D., Ramabhadran, T.V. & Greengard, P. (1992) *Proc. Natl Acad. Sci. USA* **89**, 3055–3059.

Citron, M., Oltersdorf, T., Haass, C., McConlogue, L., Hung, A.Y., Seubert, P., Vigo, P.C., Lieberburg, I. & Selkoe, D.J. (1992) *Nature* **360**, 672–674.

Clarke, N.A., Procter, A.W., Webster, M.T., Francis, P.T., Hodgkiss, A.D. & Bowen, D.M. (1993a) *Biochem. Soc. Trans.* **21**, 2425.

Clarke, N.A., Webster, M.-T., Francis, P.T., Proctor, A.W., Hodgkiss, A.D. & Bowen, D.M. (1993b) *Neurodegeneration* **2**, 243–248.

Corder, E.H., Saunders, A.M., Strittmatter, W.J., Schmechel, D.E., Gaskell, P.C., Small, G.W., Roses, A.D., Haines, J.L. & Pericak-Vance, M.A. (1993) *Science* **261**, 921–923.

da Cruz e Silva, O.A.B., Iverfeldt, K., Oltersdorf, T., Sinha, S., Lieberburg, I., Ramabhadran, T.V., Suzuki, T., Sisodia, S.S., Gandy, S. & Greengard, P. (1993) *Neuroscience* **57**, 873–877.

da Cruz e Silva, E.F., da Cruze e Silva, O.A.B., Zaia, C.T.B.V. & Greengard, P. (1995) *Molecular Medicine*, **1**, 535–541.

Daigle, I. & Li, C. (1993) *Proc. Natl Acad. Sci. USA* **90**, 12045–12049.

Davies, T.A., Fine, R.E., Johnson, R.J., Levesque, C.A., Rathbun, W.H., Seetoo, K.F., Smith,

S.J., Strohmeier, G., Volicer, L., Devla, L. *et al.* (1993) *Biochem. Biophys. Res. Commun.* **194**, 537–543.
Donnelly, R.J., Friedhoff, A.J., Beer, B., Blume, A.J. & Vitek, M.P. (1990) *Cell. Mol. Neurobiol.* **10**, 485–495.
Emmerling, M.R., Moore, C.J., Doyle, P.D., Carroll, R.T. & Davis, R.E. (1993) *Biochem. Biophys. Res. Commun.* **197**, 292–297.
Fukushima, D., Konishi, M., Maruyama, K., Miyamoto, T., Ishiura, S. & Suzuki, K. (1993) *Biochem. Biophys. Res. Commun.* **194**, 202–207.
Fukuyama, R., Chandrasekaran, K. & Rapoport, S.I. (1993) *Brain Res. Mol. Brain Res.* **17**, 17–22.
Gabuzda, D., Busciglio, J. & Yankner, B. (1993) *J. Neurochem.* **61**, 2326–2329.
Gandy, S., Czernik, A.J. & Greengard, P. (1988) *Proc. Natl Acad. Sci. USA* **85**, 6218–6221.
Gardella, J.E., Ghiso, J., Gorgone, G.A., Marratta, D., Kaplan, A.P., Frangione, B. & Gorevic, P.D. (1990) *Biochem. Biophys. Res Commun.* **173**, 1292–1298.
Gillespie, S.L., Golde, T.E., & Younkin, S.G. (1992) *Biochem. Bioiphys. Res. Commun.* **187**, 1285–1290.
Giulian, D., Woodward, J., Young, D.G., Krebs, J.F. & Lachman, L.B. (1988) *J. Neurosci.* **8**, 2485–2490.
Glenner, G.G. & Wong, C.W. (1984) *Biochem. Biophys. Res. Commun.* **120**, 885–890.
Goldgaber, D., Lerman, M.I., McBride, W.O., Saffiotti, U. & Gajdusek, D.C. (1987) *J. Neural Transm. Suppl.* **24**, 23–28.
Goldgaber, D., Harris, H.W., Hla, T., Maciag, T., Donnelly, R.J., Jacobsen, J.S., Vitek, M.P. & Gajdusek, D.C. (1989) *Proc. Natl Acad. Sci. USA* **86**, 7606–7610.
Gray, C.W. & Patel, A.J. (1993a) *Brain Res. Mol. Brain Res.* **19**, 251–256.
Grey, C.W. & Patel, A.J. (1993b) *Neuroreport* **4**, 811–814.
Griffin, W.S., Stanley, L.C., Ling, C., White, L., MacLeod, V., Perrot, L.J., White, C.L. & Araoz, C. (1989) *Proc. Natl Acad. Sci. USA* **86**, 7611–7615.
Griffin, W.S.T., Sheng, J.G., Gentleman, S.M., Graham, D.I., Mrak, R.E. & Roberts, G.W. (1994) *Neurosci. Lett.* **176**, 133–136.
Haass, C., Hung, A.Y., Schlossmacher, M.G., Teplow, D.B. & Selkoe, D.J. (1993) *J. Biol. Chem.* **268**, 3021–3024.
Hung, A.Y. & Selkoe, D.J. (1994) *EMBO J.* **13**, 534–542.
Hung A.Y., Haass, C., Nitsch, R.M., Qiu, W.Q., Citron, M., Wurtman, R.J., Growdon, J.H. & Selkoe, D.J. (1993) *J. Biol. Chem.* **268**, 22959–22962.
Hunt, H.H., Thinakaran, G., Von Koch, C., Lo, A.C.Y., Tanzi, R.E. & Sisodia, S.S. (1994) *J. Biol. Chem.* **269**, 2637–2644.
Iverfeldt, K., Walaas, S.I. & Greengard, P. (1993) *Proc. Natl Acad. Sci. USA* **90**, 4146–4150.
Jaffe, A.B., Toran-Alleraud, C.D., Greengard, P., & Goudy, S.E. (1994) *J. Biol. Chem* **269**, 13065–13068.
Kalaria, R.N., Bhatti, S.U., Palatinsky, E.A., Pennington, D.H., Shelton, E.R., Chan, H.W., Perry, G. & Lust, W.D. (1993) *Neuroreport* **4**, 211–214.
Kang, J., Lemaire, H.G., Unterbeck, A., Salbaum, J.M., Masters, C.L., Grzeschik, K.H., Multhaup, G., Beyreuther, K. & Muller-Hill, B. (1987) *Nature* **325**, 733–736.
Kawarabayashi, T., Shoji, M., Harigaya, Y., Yamaguchi, H. & Hirai, S. (1991) *Brain Res.* **563**, 334–338.
Kim, K.S., Wen, G.Y., Bancher, C., Chen, C.J.M., Sapienza, V.J., Hong, H. & Wisniewski, H.M. (1990) *Neurosci. Res. Commun.* **7**, 113–122.
Kitaguchi, N., Takahashi, Y., Tokushima, Y., Shiojiri, S. & Ito, H. (1988) *Nature* **331**, 530–532.
Koo, E.H., Sisodia, S.S., Archer, D.R., Martin, L.J., Weidemann, A., Beyreuther, K., Fischer, P., Masters, C.L. & Price, D.L. (1990) *Proc. Natl Acad. Sci. USA* **87**, 1561–1565.
La Du, M.J., Pederson, T.M., Frail, D.E., Reardon, C.A., Getz, G.S., & Falduto, M.T. (1995) *J. Biol. Chem.* **270**, 9039–9042.
Lamb, B.T., Sisodia, S.S., Lawler, A.M., Slunt, H.H., Kitt, C.A., Kearns, W.G., Pearson, P.L., Price, D.L. & Gearhart, J.D. (1993) *Nature Genet.* **5**, 22–30.

Loffler, J. & Huber, G. (1993) *Biochem. Biophys. Res. Commun.* **195**, 97–103.
Masters, C.L., Simms, G., Weinman, N.A., Multhaup, G., McDonald, B.L. & Beyreuther, K. (1985) *Proc. Natl Acad. Sci. USA* **82**, 4245–4249.
Mayeux, R., Ottman, R., Tang, M.X., Noboa, B.L., Marder, K., Gurland, B. & Stern Y. (1993) *Ann. Neurol.* **33**, 494–501.
McAreavey, M.J., Ballinger, B.R. & Fenton, G.W. (1992) *Elipepsia* **33**, 657–660.
Mobley, W.C., Neve, R.L., Prusiner, S.B. & McKinley, M.P. (1988) *Proc. Natl Acad. Sci. USA* **85**, 9811–9815.
Morin, P.J., Abraham, C.R., Amaratunga, A., Johnson, R.J., Huber, G., Sandell, J.H. & Fine, R.E. (1993) *J. Neurochem.* **61**, 464–473.
Mortimer, J.A., van Duijn, C.M., Chandra, V., Fratiglioni, L., Graves, A.B., Heyman, A., Jorm, A.F., Kokmen, E., Kondo, K., Rocca, W.A. (1991) *Int. J. Epidemiol.* **20**, 528–535.
Mullan, M., Crawford, F., Axelman, K., Houlden, H., Lilius, L., Winblad, B. & Lannfelt, L. (1992) *Nat. Genet.* **1**, 345–347.
Nitsch, R.M., Slack, B.E., Wurtman, R.J. & Growdon, J.H. (1992) *Science* **258**, 304–307.
Nitsch, R.M., Farber, S.A., Growdon, J.H. & Wurtman, R.J. (1993) *Proc. Natl Acad. Sci. USA* **90**, 5191–5193.
Nordstedt, C., Gandy, S.E., Alafuzoff, I., Caporaso, G.L., Iverfeldt, K., Grebb, J.A., Winblad, B. & Greengard, P. (1991) *Proc. Natl Acad. Sci. USA* **88**, 8910–8914.
Ohyagi, Y. & Tabira, T. (1993) *Brain Res. Mol. Brain Res.* **18**, 127–132.
Okado, H. & Okamoto, H. (1992) *Biochem. Biophys. Res. Commun.* **189**, 1561–1568.
Otsuka, N., Tomonaga, M. & Ikeda, K. (1991) *Brain Res.* **568**, 335–338.
Palmert, M.R., Usiak, M., Mayeux, R., Raskind, M., Tourtellotte, W.W. & Younkin, S.G. (1990) *Neurology* **40**, 1028–1034.
Pandiella, A. & Massague, J. (1991) *J. Biol. Chem.* **266**, 5769–5773.
Pearson, B.E. & Choi, T.K. (1993) *Proc. Natl Acad. Sci. USA* **90**, 10578–10582.
Ponte, P., Gonzalez-DeWhitt, P., Schilling, J., Miller, J., Hsu, D., Greenberg, B., Davis, K., Wallace, W., Lieberburg, I., Fuller, F. & Cordell, B. (1988) *Nature* **331**, 525–527.
Querforth, H.W. & Selkoe, D.J. (1994) *Biochemistry* **33**, 4550–4561.
Refolo, L.M., Salton, S.R., Anderson, J.P., Mehta, P. & Robakis, N.K. (1989) *Biochem. Biophys. Res Commun.* **164**, 664–670.
Robakis, N.K., Ramakrishna, N., Wolfe, G. & Wisniewski, H.M. (1987) *Proc. Natl Acad. Sci. USA* **84**, 4190–4194.
Roberts, G.W., Allsop, D. & Bruton, C. (1990) *J. Neurol. Neurosurg. Psychiatry* **53**, 373–378.
Rosen, D.R., Martin-Morris, L., Luo, L.Q. & White, K. (1989) *Proc. Natl Acad. Sci. USA* **86**, 2478–2482.
Schlossmacher, M.G., Ostaszewski, B.L., Hecker, L.I., Celi, A., Haass, C., Chin, D., Lieberburg, I., Furie, B.C., Furie, B. & Selkoe, D.J. (1992) *Neurobiol. Aging* **13**, 421–434.
Schmechel, D.E., Saunders, A.M., Strittmatter, W.J., Crain, B.J., Hulette, C.M., Joo, S.H., Pericak, V.M.A., Goldgaber, D. & Roses, A.D. (1993) *Proc. Natl Acad. Sci. USA* **90**, 9649–9653.
Schubert, D., Jin, L.W., Saitoh, T. & Cole, G. (1989) *Neuron* **3**, 689–694.
Sheng, J.G., Boop, F.A., Mrak, R.E. & Griffin, W.S.T. (1994a) *J. Neurochem.* **63**, 1872–1879.
Sheng, J.G., Wu, J.H., Skinner, R.A., Stanley, L.C., Wall, P.T., Mrak, R.E. & Griffin, W.S.T. (1994b) (submitted).
Siman, R., Card, J.P., Nelson, R.B. & Davis, L.G. (1989) *Neuron* **3**, 275–285.
Sisodia, S.S., Koo, E.H., Hoffman, P.N., Perry, G. & Price, D.L. (1993). *J. Neurosci.* **13**, 3136–3142.
Slack, B.E., Nitsch, R.M., Livneh, E., Kunz, G.M.J., Eldar, H. & Wurtman, R.J. (1993) *Ann. N.York Acad. Sci.* **695**, 128–131.
Smith, R.P., Higuchi, D.A. & Broze, G.J. (1990) *Science* **248**, 1126–1128.
Sparks, D.L., Liu, H., Scheff, S.W., Coyne, C.M. & Hunsacker, J.C. 3. (1993) *J. Neuropathol. Exp. Neurol.* **52**, 135–142.
Sprecher, C.A., Grant, F.J., Grimm, G., O'Hara, P.J., Norris, F., Norris, K. & Foster, D.C. (1993) *Biochemistry* **32**, 4481–4486.

Stanley, L.C., Mrak, R.E., Woody, R.C., Perrot, L.J., Zhang, S.-X., Marshak, D.R., Nelson, S.J. & Griffin, W.S.T. (1994) *J. Neuropathol. Exp. Neurol.* **53** 231–238.
Strittmatter, W.J., Saunders, A.M., Schmechel, D., Pericak, V.M., Enghild, J., Salvesen, G.S. & Roses, A.D. (1993) *Proc. Natl Acad. Sci. USA* **90**, 1977–1981.
Suzuki, T., Nairn, A.C., Gandy, S.E. & Greengard, P. (1992) *Neuroscience* **48**, 755–761.
Suzuki, T., Oishi, M., Marshak, D.R., Czernik, A.J., Nairn, A.C. & Greengard, P. (1994) *EMBO J.* **13**, 1114–1122.
Tanzi, R.E., Gusella, J.F., Watkins, P.C., Bruns, G.A., St George-Hyslop, P., Van Keuren, M.L., Patterson, D., Pagan, S., Kurnit, D.M. & Neve, R.L. (1987) *Science* **235**, 880–884.
Tanzi, R.E., McClatchey, A.I., Lamperti, E.D., Villa-Komaroff, L., Gusella, J.F. & Neve, R.L. (1988) *Nature* **331**, 528–530.
Tomlinson, B.E. & Corsellis, J.A.N. (1984) In *Greenfield's Neuropathology* (eds Adams, J.H., Corsellis, J.A.N. & Duchen, L.W.), pp. 951–1025. London, Arnold.
van Duijn, C.M., Tanja, T.A., Haaxma, R., Schulte, W., Saan, R.J., Lameris, A.J., Antonides, H.G. & Hofman, A. (1992) *Am. J. Epidemiol.* **135**, 775–782.
Van Nostrand, W.E., Schmaier, A.H., Farrow, J.S. & Cunningham, D.D. (1990) *Science* **248**, 745–748.
Van Nostrand, W.E., Wagner, S.L., Shankle, W.R., Farrow, J.S., Dick, M., Rozemuller, J.M., Kuiper, M.A., Wolters, E.C., Zimmerman, J., Cotman, C.W. *et al.* (1992) *Proc. Natl Acad. Sci. USA* **89**, 2551–2555.
Vaughan, P.J. & Cunningham, D.D. (1993) *J. Biol. Chem.* **268**, 3720–3727.
Wallace, W., Ahlers, S.T., Gotlib, J., Bragin, V., Sugar, J., Gluck, R., Shea, P.A., Davis, K.L., Haroutunin, V. (1993) *Proc. Natl. Acad. Sci. USA* **90**, 8712–8716.
Wasco, W., Bupp, K., Magendantz, M., Gusella, J.F., Tanzi, R.E. & Solomon, F. (1992) *Proc. Natl Acad. Sci. USA* **89**, 10758–10762.
Wasco, W., Gurubhagavatula, S., Paradis, M.D., Romano, D.M., Sisodia, S.S., Hyman, B.T., Neve, R.L. & Tanzi, R.E. (1993) *Nat. Genet.* **5**, 95–100.
Weidemann, A., Konig, G., Bunke, D., Fischer, P., Salbaum, J.M., Masters, C.L. & Beyreuther, K. (1989) *Cell* **57**, 115–126.
Yan, Y.C., Bai, Y., Wang, L.F., Miao, S.Y. & Koide, S.S. (1990) *Proc. Natl Acad. Sci. USA* **87**, 2405–2408.

CHAPTER 7
PROCESSING OF THE AMYLOID β-PROTEIN PRECURSOR

Todd E. Golde and Steven G. Younkin

Division of Neuropathology, Institute of Pathology, Case Western Reserve University, Cleveland, Ohio 44106, USA

Table of Contents

7.1 Amyloid and Alzheimer's disease

Numerous proteins or fragments of proteins including serum amyloid A, immunoglobulin light chains, transthyretin, cystatin C, pancreatic islet cell amyloid peptide, calcitonin, β2-microglobulin, apolipoprotein A-1, prion protein and the amyloid β-protein (Aβ) are capable of forming amyloid deposits *in vivo* (Glenner, 1983; Cohen and Connors, 1987). The amyloid formed by these proteins is composed of straight fibrils 5–10 nm in diameter that are deposited extracellularly and these fibrils are insoluble in all but the harshest detergents. Histopathologically, amyloid can be identified by Congo red or Thioflavin S staining because the protein within amyloid fibrils is invariably in a β-pleated sheet conformation. A number of pathological mechanisms for abnormal deposition of potentially amyloidogenic proteins have been described, including mutations resulting in variant proteins with enhanced fibrillogenic potential, overproduction of a normal protein and abnormal processing of a normal precursor resulting in a proteolytically derived fragment that aggregates into amyloid fibrils (Cohen and Connors, 1987).

Although deposition of the Aβ as amyloid in the brains of individuals with Alzheimer's disease (AD) is a characteristic and diagnostic feature of AD, only recently has strong evidence for a causal role of Aβ deposition in AD been obtained. Because

Pathobiology of Alzheimer's Disease
ISBN 0-12-286965-5

no animal models of this disease exist, this evidence has come from a convergence of genetic linkage studies showing that mutations in the amyloid β-protein precursor (APP) can cause AD and analyses showing that the disease-producing mutations alter APP processing in a way that is amyloidogenic. These studies and their implications for the role of the Aβ in AD are reviewed in this chapter.

7.2 Aβ deposition in Alzheimer's disease

In patient's with AD, large numbers of senile plaques are found throughout the cerebral neocortex and hippocampus. These plaques are present in small numbers in the brains of some aged mammals and normal elderly individuals, but they are present in large numbers only in AD (Roth *et al.*, 1966; Struble *et al.*, 1982; Terry, 1985; Selkoe *et al.*, 1987; Terry *et al.*, 1987; Daigle and Li, 1993). Each classic senile plaque is a complex lesion consisting of a spherical cluster of altered neurites surrounding a central, extracellular amyloid core (Terry, 1985). Activated microglial cells envelop the amyloid core in each classic plaque and there are reactive astrocytes at the periphery of each plaque (Wisniewski *et al.*, 1981; Dickson *et al.*, 1988; Cras *et al.*, 1990). Additional extracellular amyloid is scattered among the dystrophic neurites in classic plaques and, in most cases of AD, amyloid fibrils are also found in the walls of cerebral vessels (Glenner, 1983).

Although numerous proteins are associated with the amyloid deposits in AD (Selkoe, 1991), the principal proteinaceous component of both plaque core and vascular amyloid is a ~4 kDa polypeptide referred to as the amyloid β-protein (Glenner and Wong, 1984; Masters *et al.*, 1985; Selkoe, 1986; Kang *et al.*, 1987; Prelli *et al.*, 1988). More recently, the development of antibodies against the Aβ sequence has enabled the detection of diffuse immunoreactive Aβ deposits in the AD brain (Yamaguchi *et al.*, 1988, 1989; Mashiah, 1990). In contrast to the Aβ deposited in compact senile plaques, these deposits are not detected with Congo red or other dyes, such as Thioflavin S, that bind amyloid, nor are they accompanied by pathological neuritic or glial reaction. These lesions, which are referred to as diffuse plaques, have been thought to be early precursor lesions that eventually mature into classic senile plaques, but it remains possible that diffuse plaques are distinct nonpathological lesions. Several groups have examined the insoluble Aβ that remains after the AD brain is extracted with high concentration sodium dodecyl sulphate (SDS) (Mori *et al.*, 1992; Miller *et al.*, 1993; Roher *et al.*, 1993a, b; Gowing *et al.*, 1994). Recent reports using this approach, particularly those of Roher and his colleagues (Miller *et al.*, 1993; Roher *et al.*, 1993a, b; Gowing *et al.*, 1994), indicate that the SDS-insoluble amyloid in senile plaque cores is primarily Aβ1–42 (Miller *et al.*, 1993; Roher *et al.*, 1993a), that diffuse plaques are primarily Aβ17–42 (Gowing *et al.*, 1994), and that vascular amyloid is a mixture of 25–42% Aβ1–40 and 58–75% Aβ1–42 (Roher *et al.*, 1993b). Thus species ending at Aβ42 may be critically important in the pathogenesis of AD.

7.3 The APP

7.3.1 The APP resembles a cell surface receptor

Cloning of the gene encoding the Aβ revealed that the 43 residue Aβ is an internal peptide within a larger precursor protein referred to as the amyloid β-protein precursor (Figure 1) (Goldgaber *et al.*, 1987; Kang *et al.*, 1987; Robakis *et al.*, 1987; Tanzi *et al.*, 1987). The single copy APP gene is located on chromosome 21 and generates numerous APP isoforms through alternative splicing of several exons. The 43 residue Aβ begins 99 residues from the carboxyl terminus of the 677–770 residue APPs, proteins whose primary structure resembles many other cell surface receptors, such as the (epidermal growth factor) EGF receptor, transferrin receptor and many cytokine receptors (see Figure 1) (Kang *et al.*, 1987). The internal location of the Aβ within the APP indicates that at least two proteolytic cleavages are necessary to release it. One of these on the amino side of the Aβ would occur near the carboxyl end of the large intraluminal/extracellular domain, the other on the carboxyl side of the Aβ would appear to involve an unprecedented cleavage within the single transmembrane domain. Neither putative cleavage site necessary to generate Aβ is a consensus sequence for cleavage by a known protease.

7.3.2 APP isoforms

Multiple isoforms of the APP arise through alternative splicing of APP mRNAs. Two minor isoforms encode truncated APPs of 365 (Jacobsen *et al.*, 1991) and 563 amino acids (De Sauvage and Octave 1989) that will not be considered further because they do not contain the Aβ. Eight other isoforms arise through alternative splicing of exons 7, 8 and 15 (Goldgaber *et al.*,1987; Kang *et al.*, 1987; Kitaguchi *et al.*, 1988; Ponte *et al.*, 1988; Tanzi *et al.*, 1988; Golde *et al.*, 1990; Konig *et al.*, 1992). These isoforms encode proteins of 677–770 amino acids that are referred to by the number of amino acids encoded (e.g. APP_{695} refers to the isoform lacking exons 7 and 8, but containing exon 15). Exon 7 encodes a 56 amino acid domain that is referred to as the Kunitz protease inhibitor (KPI) domain because it is homologous to the Kunitz family of serine protease inhibitors, whereas exon 8 encodes a 19 amino acid domain with homology to the MRC OX-2 antigen found on neurons and thymocytes (Weidemann *et al.*, 1989). Although the KPI domain may have functional significance, no role in control of APP processing has been established. The recently described alternative splicing of exon 15 (Konig *et al.*, 1992) is of considerable interest, because the inclusion or exclusion of the 18 amino acids encoded by this exon, which are located 20 residues amino to the Aβ, could significantly influence processing at or near the Aβ amino terminus. To date, however, the influence of exon 15 on APP processing remains unknown.

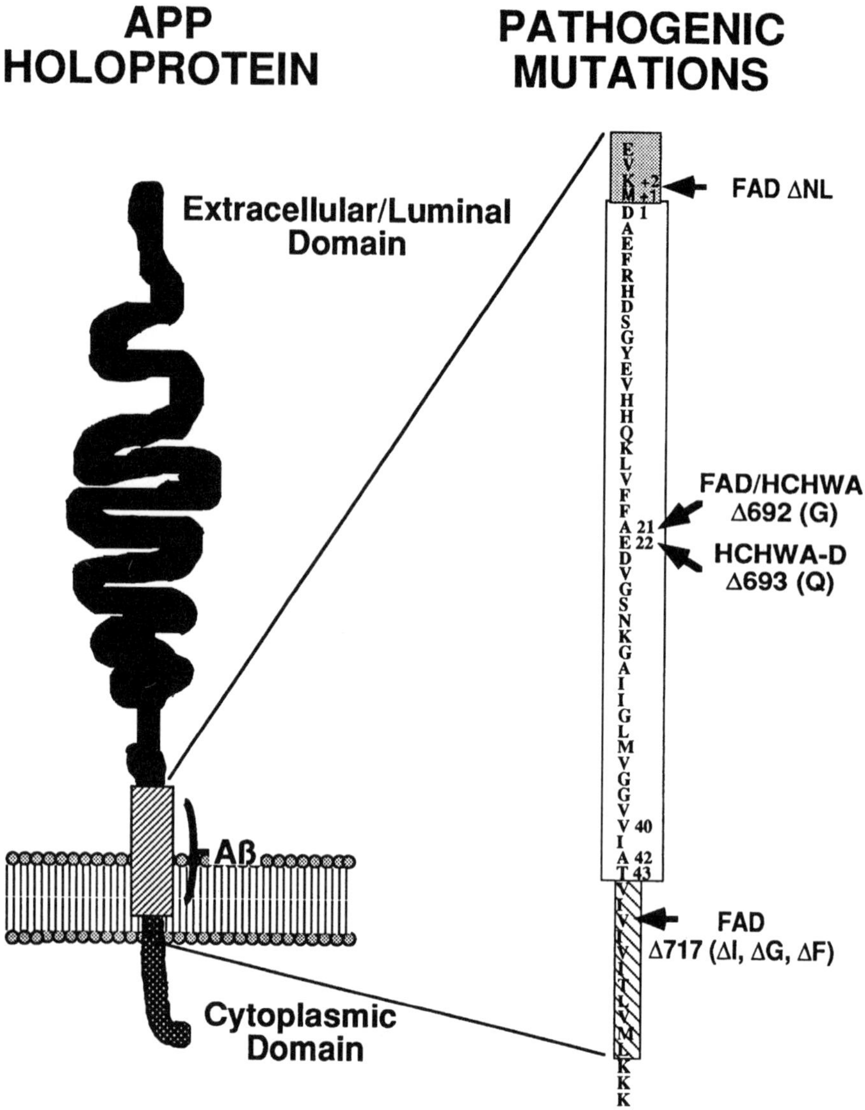

Figure 1 Structure of wild-type and FAD-linked mutant APPs. The primary structure of the APP resembles a transmembrane receptor with a single transmembrane domain (Kang *et al.*, 1987). The 43-residue Aβ sequence begins 99 residues from the carboxyl terminus of the APP and it extends from the extracellular/intraluminal region (28 residues) into the transmembrane region (15 residues). The expansion of the Aβ region on the right shows the position of the known pathogenic mutations within the APP sequence. The Aβ sequence is enclosed by an open box. The hatched box represents the remainder of the transmembrane domain followed by a lysine triplet that is thought to anchor the APP in the membrane.

7.3.3 APP expression

The APP is expressed in numerous tissues with the highest levels seen in the brain and kidney (Goedert, 1987; Neve *et al.*, 1988; Golde *et al.*, 1990; Koo *et al.*, 1990). Expression of the various isoforms is tightly regulated at both a tissue and cellular level. Peripheral tissues express mostly KPI-containing forms and, depending on the tissue examined, from 20% to 50% of these forms lack the exon 15 domain (Sandbrink *et al.*, 1993, 1994). The brain expresses the largest amount of the KPI-free APP_{695} form. At a cellular level, APP_{695} is the major isoform only in neurons, whereas glial cells express significant amounts of KPI-containing isoforms. Recent data indicate that neuronal cells express minor amounts of exon 15-free forms (>5% of total) whereas other cells within the brain express a significant amount of the KPI-containing, exon 15-free isoforms (Sandbrink *et al.*, 1993, 1994).

In other amyloidoses, the source of the amyloid deposited is a circulating intravascular protein (Cohen and Connors, 1987). The widespread expression of the APP throughout the body indicates that the Aβ deposited could, in principle, arise from a number of sources including intravascular APP. However, given the high level of APP expression in brain, it is likely that the cerebral amyloid deposited in AD is derived primarily from an intracerebral source.

7.4 Genetic evidence for a causal role of amyloid deposition in Alzheimer's disease

7.4.1 Trisomy 21 and familial AD

The pathology of AD is complex. In addition to senile plaques and amyloid deposition, there is neuronal loss in many populations. In the perikarya of neurons in these degenerating neuronal populations, there are intracellular neurofibrillary tangles containing highly insoluble paired helical filaments, and paired helical filaments are also present in the altered neurites found in senile plaques. In hippocampal pyramidal cells, two other classic intracellular lesions, granulovacuolar degeneration and Hirano bodies, are observed. The central question with respect to amyloid deposition in AD has been whether amyloid deposition triggers the complex pathology observed in AD or is an end-stage product of that pathology. The genetic forms of AD have been particularly helpful in resolving this issue. If amyloid deposition triggers the development of AD pathology, then the genetic defects that produce AD should be related to amyloid deposition. This has, so far, proven to be the case.

Individuals with Down's syndrome (DS, trisomy 21) who are over the age of 40 invariably develop central nervous system pathology that is essentially identical to that seen in AD (Ropper and Williams, 1980; Mann *et al.*, 1986). This observation is significant because it indicates that increased dosage of one or more of the loci on chromosome 21 is sufficient to cause AD. The finding that the APP gene is located on chromosome 21 immediately indicates that the APP gene is likely to be the locus (or at least one of the loci) that is responsible for the AD pathology that develops in DS. This is supported by the observation that, in DS brains, deposits labelled with antisera to the Aβ are observed before other aspects of AD pathology develop.

In rare families, Alzheimer's disease is inherited as an autosomal dominant trait. Analysis of a large number of familial Alzheimer's disease (FAD) kindreds (St George-Hyslop *et al.*, 1990) has shown them to be genetically heterogeneous. Some families with early onset of symptoms show linkage to chromosome 21 (Goate *et al.*, 1989; St George-Hyslop *et al.*, 1990) whereas other early onset families (Schellenberg *et al.*, 1988; St George-Hyslop *et al.*, 1990) and late onset families (Pericak-Vance *et al.*, 1988; St George-Hyslop *et al.*, 1990) do not. A number of FAD kindreds have now been identified in which point mutations at APP_{717} (resulting in substitution of isoleucine, phenylalanine or glycine for valine 717 in APP_{770}) cosegregate with the disease (Chartier-Harlin *et al.*, 1991; Goate *et al.*, 1991; Hardy *et al.*, 1991; Murrell *et al.*, 1991; Naruse *et al.*, 1991; Yoshioka *et al.*, 1991). These mutations have been identified in many unrelated families on different continents and they have not been detected in any controls despite exhaustive analysis. Thus there is excellent evidence that this mutation in the APP gene causes AD. Recently, another APP mutation has been shown to cosegregate with the AD phenotype in a large Swedish kindred (Mullan, 1992). In this case, the defect is a double mutation that converts the lysine-methionine located immediately amino to Aβ1 (lys670-met671 in APP_{770}) to asparagine-leucine. The location of these mutations in close proximity to the amino and carboxyl ends of Aβ (Figure 1) immediately suggests that they cause AD by altering APP processing in a way that is amyloidogenic.

On the basis of this genetic evidence, it is reasonable to propose, as a working hypothesis, that AD is a heterogeneous disorder in which multiple initiating mechanisms alter APP processing in a way that results in amyloid deposition, which, in turn, produces the complex pathology that characterizes this disorder. Recently mutations on chromosome 14 (Mullan *et al.*, 1992; St George-Hyslop *et al.*, 1992; Schellenberg *et al.*, 1992; Van Broeckhoven *et al.*, 1992) have been linked to AD, and it has been shown that the ApoE4 allele, which is located on chromosome 19, substantially increases the risk that AD will develop (Pericak-Vance *et al.*, 1991; Borgaonkar *et al.*, 1993; Corder *et al.*, 1993; Hardy *et al.*, 1993; Noguchi *et al.*, 1993; Payami *et al.*, 1993; Poirier *et al.*, 1993; Saunders *et al.*, 1993a, b; Schmechel *et al.*, 1993; Strittmater *et al.*, 1993a, b). Thes mutations are obviously not in the APP gene, but the affected genes may well encode proteins (e.g. proteases or protease inhibitors) that alter APP processing in a way that is amyloidogenic. Identifying these and other mutated genes that are linked to FAD and determining whether they alter APP processing in a way that is amyloidogenic is, therefore, an important way to continue to test the hypothesis that amyloid deposition plays a central role in the development of AD.

7.5 APP processing

7.5.1 Full-length APP is truncated at its carboxyl terminus to generate large secreted derivatives

Studies in a number of laboratories have: (i) identified a set of ~110–135 kDa membrane-associated proteins that represent full-length forms of the APP (Autilio-Gambetti *et al.*, 1988; Schubert *et al.*, 1988; Selkoe *et al.*, 1988; Palmert *et al.*, 1989a; Takio *et al.*, 1989; Oltersdorf *et al.*, 1990); and (ii) shown that, in human brain and cultured cells, these membrane-associated forms are processed into large, secreted, amino-terminal derivatives, referred to as secreted APP (sβAPP), that are readily detected in culture medium or human cerebrospinal fluid (CSF) (Schubert *et al.*, 1988, 1989; Palmert *et al.*, 1989b; Wiedemann *et al.*, 1989; Oltersdorf *et al.*, 1990). This secretion is not isoform specific and appears to occur quite rapidly in cultured cells with the half-life for turnover of newly synthesized APP estimated to be ~20–30 min. In addition, studies on cultured PC12 cells indicate that secretion can be a major fate of the APP with up to ~50% of newly synthesized APP cleaved and secreted into the medium (Wiedemann *et al.*, 1989).

Using cultured cells, several groups have established that secretion of newly synthesized APP occurs following extensive post-translational modification. During maturation, the APP is rapidly N-glycosylated then O-glycosylated and tyrosine sulphated (Wiedemann *et al.*, 1989; Oltersdorf *et al.*, 1990). Subsequently the fully glycosylated, sulphated APP holoprotein is cleaved releasing sAPP into the medium and leaving behind a small cell-associated carboxyl-terminal fragment (CTF) (Oltersdorf *et al.*, 1990).

Activation of protein kinase C (PKC) has a marked influence on APP processing in many cell types as reviewed elsewhere in this volume. In the cytoplasmic domain of the APP, there are several putative sites for phosphorylation by PKC (Gandy *et al.*, 1988, 1992; Wiedemann *et al.*, 1989; Oltersdorf *et al.*, 1990; Caporaso *et al.*, 1992b; Hung *et al.*, 1993), but it has not been definitively established that these sites are phosphorylated by PKC under physiological conditions. Hung and Selkoe have recently shown that serine phosphorylation in the APP ectodomain does occur *in vivo*, but they find no evidence that the phosphorylation state of the APP influences its processing (Hung *et al.*, 1993). Rather they suggest that the activation of PKC is likely to alter APP processing either by acting on proteases that cleave the APP or by altering its cellular targeting.

7.5.2 The α-secretory pathway

The first evidence that sAPP is produced by cleavage within the Aβ sequence was obtained by Sisodia *et al.* (1990). This group examined sAPP in CHO cells transiently transfected with a series of expression constructs in which various regions of the APP were deleted. These constructs produced sAPP of appropriate size, but secretion ceased when a critical region containing the amino portion of the Aβ was deleted. On this basis the authors concluded that, in this system, sAPP is produced by cleavage within the Aβ sequence. Subsequently, Esch *et al.* (1990), using human embryonic kidney (293) cells transfected with full-length APP_{695} or APP_{751} expression constructs definitively established that cleavage within the Aβ sequence provides sAPP. Purifying the large sAPP derivatives and a membrane-associated ~10 kDa CTF described initially by Oltersdorf *et al.* (1990), Esch *et al.* (1990) showed by direct sequencing that the soluble derivative ends at Aβ15 and that the ~10 kDa CTF begins at Aβ17. Thus, in this system, cleavage producing sAPP occurred at Aβ16 with subsequent removal of the Aβ16 lys residue from either the carboxyl end of sAPP or the amino end of the CTF by an exopeptidase. Subsequently, characterization of secreted derivatives in other cultured mammalian cells has enabled identification of sAPP derivatives ending at both Aβ15 and Aβ16 (Anderson *et al.*, 1991; Wang *et al.*, 1991). Thus it appears that the cleavage generating the APP occurs on the carboxyl side of the lys at Aβ16 with a carboxyl-peptidase subsequently removing the lys at Aβ16 from the secreted derivative. Characterization of sAPP in the brain and CSF from both AD and control patients by both biochemical and immunocytochemical methods has established that similar processing occurs *in vivo* (Palmert *et al.*, 1989b; Pasternack *et al.*, 1992), thereby validating the use of cell culture models for studying APP processing.

The protease, designated α-secretase, that cleaves the APP at Aβ16 to generate a large secreted derivative ending at Aβ15 or 16 (referred to specifically as sAPPα) and a cognate small CTF, has not been identified. Recent work (Maruyama *et al.*, 1991; Wang *et al.*, 1991; Zhong *et al.*, 1994) utilizing site-specific mutagenesis to clarify the sequence requirements for proteolytic recognition indicates that cleavage by α-secretase has relatively little sequence specificity. Only a very limited number of mutations alter cleavage, and normal levels of cleavage are seen even with fairly extensive modification of the primary amino-acid sequence at or near Aβ16. Since most known endoproteolytic cleavage events are sequence dependent, the relative nonspecificity of α-secretase indicates that it may be a novel type of endoprotease with a cleavage specificity regulated more by secondary structure of the substrate and distance from the membrane than primary sequence. Alternatively, α-secretase cleavage could represent the combined activity of multiple proteases acting at or near the same site (Zhong *et al.*, 1994).

Cleavage of APP in the α-secretory pathway can apparently occur on the cell surface or intracellularly (Haass *et al.*, 1992a; Sambamurti *et al.*, 1992; Sisodia, 1992); however, the precise location of intracellular processing is not known. As shown in Figure 2, processing of the APP in the α-secretory pathway cuts the Aβ in two, thereby preventing formation of the Aβ that is deposited as amyloid in plaque cores. For this reason, the α-secretory pathway is currently thought to be a pathway that prevents amyloid deposition.

7.5.3 Carboxyl-terminal fragments of the APP (CTFs)

Since the ubiquitous processing that occurs in the α-secretory pathway precludes Aβ generation, it was originally thought that Aβ might be generated only in association

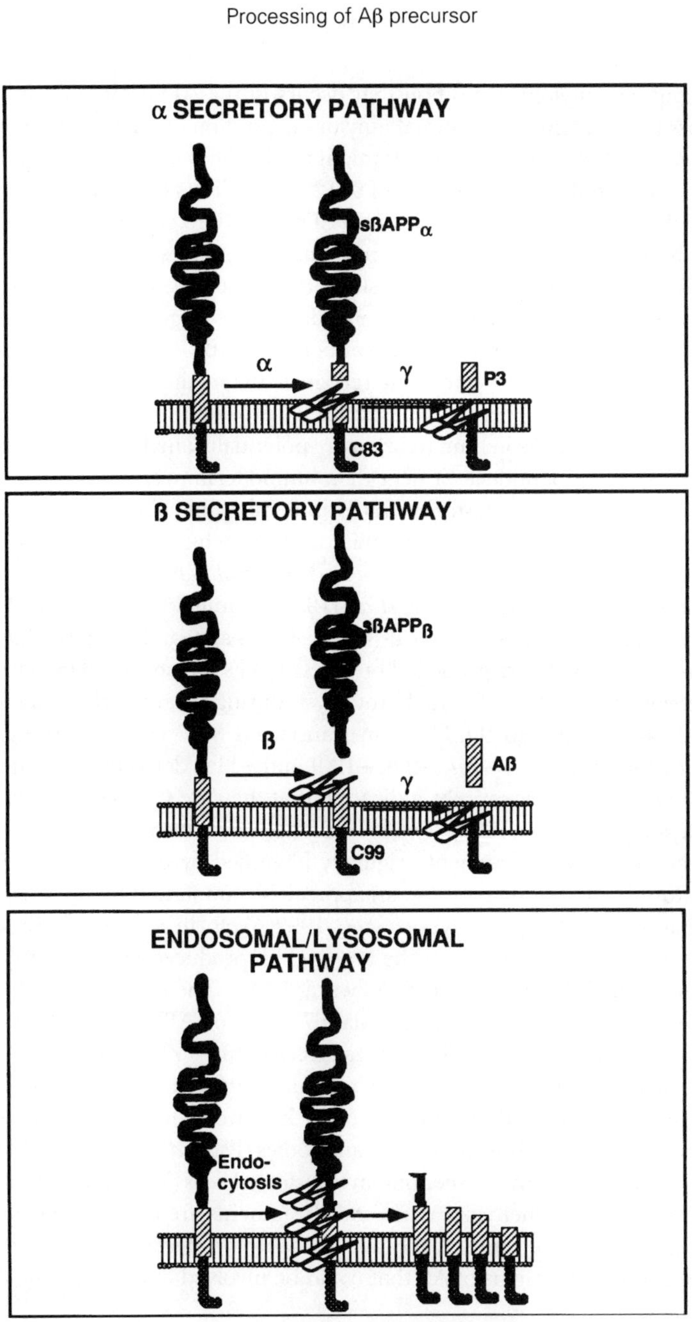

Figure 2 Pathways processing the APP. The upper panel shows the cleavage by α- and γ-secretase that is postulated to release P3, the middle panel shows the cleavage by β and γ secretase that is postulated to release 4 kDa Aβ, and the lower panel shows the complex set of carboxyl-terminal APP derivatives that may be produced by endosomal/lysosomal processing.

with the extensive amyloid deposition that occurs in the AD brain and, to a lesser extent, in association with the minimal amyloid deposition often seen in the normal elderly brain. The first evidence that this might not be the case was provided by Estus *et al.* (1992) who identified a complex set of 8–12 kDa CTFs of the APP in both human brain and cultured cells. By using Tris-tricine SDS/PAGE to compare the ~8–12 kDa CTFs from human brain with a synthetic protein, C100, corresponding to the last 100 amino acids of the APP, Estus *et al.* (1992) showed that the ~11.8 and ~11.4 kDa CTFs in human brain are potentially amyloidogenic derivatives large enough to contain the entire Aβ. In addition, they showed that antibodies to Aβ1–15 specifically recognize the ~11.8 and ~11.4 kDa CTFs but fail to recognize the smaller (~10.9, ~9.6 and ~8.7 kDa) proteins in the 8–12 kDa set. These findings provided strong evidence that the ~11.8 and ~11.4 kDa CTFs in human brain are potentially amyloidogenic derivatives that contain intact Aβ peptide at or near their amino terminus.

In cultured human neuroblastoma (M17) and embryonic kidney (293) cells, there are ~8.7, ~9.6, ~10.9 and ~11.4 CTFs similar to those in human brain, but the ~11.8 kDa derivative does not appear to be present (Estus *et al.* 1992; Cheung *et al.*, 1994). Using a pulse chase paradigm, Estus *et al.* (1992) examined transfected M17 cells expressing APP_{695} and showed that the 8–12 kDa CTFs appear during the chase interval when full-length APP is degraded. Thus these CTFs are produced by normal cellular processing and not by artifactual proteolysis during isolation. Recently, Cheung *et al.* (1994) radiosequenced the CTFs in transfected M17 cells expressing APP_{695} cDNA. They found that the ~8.7, ~9.6, ~10.9 and ~11.4 derivatives begin at Aβ1, Aβ4, Aβ10 and Aβ17, respectively, indicating that they are CTFs of 99, 95, 89 and 82 amino acids (Estus *et al.*, 1992). The 8.7 kDa derivative beginning at Aβ17, which is the α-secretase-derived fragment originally identified by Esch *et al.*, is the major CTF in cultured cells, but there is also an appreciable amount of the 11.4 kDa derivative that has intact Aβ peptide (Knops *et al.*, 1992; Wolozin *et al.*, 1992; Siman *et al.*, 1993). The amino termini of two of the larger CTF produced by FAD lymphocytes have been identified (Matsumoto and Fujiwara, 1993). These fragments of ~15 and ~16 kDa begin respectively at gly582 and ala567 (based on APP_{695}), and therefore correspond to CTF of 113 and 128 amino acids. Recently, Zhong *et al.* (1994) have reported that the major CTFs in Down's syndrome fibroblasts begin at positions Aβ19, 22 and 25. Even smaller CTF in the 6–7 kDa range have also been identified by their reactivity with carboxyl-terminal antibodies (Estus *et al.*, 1992; Wolozin *et al.*, 1992) and, in some cases, by augmention in transfected cells (T.E. Golde, unpublished observations). Based on their size alone, these fragments are unlikely to contain any Aβ peptide. Thus the 6–7 kDa CTF are probably produced by cleavage(s) at or beyond the carboxyl terminus of Aβ that could be involved in releasing the Aβ from its precursor.

The finding that the APP is normally cleaved at multiple sites within or near the Aβ established that APP processing is complex and suggested that it might involve several pathways. Analysis of the APP sequence showed that the cytoplasmic domain contains a consensus sequence, Asn-Pro-Thr-Tyr, for endocytosis via clathrin-coated pits similar to that found in many cell surface receptors with single membrane spanning domains and short cytoplasmic tails (Chen *et al.*, 1990). The first evidence that the APP is, in fact, processed within the endosomal/lysosomal system was obtained by treating cells expressing APP with classical inhibitors of endosomal/lysosomal processing. This treatment led to accumulation of the 8–12 kDa CTFs as well as larger CTFs and holoprotein (Cole *et al.*, 1989; Caporaso *et al.*, 1992a; Golde *et al.*, 1992; Haass *et al.*, 1992a; Knops *et al.*, 1992; Siman *et al.*, 1993). Subsequently, cell surface labelling of APP with biotin and anti-βAPP antibodies showed that cell surface APP is internalized via clathrin-coated pits (Haass *et al.*, 1992a), and that deletion of the endocytosis signal results in decreased internalization and increased secretion of APP (Wang *et al.*, 1991; Haass *et al.*, 1992a). Finally, direct purification of late-endosomes/lysosomes has shown unequivocally that lysosomes contain full-length APP and CTFs including multiple forms large enough to contain the entire Aβ (Haass *et al.*, 1992a). Lysosomes isolated from cells treated with leupeptin showed a marked accumulation of these CTFs providing strong evidence that a lysosomal thiol protease either directly degrades these CTFs or activates proteases that degrade them.

In short-term pulse-chase experiments examining transfected 293 cells expressing APP_{695}, Golde *et al.* (1992) found that both leupeptin, a potent inhibotor of thiol proteases, and ammonium chloride, a general inhibitor of pH-dependent proteolysis, initially inhibited production and then caused accumulation of the larger potentially amyloidogenic CTF, although neither drug had a detectable effect on APP secretion. These data indicate that potentially amyloidogenic CTF may not only be degraded but also produced in the endosomal/lysosomal pathway, although they do not preclude the possibility that potentially amyloidogenic CTF are produced outside this system.

Overall, the studies summarized above show that normal APP processing produces a complex set of 8–12 kDa CTFs that includes potentially amyloidogenic forms with the entire Aβ at or near their amino terminus, larger less well-characterized CTFs of 14–30 kDa, and smaller ~6 kDa CTFs. The exact role that CTFs play in the pathogenesis of AD is not well understood. A survey comparing CTFs in sporadic AD and control brains failed to show elevated amounts of Aβ-bearing CTFs (Estus *et al.*, 1992), so there is no evidence that sporadic AD develops in association with an elevation of these fragments. Comparison of human brain with various peripheral tissues did show, however, that brain has roughly equivalent amounts of the various 8–12 kDa derivatives, whereas peripheral tissues (liver, kidney, small intestine and muscle) have relatively low amounts of the larger potentially amyloidogenic forms (Estus *et al.*, 1992). Thus, the selective deposition of amyloid in ageing brain may be related to the comparatively large amount of the Aβ-bearing carboxyl-terminal derivatives in this tissue. Several interesting, although controversial, studies have demonstrated that overexpression of APP CTF, such as C100 in rat PC12 cells (Yankner *et al.*, 1989), mouse teratocarcinoma cell lines (Fukuchi *et al.*, 1992) and human neuroblastoma cells (Fukuchi *et al.*, 1993b) can be cytotoxic when these transfected cells are treated with agents inducing neuronal differentiation. However, a high level expression of CTF in many cell culture systems results in no apparent toxicity (T.E. Golde, unpublished observations).

7.5.4 Secretion of Aβ during normal processing of the APP

To release Aβ from Aβ-bearing CTFs, these derivatives must be cleaved within their putative membrane-spanning domains. Given the tendency of Aβ to form insoluble amyloid, it seemed that any Aβ released by this unlikely cleavage would probably be highly insoluble and therefore difficult to detect. Thus it was quite surprising when several groups developed assays for Aβ and demonstrated that soluble Aβ is normally secreted into the media by cultured cells (Seubert *et al.*, 1992; Shoji *et al.*, 1992; Busciglio *et al.*, 1993; Haass *et al.*, 1993). In these experiments, Aβ was detected either by radiolabelling cells with [^{35}S]methionine and then immunoprecipitating the radiolabelled Aβ from the media, or by immunoprecipitating unlabelled Aβ from conditioned media and then immunoblotting with a second anti-Aβ antibody. Using these paradigms, two distinct proteins of ~4 kDa and ~2–3 kDa were typically detected after separation by Tris-tricine/SDS PAGE. The ~4 kDa band comigrated with synthetic Aβ1–40 and was not pelleted by high-speed centrifugation. Remarkably, secreted 4 kDa Aβ was present at nanomolar concentrations in medium conditioned by transfected cells expressing full-length APP or Aβ-bearing CTFs. It was immediately evident that equally impressive amounts of 4 kDa Aβ are secreted *in vivo* because human CSF also contained Aβ at nanomolar concentrations.

Further characterization of the Aβ secreted by cultured cells using radioactive sequencing and mass spectrometry has shown that, like the Aβ isolated from AD amyloid, secreted Aβ is a heterogeneous mix of peptides (Haass *et al.*, 1992b; Seubert *et al.*, 1992; Shoji *et al.*, 1992; Busciglio *et al.*, 1993; Dovey *et al.*, 1993; Vigo-Pelfrey *et al.*, 1993). These Aβ peptides are catalogued in Figure 3. The major 4 kDa species in both conditioned medium and human CSF is Aβ1–40, although some Aβ1–42 is also present along with minor amounts of other peptides (e.g. Aβ1–28, Aβ1–33, Aβ1–34, Aβ3–34, Aβ1–37, Aβ1–38 and Aβ1–39). The 2–3 kDa species termed P3 begins at Aβ17 or 18. Thus it appears that the APP holoprotein is cleaved primarily at the amino terminus of the Aβ or at the α-secretase site to generate CTF of 99 or 82 amino acids that are subsequently cleaved on the carboxyl side of Aβ to generate soluble Aβ and P3 peptides. The complex array of minor Aβ peptides may be generated *de novo* by endoproteolytic cleavage of holoprotein or CTFs at many sites. Alternatively these peptides may arise from secondary processing by exopeptidases of one or several major Aβ species.

From the foregoing analyses, it appears that there are significant differences in the composition of the Aβ deposited as amyloid in the AD brain and the Aβ released into CSF or the medium of cultured cells (see Figure 3). As discussed earlier, the Aβ in plaque cores ends primarily at Aβ42, and it shows considerable amino-terminal

---

Figure 3 Known cleavage sites in the APP. All three schematics show the Aβ region of the APP. Amino termini are indicated by hatched arrows and carboxyl termini by solid arrows. The Aβ sequence begins at D1 and ends at T43. The left panel shows the sites of cleavage during secretory processing of the APP. The proteolytic activities producing cleavage at these sites are

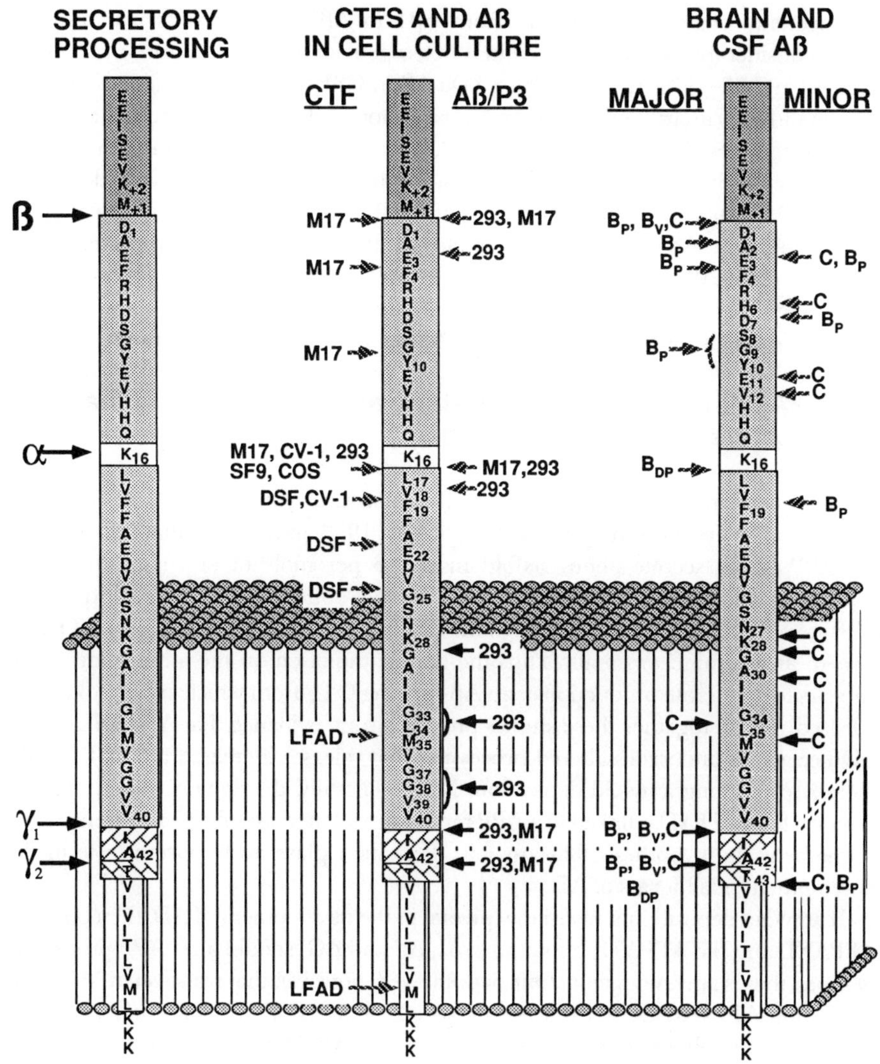

referred to as α-, β- and γ-secretase. The centre panel shows amino termini of APP carboxyl-terminal fragments (left arrows) as well as the amino and carboxyl termini of Aβ (right arrows) isolated from cell culture. M17 refers to BE2C-M17 human neuroblastoma cells, 293 to human embryonic kidney cells, CV-1 to African monkey kidney fibroblasts, DSF to Down's syndrome fibroblasts, SF9 to *Spodoptera frugiperda* cells, and LFAD to EBV transformed lymphocytes from patients with FAD. The right panel shows the amino and carboxyl termini of Aβ isolated from human brain and CSF. Arrows on the left show major cleavage sites, those on the right show minor sites. B_P indicates Aβ isolated from brain parenchyma, B_V indicates Aβ isolated from cerebral vessels, B_{DP} indicates Aβ isolated from diffuse plaques and C indicates Aβ isolated from CSF. References are cited in the text.

heterogeneity beginning anywhere from Aβ1 to Aβ11. Cerebrovascular amyloid shows considerably less amino-terminal heterogeneity and is a mixture of 25–42% Aβ1–40 and 58–75% Aβ1–42. The 4 kDa Aβ in CSF or in culture medium, on the other hand, is primarily Aβ1–40 with a small amount of Aβ1–42 and minor amounts of other peptides. Thus species ending at Aβ42, which appear to be a minor component of the Aβ that is normally secreted, constitute the bulk of the Aβ that is deposited in the AD brain. The factors that account for this highly selective accumulation of species ending at Aβ42 are currently unknown. Recently it has been reported that a P3-like peptide, Aβ17-42, is the major species present in diffuse plaques. If confirmed, this is highly significant because it suggests that cleavage by α-secretase may be involved in the production of some of the Aβ peptide that deposits in the AD brain.

7.5.5 FAD-linked APP mutants show altered processing that fosters amyloid deposition

Two independent studies using transfected human embryonic kidney (293) (Citron *et al.*, 1992) or neuroblastoma (M17) cells (Cai *et al.*, 1993) have shown that cells expressing APP_{695}ΔNL secrete about sixfold more Aβ per mole of precursor than cells expressing wild-type APP_{695}. In addition, Cai *et al.* found that M17 cells expressing APP_{695}ΔNL show a fivefold increase in the relative amount of the ~11.4 kDa CTF that has Aβ at its amino terminus. Thus the double mutation on the amino side of Aβ appears to accelerate production of an Aβ-bearing CTF from which excess Aβ is then released. Taken together these studies provide strong evidence that APPΔNL causes AD because it undergoes altered processing that releases increased amounts of Aβ, thereby promoting amyloid deposition. More generally, the demonstration that the Swedish form of FAD is linked to a APP mutation that increases Aβ production greatly strengthens the hypothesis that amyloid deposition plays an important role in the development of all forms of AD.

If amyloid deposition is invariably pivotal in the development of AD, then both the APPΔNL and the APPΔ717 mutations (ΔI, ΔF and ΔG) should alter APP processing in a way that is amyloidogenic. Our data showed, however, that transfected M17 cells expressing APP_{695}ΔI do not release increased amounts of Aβ (Cai *et al.*, 1993). To account for this observation, we proposed that the FAD-linked Δ717 mutations on the carboxyl side of Aβ (ΔI, ΔF, ΔG) shift cleavage to favour generation of longer Aβs such as Aβ1–42 or Aβ1–43. Since these longer Aβs form amyloid fibrils more rapidly than Aβ1–40 (Hilbich *et al.*, 1991; Burdick *et al.*, 1992; Jarrett and Landsbury, 1993; Jarrett *et al.*, 1993), shifting the site of cleavage could result in amyloid deposition without increasing the overall amount of Aβ produced.

To test this hypothesis, we have recently used two methods to compare the effect of the APP_{717} mutations on the relative amounts of Aβ1–40 and Aβ1–42,43 secreted by appropriately transfected M17 cells (Suzuki *et al.*, 1994). In the first method, Aβ1–40 and Aβ1–42,43 were distinguished and quantified by isolating metabolically labelled Aβ from conditioned media, digesting the Aβ with cyanogen bromide to release radiolabelled carboxyl-terminal peptides (Aβ36–40 or Aβ36–42,43), and analysing the carboxyl-terminal peptides by RPLC using a C4 column. In the second method, highly sensitive sandwich ELISAs that discriminate synthetic Aβ1–40 from Aβ1–42,43 were employed. With both methods, we invariably observed the relative amount of Aβ1–42,43 produced by APP_{717} mutants (ΔI or ΔF) to be 1.5–2.0 times that produced by control APPs. Quantification of the sandwich ELISA data showed the percentages of Aβ1–42,43 secreted by cells expressing wild-type APP_{695}, $APP_{695}\Delta I$, and $APP_{695}\Delta F$ to be 17.5±0.5%, 25.7±0.5% ($p < 0.002$) and 31.4±0.6% ($p < 0.0002$), respectively.

Although the APP_{717}-induced increase in long Aβ production is relatively modest, *in vitro* studies by Lansbury and his colleagues (Jarrett *et al.*, 1993a, b) indicate that the impact of this shift on amyloid deposition could be substantial. These studies indicate that Aβ aggregation into amyloid is a nucleation-dependent polymerization (Jarrett *et al.*, 1993a, b) and this type of seeded polymerization can be extremely concentration dependent. Numerous studies have shown that Aβ1–42 and 43 nucleate rapidly compared to Aβ1–40 (Hilbich *et al.*, 1991; Burdick *et al.*, 1992; Jarrett *et al.*, 1993a, b) and can seed aggregation of Aβ1–40 *in vitro* (Jarrett *et al.*, 1993a, b). Thus the modest increase in long Aβ produced by the Δ717 mutations could substantially increase amyloid deposition in the decades that precede the onset of clinical symptoms in patients who carry these mutations.

The complex pathology observed in AD can be caused by trisomy 21, the FAD-linked APP mutations, or by mutation of the AD3 locus on chromosome 14. In addition, there is evidence that the presence of the ApoE4 allele significantly increases the risk of AD, and it is clear that AD develops in many patients in which the aforementioned genetic factors play no role. Thus AD is clearly a heterogeneous disorder in which pathology can be initiated or accelerated by a number of genetic or environmental factors. One of the hallmarks of AD pathology is the deposition of large amounts of Aβ as amyloid. The studies of transfected cultured cells reviewed above show: (i) that the FAD-linked APPΔNL mutation, which is located immediately amino to Aβ, increases Aβ release sixfold (Citron *et al.*, 1992; Cai *et al.*, 1993); and (ii) that the APP_{717} mutations, which are located close to the carboxyl end of Aβ, enhance production of long Aβ, which is known to form insoluble amyloid fibrils more rapidly than Aβ1–40. Thus all of the FAD-linked APP mutations alter APP processing in a way that increases the likelihood of amyloid formation, and this provides strong evidence that it is by enhancing amyloid deposition that these mutations produce AD. It is likely that trisomy 21 also causes AD by enhancing amyloid deposition, since patients with trisomy 21 have an extra copy of the APP gene (Goldgaber *et al.*, 1987; Kang *et al.*, 1987; Robakis *et al.*, 1987; Tanzi *et al.*, 1987) and show increased expression of APP (Tanzi *et al.*, 1987), which is normally processed to release Aβ.

Two additional APP mutations have been described. One of these, which changes the glutamic acid at Aβ22 (residue 693 of APP_{770}) to glutamine (Levy *et al.*, 1990; von Broeckhoven *et al.*, 1990), causes hereditary cerebral haemorrhage with amyloidosis of the Dutch type (HCHWA-D), a rare autosomal dominant disorder in which vascular amyloid deposition leads invariably to stroke in the fourth and fifth decades of life (Castano *et al.*, 1988). The other, which converts the alanine at Aβ21 (residue 692 of APP_{770}) to glycine, appears to be capable of causing both early onset AD and cerebral amyloid angiopathy. Studies of Aβ1–28 and Aβ1–40 peptides with the Δ693 substitution indicate that these peptides aggregate more rapidly than the wild-type peptides (Burdick *et al.*, 1992; Fraser *et al.*, 1992; Fabian *et al.*, 1993; Sorimachi and Craik, 1994). Thus, the HCHWA-D mutation may drive cerebrovascular deposition by altering the amyloidogenicity of the Aβ peptide rather than by altering Aβ genesis. Currently the mechanism(s) that restrict Aβ deposition to the cerebrovasculature are unknown. Perhaps further characterization of the effects of the Δ692 mutation will help to clarify the mechanisms responsible for the differential deposition of Aβ in AD and HCHWA-D.

7.5.6 The β- and γ-secretory pathways

Recently, Seubert *et al.* (1993) reported the identification of a novel secreted, amino-terminal APP derivative ending at the methionine (met-1) that immediately precedes Aβ1. The approach taken in this study was to use antibodies specifically recognizing the carboxyl-terminal methionine in this derivative to distinguish it from the previously identified secreted derivative, which was specifically identified using monoclonal antibodies to the Aβ1–15 domain at its carboxyl terminus. In human CSF and in culture medium conditioned with mixed human fetal brain cultures, the novel derivative ending at met-1 was found to be a substantial part of the total secreted derivative. The report by Seubert *et al.* (1993) is important because it suggests that, in addition to the previously identified APP secretase activity (α-secretase), there is an additional activity (β-secretase) that cleaves full-length APP between met-1 and Aβ1 to produce a large, secreted derivative (sAPPβ) and a potentially amyloidogenic carboxyl-terminal derivative with Aβ at its amino terminus (see Figure 2) that would be identical to the 11.4 kDa derivative identified by Estus *et al.* (1992) and characterized by Cheung *et al.* (1994). The activity of β-secretase varies considerably with cell type. In their analysis of 293 cells expressing full-length APP, Seubert *et al.* (1993) found very little secreted derivative ending at met-1 (sAPPβ). In fact, the vast majority of the secreted derivative produced by these cells (~90%) was found to end at Aβ15 (sAPPα). Thus the sAPPβ produced by 293 cell tends, when analysed by SDS-PAGE, to get lost in the trailing edge of the more abundant sAPPα unless specific antibodies for sAPPβ are employed. This undoubtedly accounts for the failure to detect sAPPβ in early studies of transfected 293 cells (Golde *et al.*, 1992).

Recently, Cheung *et al.* (1994) have radiosequenced the Aβ and the 11.4 kDa carboxyl-terminal derivative produced by cells expressing APPΔNL. In these cells, as in cells expressing wild-type APP, both 4 kDa Aβ and the 11.4 kDa derivative begin at Aβ1. Since the ΔNL mutation is located at the β-secretase cleavage site, this suggests that it may act by enhancing β-secretase cleavage. If so, then the ΔNL mutation should increase not only the 11.4 kDa Aβ-bearing carboxyl-terminal derivative but also the sAPPβ produced by β-secretase. Using systems that are capable of distinguishing sAPPα and sAPPβ either by size or by differential detection with antibodies specifically recognizing carboxyl-terminal Aβ epitopes in sAPPα or sAPPβ, several groups have now shown that the ΔNL mutation does, in fact, markedly augment sAPPβ production. Thus analyses of the APPΔNL mutation have been highly informative. Overall, these studies indicate: (i) that the 11.4 kDa carboxyl-terminal derivative with Aβ at its amino terminus is produced, at least in part, through cleavage of the APP by β-secretase in a secretory pathway; (ii) that this 11.4 kDa derivative is an important intermediate in the generation of Aβ; and (iii) that the ΔNL mutation acts by augmenting β-secretase cleavage.

The protease(s) that release Aβ by cleaving on its carboxyl side are referred to as γ-secretase(s) (Figure 2). Because the inclusion of residues beyond Aβ40 markedly enhances amyloidogenicity, it is critically important to understand the proteolytic mechanisms that are involved in producing Aβs with differing carboxyl termini. No known endoprotease has been shown to cleave a protein within its transmembrane domain when it is still membrane associated. Thus, one of the perplexing questions regarding Aβ generation involves the mechanism by which Aβ and P3 are generated from CTFs. One possibility is that γ-secretase is a novel endoprotease that is, in fact, capable of cleaving CTFs or APP holoprotein *in situ*. Alternatively, the holoprotein or its CTFs may be released from the membrane prior to proteolyis at the γ site. This release could occur: (i) by slippage of intact holoprotein or CTFs from the membrane, a presumably rare event that might occasionally occur in the course of the complex vesicle formation and resorption events that occur during membrane trafficking; (ii) by cleavage at or near the KKK membrane stop anchor to release a small CTF and a large, Aβ-bearing amino-terminal fragment that might readily slip from the membrane environment; or (iii) by release of holoprotein or its CTFs in the course of 'autophagocytic' mechanisms that normally degrade cell membrane. Once released, holoprotein or CTFs could be cleaved by conventional endoproteases and/or exoproteases to produce the complex set of Aβs that is secreted.

No protease involved in the generation of Aβ has been rigorously identified, although an extensive literature, much of it obtained using indicator peptides, has developed reporting the isolation of candidate proteases that cleave with appropriate specificity (Ishihura *et al.*, 1989, 1990; Allsop *et al.*, 1991; Maruyama *et al.*, 1991; Small *et al.*, 1991; Tagawa *et al.*, 1991; Kojima and Omori, 1992; McDermott *et al.*, 1992; Razzaboni *et al.*, 1992; Sahasrabudhe *et al.*, 1993). The cellular compartments in which Aβ genesis occurs are also currently uncertain. Several groups using a variety of cell lines have been unable to detect intracellular Aβ (Haass *et al.*, 1992b; Busciglio *et al.*, 1993; Haass and Selkoe, 1993). Aβ secretion has, however, been shown to be blocked by both brefeldin A and monensin, inhibitors of trafficking through the Golgi apparatus (Busciglio *et al.*, 1993; Haass and Selkoe, 1993; Wertkin *et al.*, 1993). Unfortunately, the inability to detect Aβ intracellularly limits the interpretation of these studies, since Aβ may be generated intracellularly in the presence of these agents but its secretion blocked. Alternatively, as suggested by the authors, these studies may indicate that trafficking through the Golgi is necessary for Aβ production. There is evidence (Shoji *et al.*, 1992; Busciglio *et al.*, 1993; Haass and Selkoe, 1993) that Aβ secretion is inhibited by agents such as chloroquine and ammonium chloride, which act by alkalinizing compartments that are normally acidic, but interpretation of this effect is also problematic because these agents could impair various events (proteolysis, trafficking, secretion) not only in lysosomes but also in other acidic compartments such as the late trans-Golgi or endosomes. Recently, Wertkin *et al.* (1993) have detected a 4 kDa peptide in lysates of differentiated human neuroblastoma (NT2N) cells that is almost certainly Aβ, since it is recognized by specific anti-Aβ antibodies. It is likely that this peptide is intracellular and not extracellular cell-associated Aβ, since pulse chase experiments show that newly generated peptide is detected in cell lysates prior to detection in the media. It may be that further study of NT2N cells will prove particularly useful in identifying the intracellular compartments involved in Aβ production.

7.6 Implications for future research on Alzheimer's disease

If one examines the brains of demented, elderly individuals, the overwhelming majority have substantial amounts of amyloid deposited extracellularly. The central tenet of the amyloid deposition hypothesis is that amyloid deposition is a critical event in a serial cascade of age-related pathological changes that produces the dementia observed. The FAD-linked APPΔNL and APP_{717} mutations provide strong support for this hypothesis, because each of these genetic changes reliably produces AD, and each enhances amyloidogenesis by increasing secretion of total Aβ or Aβ1–42. It is now evident that there are many individuals with extensive cerebral Aβ deposition who are not demented. Given the strong genetic evidence favouring the amyloid deposition hypothesis, it is reasonable to propose that these are individuals who are at risk but have not yet progressed to dementia, either because amyloid deposition is recent or because genetic or environmental factors involved in the progression from amyloid deposition to dementia are not favourable.

If amyloid deposition is, in fact, a rate-limiting element within a serial cascade of age-related pathological changes that produces AD then, in every patient with AD, an appropriate combination of genetic and/or environmental factors has produced amyloid, and that amyloid has provoked sufficient pathology to produce dementia. From this perspective, efforts to understand AD pathogenesis, to develop an animal model, and to provide effective therapy may be viewed as elements of a comprehensive effort to understand and to learn how to control amyloid deposition and the pathology that is provoked by amyloid deposition. Although this discussion focuses primarily on the APP processing events that result in amyloid deposition, it is evident that the pathological effects of deposited amyloid are critically important. Deliberate amplification of changes postulated to be key elements in the pathological cascade induced by amyloid could, for example, be pivotal in developing an animal model for AD. Similarly, effective therapy may be provided by agents that: (1) interfere with the pathological cascade produced by amyloid; or (2) enable the nervous system to compensate for damage inflicted by amyloid-induced pathology.

With the discovery that Aβ is produced and released by normal processing of the

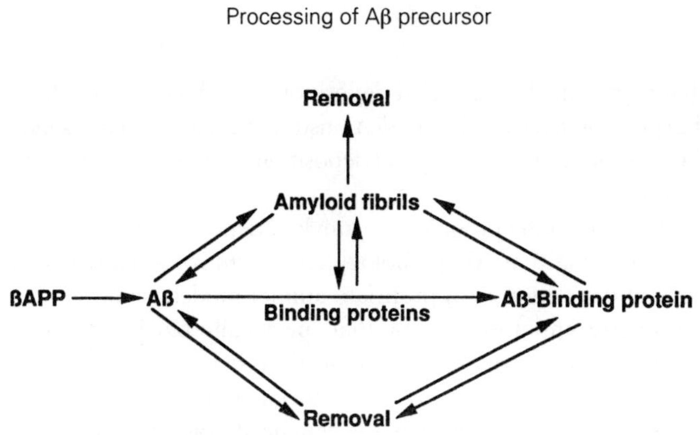

Figure 4 Processes that influence the amount of amyloid deposited in human brain. There is very little information on many of the processes likely to influence the amount of amyloid that accumulates in the ageing human brain. Thus the relative importance of the various processes illustrated is unknown.

APP, it is now clear that Aβ is an unusual secreted protein released from within the APP through cleavage by β- and γ-secretase (Figure 2). Thus in AD, as in the other known amyloidoses, amyloid deposition is a process in which a secreted protein forms insoluble fibrils that deposit extracellularly. The rate at which amyloid accumulates in the brain will depend on four processes: (i) the rate at which the various Aβ peptides are secreted; (ii) the rate at which these secreted Aβ peptides are removed (together these two rates determine Aβ concentration); (iii) the rate at which Aβ fibrils form at any prevailing concentration of Aβ; and (iv) the rate at which deposited amyloid fibrils are removed. Recently, it has become apparent that a number of extracellular proteins can form complexes with Aβ (Ghiso *et al.*, 1993; Schwarzmann *et al.*, 1993). Among the various proteins that may bind significantly to Aβ are apolipoprotein E, apolipoprotein J, heparin sulphate proteoglycans, α_1-antichymotrypsin, and transthyretin. These Aβ-binding proteins are potentially very important not only because they will influence the fraction of Aβ that is free but also because they may substantially influence removal of secreted Aβ, the rate at which amyloid fibrils form, or the rate at which amyloid fibrils are removed (Figure 4).

7.6.1 Implications for AD pathogenesis

At autopsy, there is some Aβ deposited in the brain of almost everyone over 65 indicating that the conditions needed for minimal amyloid deposition are commonplace in the elderly human brain. The realization that the incidence of AD may be as high as 50% in individuals over the age of 85 indicates that, with sufficient ageing, an extraordinary fraction of the population may deposit enough amyloid to produce dementia. Thus it appears that there is a strong tendency toward amyloid deposition in the ageing human brain and that relatively subtle differences in the factors that promote amyloid deposition may decide whether or not substantial amounts of amyloid are deposited. Since there appears to be considerable interindividual variation in the pathology that is provoked by amyloid deposition, the challenge for those favouring the amyloid deposition hypothesis is to demonstrate that all AD can be accounted for in terms of factors that influence amyloid deposition or pathology that is provoked by amyloid deposition.

It is likely that compelling proof for the amyloid deposition hypothesis will be provided either by showing that AD pathology and cognitive impairment can be provoked by amyloid deposition in appropriate animal models or by showing that AD can be halted or prevented by therapy that specifically blocks amyloid formation. These possibilities are discussed in separate sections below. Until such proof is available, the critical issues with respect to pathogenesis are to determine whether or not amyloid deposition is enhanced by all of the genetic alterations that promote AD and to determine if factors that enhance amyloid deposition are major risk factors for the development of 'sporadic' AD. These issues are discussed below in terms of the key factors regulating amyloid deposition (Figure 4).

7.6.1.1 Formation of secreted Aβ

It appears that the FAD-linked APP mutations enhance amyloid deposition by increasing the rate at which total Aβ or Aβ1–42 are secreted. It is likely that a similar mechanism operates in trisomy 21 where there is an extra copy of the APP gene. Thus it is reasonable to postulate that other FAD-linked mutations may act by increasing Aβ secretion. Mutation within β- or γ-secretase could alter their activity in a way that increases Aβ secretion as could mutation in any protease that activates β- or γ-secretase. More generally, any mutation that alters the expression, trafficking, post-translational modification or proteolysis of the APP, β-secretase or γ-secretase could, in principle, increase secretion of total Aβ, Aβ1–42, or some other highly amyloidogenic peptide (e.g. Aβ1–43 or 44). Since the expression and processing of all three molecules (APP, β-secretase and γ-secretase) is likely to be influenced by cell–cell interaction, genetically induced amyloidogenic change could even be produced indirectly by altering cells whose only role is to influence other cells to increase secretion of Aβ. Thus the mechanisms causing specific mutations to increase Aβ secretion may be complex, impossible to predict and difficult to elucidate. Once a specific mutation is identified, it may nonetheless be possible to determine whether it increases Aβ secretion by expressing the mutant transgene in appropriate cells and analysing secreted Aβ as has been done for the FAD-linked APP mutations. As the sensitivity and specificity of assays for secreted Aβ improve, it may even be possible to demonstrate that specific mutations increase Aβ secretion prior to identification of the mutation by showing that secretion is increased in cells (e.g. fibroblasts or transformed lymphocytes) from affected individuals who harbour the mutant gene.

7.6.1.2 Removal of secreted, soluble Aβ

Together, the rates of formation and removal of secreted, soluble Aβ determine the concentration of soluble Aβ, a factor that is critically important in determining the rate of amyloid formation. Essentially nothing is known about Aβ removal in the brain. Removal may take place either through passive diffusion into CSF (which flows into the bloodstream) or through local removal within the brain by (i) extracellular proteolysis and/or (ii) cellular uptake followed by intracellular removal. It is possible that cultured cells will prove as useful for studying Aβ removal as they have been for studying Aβ secretion. If so, it may ultimately be possible to determine whether any of the FAD-linked mutations act by impairing Aβ removal. It should be noted that the mechanisms causing specific mutations to impair removal may be as complex as those described above for enhancing Aβ secretion. Thus as new FAD mutations are identified, it may not be obvious whether they act by impairing removal and it will be important to test this possibility.

Any Aβ binding protein may facilitate the uptake and removal of soluble Aβ. It is, for example, possible that ApoE acts in this manner. If so, then the ApoE4 allele might promote AD because it is less effective than other ApoE alleles in removing soluble Aβ. Similar considerations apply to other Aβ-binding proteins. Thus it will be important to determine whether or not Aβ removal is facilitated by any of the Aβ-binding proteins and, if so, to determine whether change in any of these proteins plays an important role in determining whether or not amyloid is deposited.

7.6.1.3 Other considerations

Factors that accelerate amyloid fibril formation in a concentration-independent manner may also play an important role in pathogenesis. There is, in fact, evidence that the HCHWA-D-linked mutation in Aβ accelerates fibril formation *in vitro* suggesting that accelerated fibril formation may play an important role in the deposition of cerebrovascular amyloid that characterizes this disorder. Similarly, the mixed phenotype that is apparently associated with mutation at Aβ21 may be related, in part, to accelerated fibril formation.

Because amyloid is perceived to be highly insoluble and therefore difficult to remove, scant attention has been paid to the possibility that Aβ deposits may turn over. It seems likely that Aβ deposited either as amyloid fibrils or in some other form may often be effectively removed preventing the formation of neuritic plaques and that the amyloid in classic neuritic plaques may also turn over enabling much of the pathology associated with these plaques to resolve. In fact, the poor correlation between deposited Aβ or plaque number and the severity of dementia suggests that amyloid deposition and plaque number may reach a steady state dictated by the rate at which Aβ is deposited and the rate at which it is removed, most likely by microglia ingestion or extracellular proteolysis. In this concept, Aβ is deposited, a plaque may form causing pathology, some of which is probably irreversible, and the plaque then resolves. Thus, in the AD brain, there may be an ever changing population of diffuse and neuritic plaques that is relatively constant in number and that with time leaves behind an increasing burden of irreversible pathology. If this concept is correct, then AD pathogenesis may be governed as much by the mechanism(s) removing Aβ deposits as by the mechanism(s) that cause Aβ to deposit in the first place.

The Aβ binding proteins may play a critical role in amyloid formation. These proteins could either accelerate or impair amyloid formation and they could do so either at the nucleus-forming stage or during the polymerization that occurs after nuclei are formed. Proteins that coprecipitate with Aβ or that bind to amyloid after it is formed may also play an important role in determining the rate at which Aβ deposits are removed. Thus the influence of the various ApoE alleles on the development of AD may be explained by differences in Aβ binding that influence the effect that the various alleles have on Aβ deposition or on the removal of Aβ deposits. The recent study of Snow *et al.* showing that Aβ1–40 deposits and forms amyloid in rat brain far better when it is co-injected with heparin sulfate proteoglycan (HSPG) than when it is injected alone emphasizes the potential importance of HSPG in particular and more generally of the Aβ-binding proteins as a class. From this brief discussion, it is evident that FAD could, in principle, be caused by mutation in any of the important Aβ-binding proteins or in any of the plethora of genes that may regulate the concentration of these proteins. Environmental or polygenic factors that influence the Aβ-binding proteins could play a similar role in the development of sporadic AD.

From the foregoing discussion, it is evident that a complex set of factors determines the amount of amyloid deposited at any time in the human brain and that relatively subtle changes in these factors may be pivotal in determining whether or not AD develops. In attempting to determine which factors are most important, it may be useful to begin by analysing the concentration of soluble Aβ, since one set of mechanisms (those that influence the rate of formation or removal of secreted Aβ) acts by increasing soluble Aβ concentration, whereas another set (those that influence removal of Aβ deposits or that have a concentration-independent influence on fibril formation) does not. With the development of highly sensitive and specific sandwich ELISAs, it is now straightforward to analyse CSF or conditioned medium for Aβs ending either at Aβ40 or at Aβ42,43. It may soon be possible to analyse Aβ similarly in plasma. Since plasma is far easier to obtain than CSF, the ability to measure the concentration of specific Aβs in plasma could be an especially powerful tool for evaluating the role in AD pathogenesis of factors that increase the concentration of soluble Aβ.

7.6.2 Implications for animal models of AD

To model AD in an experimental animal based on the amyloid deposition hypothesis developed above, the concentration of soluble Aβ in brain must be increased to the point that Aβ is deposited extracellularly. This deposited Aβ must then persist and trigger pathology similar to that observed in the AD brain. In an effort to develop such a model, many groups have produced transgenic animals in which there is increased expression of wild-type human APP or Aβ-bearing CTFs (Quon *et al.*, 1991; Kammescheidt *et al.*, 1992; Fukuchi *et al.*, 1993a; Lamb *et al.*, 1993; Hsiao *et al.*, 1994). So far, the amount of insoluble Aβ deposited in the brains of these animals has been unimpressive. Thus there has been little reason to expect AD pathology to develop in these animals and no convincing AD-like pathology has been observed. The foregoing discussion of amyloid deposition indicates that there are many explanations for the difficulty that has been encountered in modelling amyloid deposition in transgenic animals. These explanations fall into two categories: (i) soluble Aβ in the transgenic brain has not increased in the same way that it increases in the AD brain; and (ii) amyloid has not deposited in the transgenic brain despite an adequate increase in soluble Aβ either because the factors determining fibril formation are less favourable or because removal of Aβ deposits is faster in the rodent than in the aged human brain.

Soluble Aβ could fail to rise sufficiently if the rodent brain processes a small fraction of APP molecules to Aβ as compared to human brain, if rodent brain produces less of the critically important Aβ1–42 species, or if the rodent brain rapidly removes total soluble Aβ or the Aβ1–42 species. Although measurements of the soluble Aβ in brain is challenging, it should nonetheless be possible to assess whether or not there has been a sufficient increase by estimating the concentration of soluble Aβ in the AD and control human brain, and comparing those concentrations to the concentrations observed in control and transgenic brains. If soluble Aβ is not sufficiently elevated, rational efforts could be made to increase it to an appropriate target concentration. If soluble Aβ is rapidly removed in rodent brain, then it may be far easier to increase the level of soluble Aβ by impairing removal than by increasing secretion. If a sufficient understanding of Aβ removal can be developed, it might even be possible to produce amyloid deposition by simply blocking removal of the Aβ that is secreted endogenously.

If amyloid fails to accumulate even when there is an appropriate elevation of soluble Aβ, then efforts could be made to mimic the specific conditions in the aged human brain that favour this accumulation. This might be done by increasing expression of HSPG, ApoE4 or other Aβ binding proteins thought to influence Aβ deposition. As appropriate information develops, it may also be possible to foster amyloid accumulation by specifically blocking mechanisms involved in removing amyloid deposits.

As the appropriate analytical methods are developed and information on the factors that determine amyloid deposition accumulates, it should be possible to develop animals in which cerebral amyloid is deposited in an increasingly rational manner. Careful exploitation of the APP mutations known to cause FAD and identification of additional mutations that produce FAD should be particularly helpful in developing this animal model. In culture, cells expressing APPs that have both the ΔNL mutation and a APP_{717} mutation not only secrete more total Aβ, they also secrete a higher percentage of Aβ1–42. Thus, compared to wild-type APP, the combined mutant produces almost ten times more of the critical Aβ1–42 form. By creating transgenic animals expressing this combined mutant or another artificial construct that results in secretion of high levels of Aβ and then further engineering these animals to express other genes shown to cause FAD, it is quite reasonable to expect that an animal model for AD may ultimately be developed.

7.6.3 Implications for therapy of AD

Based on the amyloid deposition hypothesis, it is rational to develop therapies for AD aimed at: (i) blocking formation and secretion of soluble Aβ; (ii) enhancing removal of secreted, soluble Aβ; (iii) blocking the formation of amyloid fibrils from soluble Aβ; (iv) enhancing removal of fibrillar Aβ; (v) interfering with the pathological cascade induced by amyloid (e.g. by interfering with the complement activation that amyloid may engender); or (vi) developing agents (e.g. anticholinesterases such as tacrine) that help the nervous system to compensate for amyloid-induced pathology.

The strategy of blocking formation and secretion of Aβ is particularly attractive from both theoretical and practical points of view. For reasons discussed above, it is very likely that amyloid formation is highly concentration dependent. Thus relatively small reductions in Aβ concentration achieved by partial inhibition of β- or γ-secretase may be highly effective in reducing amyloid deposition. This means that effective therapy may be possible even if Aβ or the proteases that produce it are critical for normal cellular function. It is important to note that blocking amyloid formation could in time result in very substantial reductions in the total amyloid deposited and possibly in the total number of plaques present, if there are normal mechanisms at work that remove deposited Aβ permitting plaques to resolve. Thus halting amyloid deposition may halt disease progression and possibly even permit some recovery of function. By using sandwich ELISAs to measure the Aβ secreted by transfected cells, it is now straightforward to develop high throughput screens for compounds that impair Aβ secretion using a conventional 96-well format. Thus it is practical to screen the vast array of compounds available in the pharmaceutical industry for prototypic nontoxic compounds that inhibit Aβ secretion. Once such compounds are identified, it should be straightforward, using published methods, to determine whether they inhibit β- or γ-secretase or act through some other mechanism. There is, in fact, a good chance that several classes of compounds will be identified that work through different and possibly synergistic mechanisms. It will, of course, be important to determine whether these compounds block secretion of Aβ1–42, which appears to be particularly important for amyloid formation, as well as the more abundant Aβ1–40, and it should be straightforward to do so. An important practical advantage of developing compounds that block the formation and release of Aβ is that the transition to development of an effective oral drug can proceed in the absence of an animal model for AD using reduction of soluble Aβ in the CSF, plasma, or brain of appropriate animals as an endpoint.

Ideally, one would like to develop specific inhibitors of β- or γ-secretase by working directly with the isolated proteases. In this regard, it is important to point out that β- and γ-secretase are activities. At the moment, it is not known if either β- or γ-secretase activity resides in a single protease or in a set of proteases, and it is unclear whether the carboxyl-terminal heterogeneity in secreted Aβ is a result of cleavage by a single promiscuous γ-secretase, a pair of endoproteases that cleave at Aβ40 and Aβ42, or one or several endoproteases acting in concert with exoproteases that trim the carboxyl-end of secreted Aβ. In isolating proteases with β- or γ-secretase activity, it would obviously be invaluable to have an array of inhibitors with known effects in order to avoid isolating irrelevant proteases that by chance cleave the APP at its γ- and β-cleavage sites.

Based on the amyloid deposition hypothesis, one could also seek useful new drugs by developing rapid *in vitro* screens for compounds that: (i) inhibit the formation of amyloid from synthetic Aβ; (ii) inhibit the binding of Aβ to proteins such as HSPG, which are postulated to influence amyloid formation or removal; or (iii) inhibit events likely to be important in pathogenesis such as Aβ-induced complement activation. The efficacy *in vivo* of compounds identified through *in vitro* screens of this sort would either have to await the development of animal models for AD or be pursued in aggressive clinical trials. Given the recent evidence that anti-inflammatory drugs may be useful in slowing AD, it may also be worthwhile to pursue this class, which presumably acts by interfering in some way with pathology that is provoked by amyloid.

This brief description of therapeutic approaches made possible by recent advances in our understanding of amyloid deposition has focused on approaches that currently seem most likely to be fruitful. As additional advances are made, other therapeutic approaches may become more attractive. In closing, it should be pointed out that, in the search for drugs effective in preventing amyloid deposition, compounds may be identified that prove useful in developing an animal model for AD. Thus this search could uncover drugs that increase the production and secretion of Aβ or that enhance the formation of amyloid from synthetic Aβ, and these agents could be quite useful in developing an animal model for AD.

References

Allsop, D., Yamamoto, T., Kametani, F., Miyazaki, N. & Ishii, T. (1991) *Brain Res.* **551**, 1–9.
Anderson, J.P., Esch, F.S., Keim, P.S., Sambamurti, K., Lieberburg, I. & Robakis, N.K. (1991) *Neurosci. Lett.* **128**, 126–128.
Autilio-Gambetti, L., Morandi, A., Tabaton, M. *et al.* (1988) *FEBS Lett.* **241**, 94–98.
Borgaonkar, D.S., Schmidt, L.C., Martin, S.E. *et al.*(1993) *Lancet* **342**, 625.
Burdick, D., Soreghen, B., Kwon, M. *et al.* (1992) *J. Biol. Chem.* **267**, 546–554.
Busciglio, J., Gabuzda, D., Matsudaira, P. & Yankner, B. (1993) *Proc. Natl Acad. Sci. USA* **90**, 2092–2096.
Cai, X., Golde, T.E. & Younkin, S.G. (1993) *Science* **259**, 514–516.
Caporaso, G.L., Gandy, S.E., Buxbaum, J.D. & Greengard, P. (1992a) *Proc. Natl Acad. Sci. USA* **89**, 2252–2256.
Caporaso, G.L., Gandy, S.E., Buxbaum, J.D., Ramabhadran, T.V. & Greengard, P. (1992b) *Proc. Natl Acad. Sci. USA* **89**, 3055–3059.
Castano, E.M. & Frangione, B.F. (1988) *Lab. Invest.* **58**, 122–132.
Chartier-Harlin, M.-C., Crawford, F., Houlden, H. *et al.* (1991) *Nature* **353**, 844–846.
Chen, W.-J., Goldstein, J.L. & Brown, M.S. (1990) *J. Biol. Chem.* **265**, 3116–3123.
Cheung, T., Ghiso, J., Shoji, M. *et al.* (1994) *Amyloid: Int. J. Exp. Clin. Invest.* **1**, 30–38.
Citron, M., Oltersdorf, T., Haas, C. *et al.* (1992) *Nature* **360**, 672–674.
Cohen, A.S. & Connors, L.H. (1987) *J. Pathol.* **151**, 1–10.
Cole, G.M., Huynh, T.V. & Saitoh, T. (1989) *Neurochem. Res.* **14**, 933–939.
Corder, E.H., Saunders, A.M., Strittmatter, W.J. *et al.* (1993) *Science* **261**, 921–923.
Cras, P., Kawai, M., Siedlak, S. *et al.* (1990) *Am. J. Pathol.* **137**, 241–246.
Daigle, I. & Li, C. (1993) *Proc. Natl Acad. Sci. USA* **90**, 12045–12049.
De Sauvage, F. & Octave, J.-N. (1989) *Science* **245**, 651–653.
Dickson, D.W., Farlo, J., Davis, P., Crystal, H., Fuld, P. & Yen, S.C. (1988) *Am. J. Pathol.* **132**, 86–101.

Dovey, H.F., Suomesaari-Chrysler, S., Lieberburg, I., Hinha, S. & Kiem, P.S. (1993) *NeuroReport* **4**, 1039–1042.
Esch, F.S., Keim, P.S., Beattie, E.C. *et al.* (1990) *Science* **248**, 1122–1124.
Estus, S., Golde, T.E., Kunishita, T. *et al.* (1992) *Science* **255**, 726–728.
Fabian, H., Szendrei, G.I. & Mantsch, H.H., Jr. L.O. (1993) *Biochem. Biophys. Res. Commun.* **191**, 232–239.
Fraser, P.E., Nguyen, J.T., Inouye, H. *et al.* (1992) *Biochemistry* **31**, 10716–10723.
Fukuchi, K., Kamino, K., Deeb, S.S., Smith, A.C., Dang, T. & Martin, G.M. (1992) *Biochem. Biophys. Res. Commun.* **182**, 165–173.
Fukuchi, K., Ogburn, C.E., Smith, A.C., Kunkel, D.D. *et al.* (1993a) *Ann. N. York Acad. Sci.* **695**, 217–223.
Fukuchi, K., Sopher, B., Furlong, C.E., Smith, A.C., Dang, N.T. & Martin, G.M. (1993b) *Neurosci. Lett.* **154**, 145–148.
Gandy, S., Czernik, A.J. & Greengard, P. (1988) *Proc. Natl Acad. Sci. USA* **85**, 6218–6221.
Gandy, S., Bhasin, R., Ramabhadran, T. *et al.* (1992) *J. Neurochem.* **58**, 383–386.
Ghiso, J., Matsubara, E., Koudinov, A. *et al.* (1993) *Biochem. J.* **293**, 27–30.
Glenner, G.G. (1983) Alzheimer's disease: multiple cerebral amyloidosis. In *Banbury Report 15: Biological Aspects of Alzheimer's Disease* (ed. Katzman, R.) pp. 137–144. New York, Cold Spring Harbor Laboratory.
Glenner, G.G. & Wong, C.W. (1984) *Biochem. Biophys. Res. Commun.* **122**, 1131–1135.
Goate, A.M., Owen, M.J., James, L.A. *et al.* (1989) *Lancet* 352–355.
Goate, A., Chartier-Harlin, M.-C., Mullan, M. *et al.* (1991) *Nature* **349**, 704–706.
Goedert, M. (1987) *EMBO J.* **6**, 3627–3632.
Golde, T.E., Estus, S.G., Usiak, M., Younkin, L.H. & Younkin, S.G. (1990) *Neuron* **4**, 253–267.
Golde, T.E., Estus, S., Younkin, L., Selkoe, D.J. & Younkin, S.G. (1992) *Science* **255**, 728–730.
Goldgaber, D., Lerman, M.I., McBride, O.W., Saffiotti, U. & Gajdusek, D.C. (1987) *Science* **235**, 877–880.
Gowing, E., Roher, A.E., Woods, A.S. *et al.* (1994) *J. Biol. Chem.* **269**, 10987–10990.
Haass, C., Koo, E.H., Mellon, A., Hung, A.Y. & Selkoe, D.J. (1992a) *Nature* **357**, 500–503.
Haass, C., Schlossmacher, M.G., Hung, A.Y. *et al.* (1992b) *Nature* **359**, 322–325.
Haass, C. & Selkoe, D.J. (1993) *Cell* **75**, 1039–1042.
Hardy, J. *et al.* (1991) *Lancet* **337**, 1342–1343.
Hardy, J., Houlden, H., Collinge, J. *et al.* (1993) *Lancet* **342**, 737–738.
Hilbich, C., Kisters-Woike, B., Reed, J., Masters, C.L. & Beyreuther, K. (1991) *J. Mol. Biol.* **218**, 149–163.
Hsiao, K.K., Olson, K., Borchelt, D.R., Johannsdottir, R., Shen, Z., Yunis, W.S., Xu, X., Kitt, C.A., Diedrich, J., Sisodia, S.S., Carlson, G., Price, D., Clark, H.B. & Iadecola, C. (1994) *Neurobiol. Aging* **15**, S76.
Hung, A.Y., Haass, C., Nitsch, R.M. *et al.* (1993) *J. Biol. Chem.* **268**, 22959–22962.
Ishiura, S., Toshifumi, T., Tabira, T. & Sugita, H. (1989) *FEBS Lett.* **257**, 388–392.
Ishiura, S., Tsukahara, T., Tabira, T., Shimizu, T., Arahate, K. & Sugita, H. (1990) *FEBS Lett.* **260**, 131–134.
Jacobsen, J.S., Muerkel, H.A., Blume, A.J. & Vitek, M.P. (1991) *Neurobiol. Aging* **12**, 575–583.
Jarrett, J.T. & Lansbury, P.T., Jr (1993) *Cell* **73**, 1055–1058.
Jarrett, J.T., Berger, E.P. & Lansbury, P.T., Jr (1993) *Biochemistry* **32**, 4693–4697.
Kammesheidt, A., Boyce, F.M., Spanyoannis, A.F., Cumming, B.J., Ortegon, M., Cotman, C., Vaught, J.L. & Neve, R.L. (1992) *Proc. Natl Acad. Sci. USA* **89**, 10857–10861.
Kang, J., Lemaire, H.-G., Unterbeck, A. *et al.* (1987) *Nature* **325**, 733–736.
Kitaguchi, N., Takahashi, Y., Tokushima, Y., Shiojiri, S. & Ito, H. (1988) *Nature* **331**, 530–532.
Knops, J., Lieberburg, I. & Sinha, S. (1992) *J. Biol. Chem.* **267**, 16022–16024.
Kojima, S. & Omori, M. (1992) *FEBS Lett.* **304**, 57–60.
Konig, G., Monning, U., Czech, C. *et al.* (1992) *J. Biol. Chem.* **267**, 10804–10809.

Koo, E.H., Sisodia, S.S., Cork, L.C., Unterbeck, A., Bayney, R.M. & Price, D.L. (1990) *Neuron* **2**, 97–104.
Lamb, B., Sisodia, S.S., Lawlwe, A.M., Slunt, H.H., Kitt, C.A., Kearns, W.G., Pearson, P.L., Price, D.L. & Gearhart, J.D. (1993) *Nature Genet.* **5**, 22–29.
Levy, Carman, M.D., Fernandez-Madrid, I.J. *et al.* (1990) *Science* **248**, 1124–1126.
Mann, D.M.A., Yates, P.O., Marcyniuk, C. & Ravindra, C.R. (1986) *Neuropath. Appl. Neurobiol.* **12**, 447–457.
Maruyama, K., Kametami, F., Usami, M., Yamao-Harigaya, W. & Tanaka, K. (1991) 'Secretase', *Biochem. Biophys. Res. Commun.* **179**, 1670–1676.
Masliah, E., Terry, R.D., Mallory, M., Alford, M. & Hanson, L.A. (1990) *Am. J. Pathol* **137**, 1293–1297.
Masters, C.L., Simms, G., Weinman, N.A., Multhaup, G., McDonald, B.L. & Beyreuther, K. (1985) *Proc. Natl Acad. Sci. USA* **82**, 4245–4249.
Matsumoto, A. & Fujiwara, Y. (1993) *Eur. J. Biochem.* **217**, 21–27.
McDermott, J., Biggins, J. & Gibson, A. (1992) *Biochem. Soc. Trans.* **20**, 206S.
Miller, D.L., Papayannopoulos, I.A., Styles, J. *et al.* (1993) *Arch. Biochem.* **301**, 41–52.
Mori, H., Takio, K., Ogawara, M. & Selkoe, D.J. (1992) *J. Biol. Chem.* **267**, 17082–17086.
Mullan, M. (1992) *Br. Med. J.* **305**, 1108.
Mullan, M., Houlden, H., Windelspecht, M. *et al.* (1992) *Nature Genet.* **2**, 340–342.
Murrell, J., Farlow, M., Ghetti, B. & Benson, M.D. (1991) *Science* **254**, 97–99.
Naruse, S., Igarashi, S., Aoki, K. *et al.* (1991) *Lancet* **337**, 978–979.
Neve, R.L., Finch, E.A. & Dawes, L.R. (1988) *Neuron* **1**, 669–677.
Noguchi, S., Murakami, K. & Yamada, N. (1993) *Lancet* **342**, 737.
Oltersdorf, T., Ward, P.J., Henriksson, T. *et al.* (1990) *J. Biol. Chem.* **265**, 4492–4497.
Palmert, M.R., Podlisny, M.B., Witker, D.S. *et al.* (1989a) *Proc. Natl Acad. Sci. USA* **86**, 6338–6342.
Palmert, M.R., Siedlak, S., Podlisny, M.B. *et al.* (1989b) *Biochem. Biophys. Res. Commun.* **165**, 182–188.
Pasternack, J., Palmert, M., Usiak, M. *et al.* (1992) *Biochem.* **31**, 10936–10940.
Payami, H., Kaye, J., Heston, L.L., Bird, T.D. & Schellenberg, G.D. (1993) *Lancet* **342**, 738.
Pericak-Vance, M., Yamaoka, L.H., Haynes, C.S. *et al.* (1988) *Expl. Neurol.* **102**, 271–279.
Pericak-Vance, M.A., Bebout, J.L., Gaskell, P.C., Jr. *et al.* (1991) *Am. J. Hum. Genet.* **48** 1034–1050.
Poirer, J., Davignon, J., Bouthillier, D., Kogan, S., Bertrand, P. & Gauthier, S. (1993) *Lancet* **342**, 697–699.
Ponte, P., Gonzalez-DeWhitt, P., Schilling, J. *et al.* (1988) *Nature* **331**, 525–527.
Prelli, F., Castano, E., Glenner, G.G. & Frangione, B. (1988) *Neurochem.* **51**, 648–651.
Quon, D., Wang, Y., Catalano, R., Scardina, J.M., Murakami, K. & Cordell, B. (1991) *Nature* **352**, 239–241.
Razzaboni, B., Papastoitsis, G., Koo, E. & Abraham, C. (1992) *Brain Res.* **589**, 207–216.
Robakis, N.K., Wisniewski, H.M., Jenkins, E.C. *et al.* (1987) *Lancet* **14**, 384–385.
Roher, A.E., Lowenson, J.D., Clarke, S. *et al.* (1993a) *J. Biol. Chem.* **268**, 3072–3083.
Roher, A.E., Lowenson, J.D., Clarke, S. *et al.* (1993b) *J. Biol. Chem.* **90**,10836–10840
Ropper, A.H. & Williams, R.S. (1980) *Neurology* **30**, 639–644.
Roth, M., Tomlinson, B.E. & Blessed, G. (1966) *Nature* **209**, 109–110.
Sahasrabudhe, S., Brown, A., Hulmes, J. *et al.* (1993) *J. Biol. Chem.* **268**, 16699–16705.
Sambamurti, K., Shioi, J., Anderson, J.P., Pappolla, M.A. & Robakis, N.K. (1992) *J. Neurosci. Res.* **33**, 319–329.
Sandbrink, R., Banati, R., Masters, C.L., Beyreuther, K. & Konig, B. (1993) *Ann. N. York Acad. Sci.* **695**, 183–189.
Sandbrink, R., Masters, C.L. & Beyreuther, K. (1994) *J. Biol. Chem.* **269**, 15510–15517.
Saunders, A.M., Schmader, K., Breitner, J.C.S. *et al.* (1993a) *Lancet* **342**, 710–711.
Saunders, A.M., Strittmatter, W.J, Schmechel, D. *et al.* (1993b) *Neurology* **43**, 1467–1472.

Schellenberg, G., Bird, T.D., Wijsman, E.M. *et al.* (1988) *Science* **241**, 1507–1510.
Schellenberg, G.D., Bird, T.D., Wijsman, E.M. *et al.* (1992) *Science* **258**, 668–671.
Schmechel, D.E., Saunders, A.M., Strittmatter, W.J. *et al.* (1993) *Proc. Natl Acad. Sci. USA* **90**, 9649–9653.
Schubert, D., Schroeder, R., LaCorbiere, M., Saitoh, T. & Cole, G. (1988) *Science* **241**, 223–226.
Schubert, D., Jin, L.-W., Saitoh, T. & Cole, G. (1989) *Neuron* **3**, 689–694.
Schwarzmann, A.L., Gregori, L., Strittmatter, W.J. *et al.* (1993) *Abstracts Soc. Neurosci.* **19**, 860.
Selkoe, D.J. (1986) *Neurobiol. Aging* **7**, 425–432.
Selkoe, D.J. (1991) *Neuron* **6**, 487–498.
Selkoe, D.J., Bell, B.S., Podlisny, M.B., Price, D.L. & Cork, L.C. (1987) *Science* **235**, 873–876.
Selkoe, D.J., Podlisny, M.B., Joachim, C.L. *et al.* (1988) *Proc. Natl Acad. Sci. USA* **85**, 7341–7345.
Seubert, P., Vigo-Pelfrey, C. & Esch, F. *et al.* (1992) *Nature* **359**, 325–327.
Seubert, P., Oltersdorf, T., Lee, M. *et al.* (1993) *Nature* **361**, 260–263.
Shoji, M., Golde, T.E., Ghiso, J. *et al.* (1992) *Science* **258**, 126–129.
Siman, R., Mistretta, S., Durkin, J.T. *et al.* (1993) *J. Biol. Chem.* **268**, 16602–16609.
Sisodia, S.S. (1992) *Med. Sci.* **89**, 6075–6079.
Sisodia, S.S., Koo, E.H., Beyreuther, K., Unterbeck, A. & Price, D.L. (1990) *Science* **248**, 492–495.
Small, D., Moir, R., Fuller, S. *et al.* (1991) *Biochemistry* **30**, 10795–10799.
Sorimachi, K. & Craik, D.J. (1994) *Eur. J. Biochem.* **219**, 237–251.
St George-Hyslop, P., Haines, J.L., Farrer, L.A., Polinsky, R., Van Broeckhoven, C., Goate, A., McLachlan, D.R., Orr, H., Bruni, A.C., Sorbi, S., Rainero, I., Foncin, J.-F., Pollen, D., Countu, J.-M., Tupler, R., Voskresenskaya, N., Mayeux, R., Growdon, J., Fried, V.A., Myers, R.H., Nee, L., Backhovens, H., Marbin, J.-J., Rossor, M., Owen, M.J., Mullan, M., Percy, M.E., Karlinsky, K., Rich, S., Heston, L., Montesi, M., Mortilla, M., Nacmias, N. & Gusella, J.F. (1990) *Nature* **347**, 194–197.
St George-Hyslop, P., Haines, J., Rogaev, E. & Hardy, J. (1992) *Nature Genet.* **2**, 330–334.
Strittmater, W.J., Weisgraber, K.H., Huang, D.Y. *et al.* (1993a) *Proc. Natl Acad. Sci. USA* **90**, 8098–8102.
Strittmatter, W.J., Saunders, A.M., Schmechel, D. *et al.* (1993b) *Proc. Natl Acad. Sci. USA* **90**, 1977–1981.
Struble, R.G., Cork, L.C., Whitehouse, P.J. & Price, D.L. (1982) *Science* **216**, 413–415.
Suzuki, N., Cheung, T.T., Cai, X.-D. *et al.* (1994) *Science* **264**, 1336–1340.
Tagawa, K., Kunishita, T., Maruyama, K. *et al.* (1991) *Biochem. Biophys. Res. Commun.* **177**, 377–387.
Takio, K., Hasegawa, M., Titani, K. & Ihara, Y. (1989) *Biochem. Biophys. Res. Commun.* **160**, 1296–1301.
Tanzi, R.E., Gusella, J.F., Watkins, P.C. *et al.* (1987) *Science* **235**, 880–884.
Tanzi, R.E., McClatchey, A.I., Lamperti, E.D., Villa-Komaroff, L., Gusella, J.F. & Neve, R.L. (1988) *Nature* **331**, 528–530.
Terry, R.D. (1985) Alzheimer's disease. In *Textbook of Neuropathology* (eds Davis, R.L. & Robertson, D.M.) pp. 824–841. Baltimore, Williams & Wilkins.
Terry, R.D., Lawrence, L.A., DeTeresa, R., Davies, P., Tobias, H. & Katzman, R. (1987) *J. Neuropathol. Exp. Neurol.* **46**, 262–268.
Van Broeckhoven, C., Backhovens, H., Cruits, M. *et al.* (1992) *Nature Genet.* **2**, 335–339.
Van Broeckhoven, C., Haan, J., Bakker, E. *et al.* (1990) *Science* **248**, 1120–1122.
Vigo-Pelfrey, C., Lee, D., Keim, P., Lieberburg, I. & Schenk, D.B. (1993) *J. Neurochem.* **61**, 19965–19968.
Wang, R., Meschia, J.F., Cotter, R.J. & Sisodia, S.S. (1991) *J. Biol. Chem.* **266**, 16960–16964.
Wertkin, A.M., Turner, R.S., Pleasure, S.J. *et al.* (1993) *Proc. Natl Acad. Sci. USA* **90**, 9513–9517.
Wiedemann, A., Konig, G., Bunke, D. *et al.* (1989) *Cell* **57**, 115–126.
Wisniewski, H.M., Moretz, R.C. & Lossinksy, A.S. (1981) *Ann. Neurol.* **10**, 517–522.
Wolozin, B., Bacic, M., Merrill, M.J. *et al.* (1992) *J. Neurosci. Res.* **33**, 163–169.

Yamaguchi, H., Hirai, S., Morimatsu, M., Shoji, M. & Harigaya, Y. (1988) *Acta Neuropathol.* **77**, 113–119.
Yamaguchi, H., Nakazato, Y., Hirai, S., Shoji, M. & Harigaya, Y. (1989) *Acta Neuropathol.* **77**, 314–319
Yankner, B.A., Dawes, L.R., Fisher, S., Villa-Komaroff, L., Oster-Granite, M.L. & Neve, R.L. (1989) *Science* **245**, 417–420.
Yoshioka, K., Miki, T., Katsuya, T., Ogihara, T. & Sakaki, Y. (1991) *Biochem. Biophys. Res. Commun.* **178**, 1141–1146.
Zhong, Z., Higaki, J., Murakami, K. *et al.* (1994) *J. Biol. Chem.* **269**, 627–632.

CHAPTER 8

NEUROTOXICITY OF AMYLOID β-PROTEIN AND THE AMYLOID β-PROTEIN PRECURSOR

Kazuaki Yoshikawa

Department of Regulation of Macromolecular Functions, Institute for Protein Research, Osaka University, 3–2 Yamadaoka, Suita, Osaka 565, Japan

Table of Contents

8.1 Introduction

The ultimate goal of research into Alzheimer's disease (AD) is to elucidate the mechanism underlying the massive neuronal death seen in the brain with this disease. Among various pathological features of AD, deposition of amyloid fibrils in the brain is thought to be the most important hallmark. Indeed, many researchers have made intensive studies on the mechanism of amyloid deposition in their attempts to understand the pathogenesis of AD. However, it is unclear at present whether or not amyloid deposition directly induces neuronal death in AD.

Amyloid fibrils are deposited in senile plaques and along the walls of the cerebral blood vessels in AD. Several types of senile plaques are classified according to their morphological characteristics (e.g. diffuse, primitive and classical plaques). Mature senile plaques consist of amyloid deposits surrounded by dystrophic neurites (neuritic plaques). These morphological characteristics have led to speculation that the amyloid masses at the extracellular space lead to the death of adjacent neurons. In peripheral amyloidoses including primary and secondary amyloidoses, large masses of amyloid induce organ dysfunction. These observations favour the

Pathobiology of Alzheimer's Disease
ISBN 0-12-286965-5

proposition that amyloid deposition is responsible for neuronal degeneration in AD.

The principal constituent of amyloid fibrils in AD is amyloid β-protein (Aβ) (Glenner and Wong, 1984), which is derived from its precursor, the amyloid β-protein precursor (APP). There are three major species of APP: APP_{695} (Kang *et al.*, 1987), APP_{751} (Ponte *et al.*, 1988; Tanzi *et al.*, 1988), and APP_{770} (Kitaguchi *et al.*, 1988). In APP molecules, the Aβ domain is situated at the boundary between the extracellular domain and the transmembrane domain (Figure 1). Studies on the generation of Aβ from APP molecules may thus be indispensable for understanding the mechanism of amyloidogenesis and its potential neurotoxicity.

To test whether the Aβ deposition causes neuronal death in AD, neurotoxic effects of synthetic Aβ peptides have been tested both *in vitro* and *in vivo*. DNA-transfected cell models overexpressing parts of APP or full-length APP have been utilized to examine whether intracellular overproduction of APP affects the processing of APP and cell viability. In this article, the recent findings on the neurotoxic effects of Aβ and related peptides are reviewed. Furthermore, on the basis of the findings with DNA-transfected cell models, a novel hypothesis for neuronal death in AD is presented.

8.2 Neurotoxicity of Aβ and related peptides

8.2.1 *In vitro* neurotoxicity

The neurotoxicity of an Aβ-containing fragment was first demonstrated in a cultured cell system of PC12 phaeochromocytoma cells transfected with complementary DNA (cDNA) encoding a carboxyl (C)-terminal fragment of APP (Yankner *et al.*, 1989). They found that PC12 cells transfected with cDNA encoding the C-terminal fragment containing the entire Aβ region and the cytoplasmic domain showed severe degeneration when the cells were treated with nerve growth factor (NGF), an inducer of neuronal differentiation in PC12 cells. In addition, they found that conditioned medium from transfected PC12 or NIH3T3 cells was toxic to neurons in primary cultures. The toxic agent was removed by immunoabsorption with an antibody directed against Aβ. Thus, they concluded that the secreted C-terminal fragment of APP exerts the neurotoxic effect.

Using hippocampal neuronal cultures, they further demonstrated that Aβ1–40 was trophic to immature neurons at low concentrations and neurotoxic to mature neurons at high concentrations (Yankner *et al.*, 1990b). The neurotoxicity of Aβ1–40 was observed at 1–20 μM, and Aβ25–35 was found to be the biologically active core sequence that mediates both the trophic and toxic effects. The effects of Aβ were mimicked by tachykinin antagonists and completely reversed by specific tachykinin agonists such as substance P and physalaemin.

In another study, Yankner *et al.* (1990a) demonstrated that Aβ1–40 showed signif-

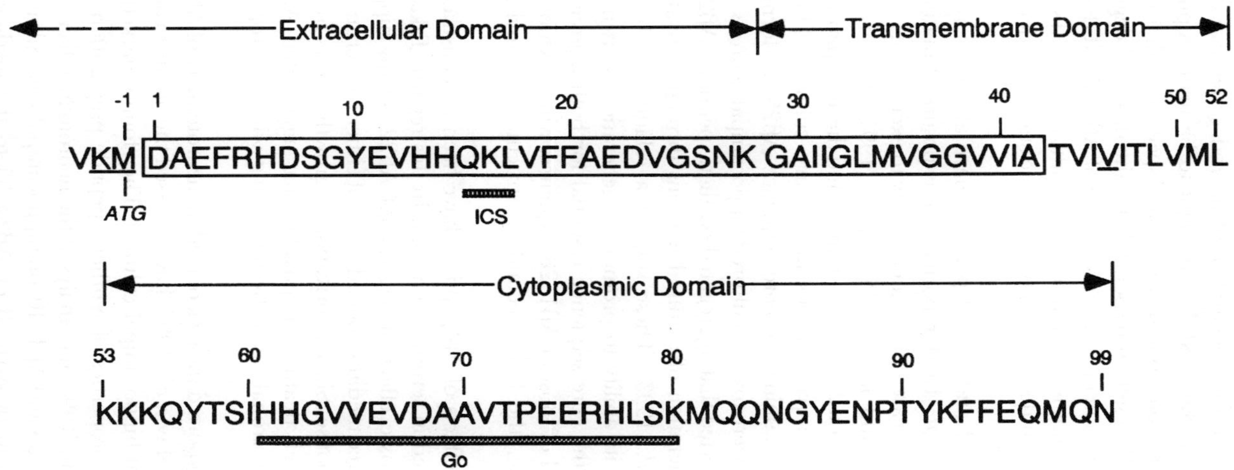

Figure 1 Amino-acid sequence of the carboxyl terminus of APP. Carboxyl-terminal amino-acid residues of APP containing the Aβ domain are represented by the single letter notation, and numbered, beginning with the first amino (N)-terminal residue of Aβ domain. The secretory cleavage occurs at residues 15–17 within the Aβ domain. The Go protein binding sequence is located in the cytoplasmic domain (61–80). Amino-acid residues mutated in familial AD cases are underlined. Met codon (ATG) at –1 can be utilized to initiate the translation of the C-terminal fragment consisting of 100 amino-acid residues (APP-C100). In the following pages, amino-acid residues are numbered as shown in this figure. ICS, internal cleavage site; Go, Go protein binding domain.

icant neurotoxicity in the presence of NGF: NGF increased the neurotoxic potency of Aβ (EC_{50} of neurotoxicity of Aβ decreased from 0.1 μM to 1 pM). Exposure of Aβ alone resulted in a marked induction of immunoreactive NGF receptors, and addition of NGF with Aβ resulted in the appearance of neurodegenerative changes in NGF-receptor positive neurons. Since NGF is thought to be neurotrophic to hippocampal and basal forebrain cholinergic neurons, they speculated that neurotoxic interaction of Aβ with the NGF system is responsible for the neuronal degeneration. Roher *et al.* (1991) also reported a neurotoxic interaction of Aβ with NGF in peripheral neurons; amyloid core proteins isolated from brains with AD showed severe toxic effects on cultured sympathetic and sensory neurons whose survival is dependent on NGF.

The neurotoxic effects of Aβ and related peptides have also been reported by Cotman and colleagues: Koh *et al.* (1990) reported that exposure of mature cortical neurons to synthetic Aβ1–42 increased their vulnerability to excitotoxins such as glutamate, *N*-methyl-D-aspartate and kainate. Pike *et al.* (1991) demonstrated that Aβ1–42 showed aggregation after a 2–4 day incubation, and the aged peptide exerted a neurotoxic effect on hippocampal neurons in primary cultures. They further reported that Aβ peptides containing the core sequence Aβ25–35 formed stable aggregates and these peptides were toxic to cultured hippocampal neurons (Pike *et al.*, 1993). Furthermore, they reported that aggregated synthetic peptides (Aβ25–35 and Aβ1–42) triggered the degeneration of cultured neurons through activation of an apoptotic pathway (Loo *et al.*, 1993). These suggest a relationship between the aggregation state of Aβ and its ability to promote degeneration. It is noteworthy, however, that Aβ25–35 (i.e. the core sequence essential for both aggregation and neurotoxicity *in vitro*) is devoid of most of Aβ1–28 sequence, which exhibits an anti-parallel β-sheet structure (i.e. the basic secondary structure of amyloid fibrils) (Kirschner *et al.*, 1987).

Most of these neurotoxic effects of Aβ peptides have been demonstrated *in vitro* using primary neurons cultured in serum-free media. In this regard, Takadera *et al.* (1993) demonstrated a cytotoxic effect of aggregated Aβ22–35 on hippocampal neurons cultured in a serum-free medium, but addition of either calf serum or insulin prevented the neurotoxic response. Since deprivation of growth factors from culture media causes an apoptotic cell death, it is possible that aggregated Aβ at high concentrations may induce neuronal death by interfering with the trophic mechanisms of some growth factors.

Several laboratories have reported the lack of evident neurotoxicity of Aβ peptides using various *in vitro* systems: May *et al.* (1992) have assessed the *in vitro* toxicity of Aβ using high-density primary rat hippocampal cultures. They found marked lot-to-lot differences in the neurotoxic properties of synthetic peptide preparations obtained from the same supplier. Malouf (1992) has analysed the neurotoxic effects of various Aβ fragments (Aβ1–28, Aβ25–35 and Aβ1–40) on hippocampal slice cultures instead of dissociated cultures; addition of Aβ1–28 or Aβ25–35 to the medium did not produce significant changes in dendritic length or the number of branches. To avoid potential access problems, they injected these peptides directly into the cultures, and found that only Aβ1–40 produced neurodegeneration around the site of injection. Mattson and Rydel (1992) found that Aβ exacerbated the ongoing neuronal death when cultures were unstable; Aβ1–40 and Aβ25–35 had a significant negative effect on neuronal survival in low cell-density rat hippocampal cultures in which considerable neuronal death occurred in control cultures. However these Aβ peptides did not affect neuronal survival in more stable high cell-density cultures where little cell death occurred in control cultures. Therefore, they assumed that Aβ peptides would not cause neuronal damage in stable healthy cultures in which there is no ongoing neuronal death.

We tested the neurotoxic effect of Aβ25–35 on hippocampal neurons which were prepared from a rat fetus at embryonic day 18 and cultured for 4 days. We found no neurotoxic effects of Aβ25–35, tested up to 100 μM, on the hippocampal neurons cultured in either serum-free or serum-supplemented medium (10% fetal calf serum). We thus concluded that Aβ25–35 had no toxic effect on the 'healthy neurons' (unpublished observations).

Taken together, it is suggested that the neurotoxic effects of Aβ peptides are dependent on various factors, including cultured neurons (e.g. derivation, maturity and cell density), synthetic peptides, vehicles and culture media (e.g. serum and growth factors).

8.2.2 *In vivo* neurotoxicity

Kowall *et al.* (1991) reported that injection of Aβ into the adult rat cerebral cortex caused profound neurodegenerative changes including neuronal loss and degenerating neurites, and that co-administration of substance P prevented the neurotoxic effects. In these experiments, they dissolved Aβ in a vehicle of 35% acetonitrile/ 0.1% trifluoroacetic acid, and confirmed that the vehicle alone showed no neurotoxicity. Kowall *et al.* (1992) later made an additional report that injections of Aβ1–40 and Aβ25–35 produced localized necrosis surrounded by a zone of neuronal loss and gliosis in rat and monkey cerebral cortex. In contrast, Waite *et al.* (1992) reported that the 35% acetonitrile solvent alone showed gross toxicity, which was markedly enhanced by Aβ, while Aβ dissolved in a cyclodextrin solution produced no marked toxic effects. Rush *et al.* (1992) found that Aβ1–40 or Aβ25–35 in a vehicle of 10% dimethyl sulphoxide induced tissue loss and neurodegeneration in rat but failed to replicate the antagonism of the toxicity with tachykinin agonists.

Several laboratories have reported the lack of *in vivo* neurotoxicity in experimental animals. Podlisny *et al.* (1992) reported that synthetic Aβ1–40 injected in monkey cerebral cortex failed to produce specific neurotoxicity after 2 weeks or 3 months. Games *et al.* (1992) observed the lack of Alzheimer pathology after injecting or infusing Aβ1–40, Aβ1–38 and Aβ25–35 into adult rat brain. Stein-Behrens *et al.* (1992) did not observe any evidence of consistent damage as tested by injecting Aβ25–35 into the rat hippocamus *in vivo*. Clemens and Stephenson (1992) evaluated the *in vivo* effects of Aβ peptides using lipid matrix implants to prolong tissue exposure, but found no cytopathology characteristics of AD in implanted young or old rat brain. Thus, it seems very difficult to produce clear neuropathological changes resembling those seen in AD by injecting Aβ peptides directly into the brains of experimental animals.

8.2.3 Amyloid deposition and neurotoxicity in the human brain

Diffuse plaques consist of amorphous aggregates with a small amount of amyloid fibrils, but this type of senile plaque is not accompanied by degenerative changes in neurons and neurites. These findings suggest that the aggregations of Aβ accompanying nascent amyloid fibrils are nontoxic or biologically inert towards adjacent neurons *in vivo*.

Amyloid angiopathy in the brain with AD is normally accompanied by very few dystrophic neurites. Furthermore, hereditary cerebral haemorrhage with amyloidosis-Dutch type (HCHWA-D) is characterized by the deposition of Aβ in the meningeal and cortical arterioles. In HCHWA-D patients, a point mutation (G to C) was found in the APP gene causing a single amino-acid substitution (glu to gln) at position 22 of Aβ itself (Levy *et al.*, 1990). In HCHWA-D, neurofibrillary tangles and dystrophic neurites have never been found (Haan *et al.*, 1991), suggesting that the deposition of Aβ is unrelated to neuritic reactions. These observations from human brains raise the possibility that Aβ deposition *per se* causes no neuronal degeneration *in vivo*.

8.3 Neurotoxicity of carboxyl terminal fragments of APP

8.3.1 Amyloidogenic property of carboxyl-terminal fragments of APP

In senile plaques, Aβ forms the amyloid structure consisting of single helical fibres. A high concentration of Aβ1–28 forms a fibril-like structure *in vitro* (Kirschner *et al.*, 1987). A C-terminal fragment of APP (APP-CT, equivalent to APP-C100) tends to self-aggregate in a cell-free translation system (Dyrks *et al.*, 1988), suggesting that C-terminal fragments of APP including Aβ domain and cytoplasmic domain are capable of aggregation.

A transient transfection of APP-C100 (i.e. C-terminal 100 amino-acid residues containing the entire Aβ domain and cytoplasmic domain) into COS cells led to an accumulation of inclusion-like structures near nuclei (Maruyama *et al.*, 1990). Electron microscopic observations revealed that the inclusions consisted of single helical fibres having varying diameters (8–22 nm) (Figure 2). These fibres contained the C-terminal sequences of APP, suggesting that APP-C100 forms an amyloid-like structure without being further processed to Aβ. Therefore, it is conceivable that a single cleavage near the N-terminus of Aβ produces an APP-C100-like fragment, which eventually forms an initial type of amyloid fibrils ('primordial fibrils') (Yoshikawa *et al.*, 1991). Double missense mutations have been found at codon 670 (lys to asn) and codon 671

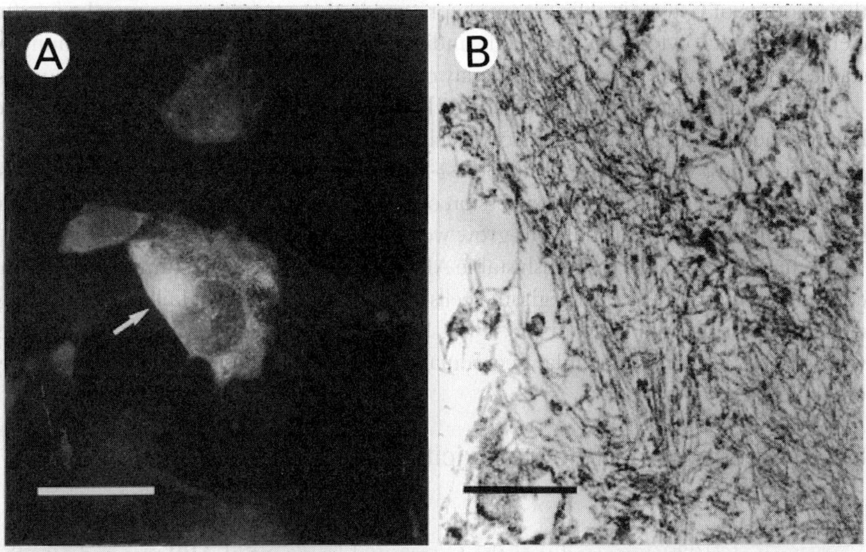

Figure 2 Accumulation of amyloid-like fibrils in COS-1 cells transfected with APP-C100. COS-1 cells were transfected with cDNA encoding APP-C100 without a signal sequence (hβ) and examined 3 days after transfection. (A) Fluorescent micrograph of APP-C100 cDNA transfected cells. Transfectants were stained with an anti-C-terminal antibody followed by indirect immunofluorescent cytochemistry. Note the intensely immunoreactive inclusion-like deposit (arrow) in the perinuclear region. Scale bar = 50 μm. (B) Electron micrograph of hβ-transfected COS cells. An ultrathin section of the cell having the inclusion-like deposit was stained with lead citrate and uranyl acetate, and observed under a transmission electron microscope. Note the accumulation of single helical fibres, which resembles β amyloid fibrils. Scale bar = 0.5 μm.

(met to leu) in a Swedish family (see Figure 1) (Mullan *et al.*, 1992). This type of mutation is valuable to correlate the abnormal cleavage near the N-terminus of Aβ with amyloid-associated pathogenesis of AD.

8.3.2 Neurotoxicity of C-terminal fragments of APP

By transfecting DNA into cultured cells, C-terminal fragments of APP have been overexpressed in cell lines which differentiate into neurons in response to chemical inducers. As described previously, overexpression of the C-terminal fragment of APP in PC12 cells induced degeneration when treated with NGF (Yankner *et al.*, 1989). Fukuchi *et al.* (1992b) reported a similar finding: P19 cells (i.e. a mouse embryonal carcinoma cell line) transfected with cDNA encoding APP-C100 with a signal sequence showed degenerative changes when the cells were differentiated into neural cells by retinoic acid treatment. These stable PC12 and P19 transfectants of C-terminal fragments of APP showed cell death only when differentiated into neurons. Therefore, the amounts of C-terminal fragments expressed in these transfectants may be insufficient to produce cytotoxicity towards undifferentiated cells (otherwise the cloning of stable transfectants is impossible) but may exert toxic effects on differentiated neurons.

Fukuchi *et al.* (1992a) transfected a cDNA sequence encoding APP-C100 into heterogeneous human neuroblastoma cell lines (SK-N-SH), and isolated six independent clones, which were all fibroblast-like cells having nonneuronal phenotypes. These results suggest that overexpression of C-terminal fragments is toxic to neurons, so that only the nonneuronal cells grow well to form colonies.

We have attempted to establish stable APP-C100 cDNA transfectants using various cell lines including PC12, P19, glioma cells and NIH3T3 cells, but failed to obtain any clones expressing high amounts of APP-C100 (unpublished observations). Therefore, we assumed that an intracellular accumulation of APP-C100 is extremely toxic to any type of cells.

8.3.3 Mechanism of neurotoxicity of C-terminal fragments of APP

Yankner *et al.* (1989) reported that the C-terminal fragment of APP secreted into the conditioned medium caused neuronal death, although the construct they used lacks a signal sequence for secretion. We made a similar construct (hβ) encoding APP-C100 without a signal sequence, but no secretion of APP-C100 from the transfected cells into the medium was detected (Maruyama *et al.*, 1990). Another construct (shβ) encoding APP-C100 with a signal sequence at the N-terminus was found to be successfully integrated into the membrane of COS cells and secreted into the medium. Neither the conditioned media of these hβ– and shβ– transfectants nor the cell extracts showed neurotoxicity when added in primary cultured hippocampal neurons (unpublished observations). Fukuchi *et al.* (1992b) reported a similar finding that the conditioned media of C-terminal cDNA-transfectants showed no neuronal degeneration when added in the culture medium. Furthermore, we have examined the *in vitro* neurotoxicity of APP-C100 purified from cDNA-transfected bacterial cells. The APP-C100 preparation showed no toxic effect on hippocampal neurons in primary cultures (Figure 3). These findings together suggest that APP-C100 or similar C terminal fragments have no degenerative effects on primary cultured neurons 'from outside'. Thus, it seems reasonable to conclude that C-terminal fragments exert their cytotoxic effects 'from inside'.

8.4 Neurotoxicity of abnormally metabolized APP fragments

8.4.1 An embryonal carcinoma cell model

P19 embryonal carcinoma cells differentiate into neurons and glial cells when treated with retinoic acid. The neural differentiation of P19 cells causes a marked increase in

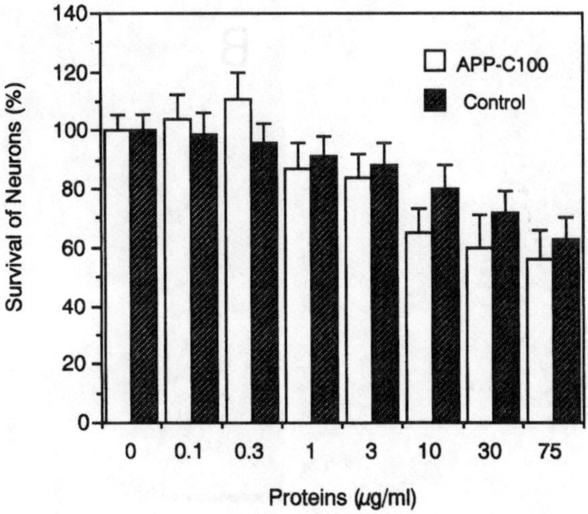

Figure 3 Lack of *in vitro* neurotoxicity of purified APP-C100. APP-C100 expressed in *E. coli* using a prokaryotic expression system was separated by a preparative sodium dodecyl sulphate (SDS)-polyacrylamide gel electrophoresis, treated with acetone, and dissolved in betaine. Hippocampal neurons, prepared from rat fetus at embryonic day 18, were cultured for 4 days, and the purified APP-C100 was added in the cultures (a final concentration of betaine was 0.8%). As a control, proteins having a molecular size similar to that of APP-C100 (about 16 kDa) were prepared from untransfected bacteria and added in the cultures (Control). No statistically significant difference was noted between APP-C100-treated and control groups.

APP gene expression (Yoshikawa *et al.*, 1990). We transfected full-length APP cDNA into undifferentiated P19 cells using a mammalian expression vector and examined whether overexpression of APP produces abnormal metabolism that may affect the cell viability after neural differentiation (Yoshikawa *et al.*, 1992). One hundred stable clones were randomly selected and the clones expressing higher APP immunoreactivities were recloned by a limiting dilution method. For each APP_{695}- and APP_{770}-cDNA transfectant, four clones that stably express high levels of APP were established. When these cells were neurally differentiated by retinoic acid treatment, differentiated post-mitotic neurons showed a severe degeneration, and disappeared in a few days (Figure 4B). Most of the surviving cells in the mixed cultures were degenerating neurons and nonneuronal cells (presumably astrocyte precursors), and some had intensely APP immunoreactive materials in lysosome-like structures (Figure 4D). Western blot analysis revealed that the differentiated P19 cells contained degraded 12–100 kDa C-terminal fragments (Figure 5A). Among them, many bands corresponding to 16.5–35 kDa were larger than APP-C100 (16 kDa), suggesting that most of these N-terminal truncated C-terminal fragments are potentially amyloidogenic. These results suggest that the overexpression of APP induces degeneration of post-mitotic neurons and concomitantly generates the abnormally processed C-terminal fragments.

Neuronal death was always seen in APP cDNA transfected P19 cells having low

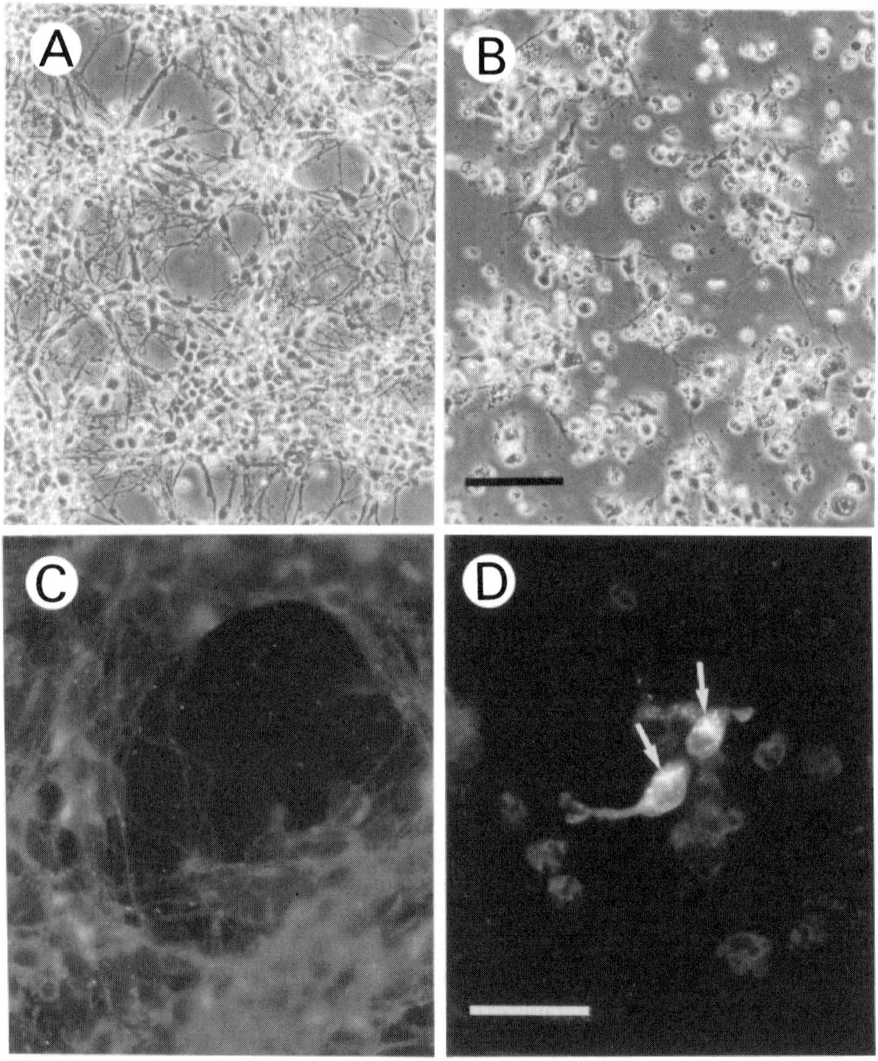

Figure 4 Degeneration of post-mitotic neurons derived from P19 embryonal carcinoma cells transfected with human full-length APP cDNA. P19 embryonal carcinoma cells were stably transfected with cDNA encoding full-length human APP_{695} and neurally differentiated by retinoic acid treatment for 3 days. The cells were examined 4 days after the treatment. (A,C) Control cells transfected with the empty vector. (B,D) Cells transfected with full-length human APP_{695} cDNA. (A,B) Phase-contrast micrography. (C,D) Fluorescent immunocytochemistry using the anti-C-terminal antibody. APP_{695} cDNA transfectants show severe degenerative changes (B), while control transfectants differentiate normally into neurons (A). Note the accumulation of APP-C-terminal fragments in lysosome-like organelles (arrows in D). Scale bars = 100 μm (for A,B) and 50 μm (for C,D).

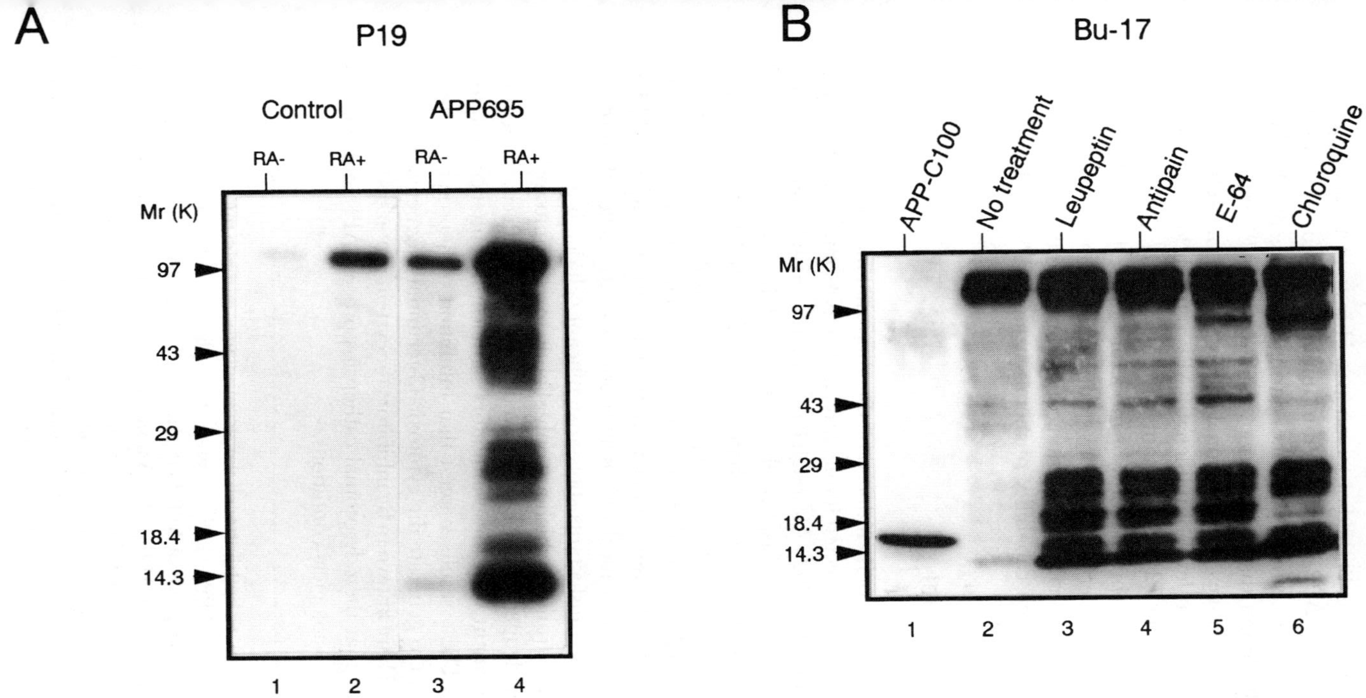

Figure 5 Generation of APP C-terminal fragments in P19 cells and Bu-17 glioma cells transfected with full-length APP cDNA. APP derivatives in total cell lysates of the transfectants were analysed by Western blotting using the anti-C-terminal antibody. (A) APP_{695} cDNA transfected P19 cells. The control cells (transfected with the empty vector) (Control, lanes 1 and 2) and APP_{695} transfectants (APP_{695}, lanes 3 and 4) were treated with (RA+, lanes 2 and 4) or without (RA-, lanes 1 and 3) retinoic acid for 3 days, incubated for 4 days, and collected. Note the generation of the C-terminal fragments with various sizes in the retinoic acid-treated APP_{695} cDNA transfectants. (B) APP_{770} cDNA transfected Bu-17 glioma cells. The transfectants were cultured in the presence of various protease inhibitors for 12 h (for chloroquine) or 24 h (for other compounds). Note the generation of the C-terminal fragments, similar to those found in the P19 transfectants, in the cDNA transfected glioma cells treated with protease inhibitors. Lane 1, APP-C100 (COS cells); lane 2, no treatment; lane 3, leupeptin (100 μg/ml); lane 4, antipain (100 μg/ml); lane 5, E-64 (6 μM); lane 6, chloroquine (100 μM).

passage numbers, and only a few transfectant lines no longer presented neuronal death after repeating subcultures (unpublished observations). This suggests that overexpression or ectopic expression of APP is toxic even to undifferentiated cells and the cell population expressing high levels of APP reduces during the course of subculturing.

Yankner *et al.* (1989) found that transfection of full-length APP into PC12 cells induced no degenerative changes after neuronal differentiation by NGF treatment. We noticed that the expression level of full-length APP is much lower in the PC12 transfectants than those in our P19 transfectants (Yoshikawa *et al.*, 1992). We found that stable APP transfectants expressing low or moderate levels of APP showed little or no degeneration of P19-derived neurons (unpublished observations). These observations suggest that neuronal degeneration can take place only when APP levels are high enough to exceed the capability of metabolizing APP completely in the neurons.

8.4.2 A glioma cell model

To examine whether the cytotoxicity induced by the abnormal metabolism of APP is specific to neurons, we then transfected full-length APP cDNAs into a human glioma cell line (Bu-17) (Hayashi *et al.*, 1992). Although large amounts of full-length APP were present in these stable transfectants, smaller C-terminal fragments were barely detected. When these cells were treated with protease inhibitors (leupeptin, antipain and E-64) or the lysosomotropic agent chloroquine, aberrantly processed C-terminal fragments of APP (potentially amyloidogenic C-terminal fragments) were generated (Figure 5B). When APP cDNA transfectants were treated with chloroquine, APP C-terminal immunoreactivity was localized in lysosome-like organelles (Figure 6A). Furthermore, APP cDNA transfectants showed significant cell death in response to chloroquine at low concentrations which are only slightly toxic to untransfected controls (Figure 6B). Neither leupeptin nor E-64 showed degenerative effects on the transfectants but the pretreatment with these inhibitors significantly enhanced the toxic effects of chloroquine. These results suggest that inhibition of lysosomal proteases impairs APP metabolism to generate the cytotoxic C-terminal fragments. Since leupeptin, antipain and E-64 are known to be inhibitors for cystein proteases (e.g. cathepsin B and L), it is likely that these lysosomal cathepsins process APP into nontoxic small fragments under physiological conditions.

Figure 6 Chloroquine induces both intracellular accumulation of C-terminal fragments of APP and degeneration of APP cDNA transfected glioma cells. (A) Fluorescent immunocytochemistry for C-terminus of APP. Left panel, APP_{770} cDNA transfectants without chloroquine treatment. Right panel, APP_{770} cDNA transfectant treated with chloroquine at 150 μM for 24 h. Note that a fine reticular distribution of the immunoreactivity changes into the granular accumulations in lysosome-like organelles in response to chloroquine treatment. (B) Dose–response curves of chloroquine-induced cytotoxicity in APP cDNA transfected glioma cells. The untransfected cells (untransfected), APP_{695} cDNA transfectants (APP695), and APP_{770} cDNA transfectants (APP770) were grown to a near confluent density, and treated with various

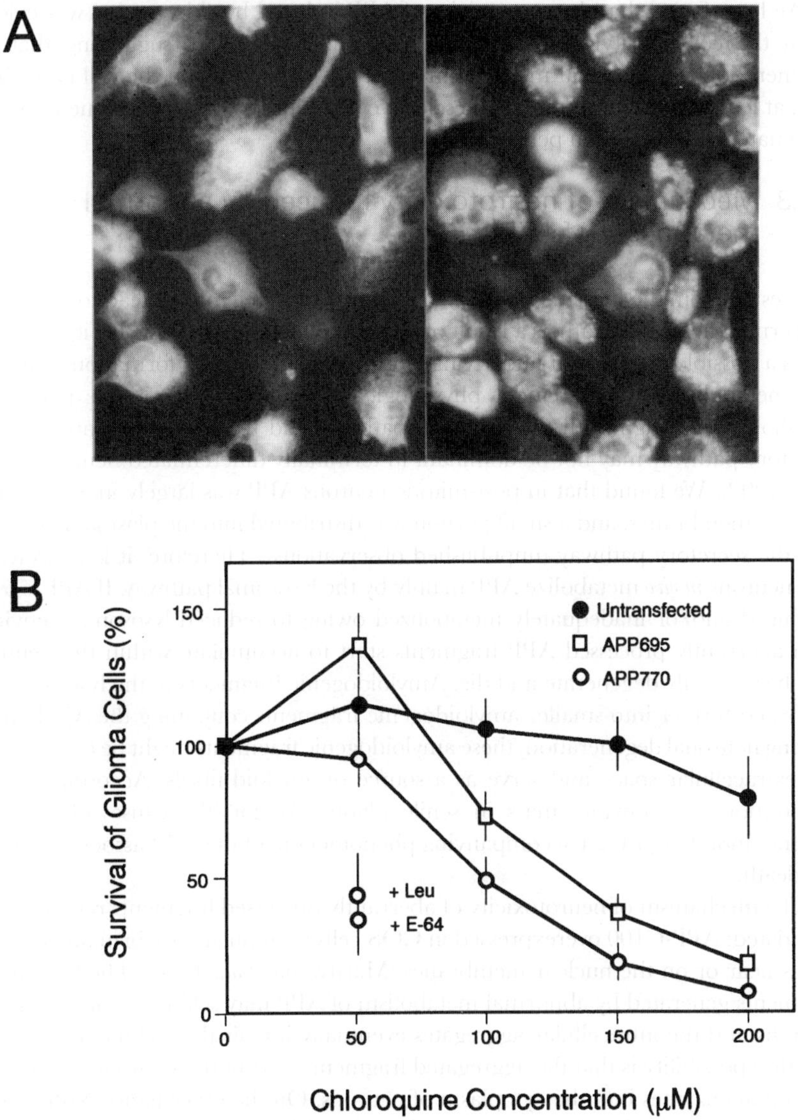

concentrations of chloroquine for 3 days. The numbers of surviving cells were determined by trypan blue exclusion test. Note that the numbers of viable APP cDNA transfectants significantly decreased in response to chloroquine treatment. In addition, APP_{770} cDNA transfectants were pretreated with leupeptin (100 μg/ml) (+Leu) or E-64 (50 μM)(+E-64) for 24 h and then treated with chloroquine (50 μM) for 3 days. Note the significant increase in susceptibility of the pretreated group to chloroquine. Normalized survival rates are expressed as the mean ± SEM of four experiments.

We have found that the cytotoxicity of APP induced by chloroquine was evident when the cell density was confluent (Hayashi *et al.*, 1992), suggesting that cell degeneration by abnormal metabolism of APP is dependent on the cell cycle. This may, at least in part, account for the selective death of P19 cell-derived neurons that terminate the cell division permanently.

8.4.3 Mechanism of neurotoxicity induced by abnormal metabolism of APP

The results from the full-length APP cDNA transfected cell models lead to a new idea concerning neuronal death as a result of abnormal metabolism of APP. Figure 7 illustrates a possible mechanism underlying neuronal death and Aβ formation by abnormal metabolism of APP. Under physiological conditions, APP in the neurons is metabolized through both the secretory pathway and the lysosomal pathway. The secretory pathway may not be dominant in terminally differentiated neurons (Hung *et al.*, 1992). We found that in post-mitotic neurons APP was largely integrated into internal membranes, and a small portion was distributed into the plasma membrane and the secretory pathway (unpublished observations). Therefore, it is conceivable that neurons *in vivo* metabolize APP mainly by the lysosomal pathway. If APP is overproduced and/or inadequately metabolized owing to reduced lysosomal activities, then aberrantly processed APP fragments start to accumulate within the neurons, which eventually degenerate and die. Amyloidogenic fragments in the lysosomes are processed further into smaller amyloidogenic fragments containing the Aβ domain. During neuronal degeneration, these amyloidogenic fragments might be expelled into the extracellular space and serve as a source of amyloid fibrils. According to this proposition, the amyloid mass in senile plaques is not 'the cause' of neuronal degeneration but just 'an accompanying phenomenon' of the APP-associated neuronal death.

The mechanism of neurotoxicity of aberrantly processed fragments remains to be elucidated: APP-C100 overexpressed in COS cells accumulates to form amyloid-like fibrils near or on the nuclear membranes (Maruyama *et al.*, 1990). The C-terminal fragments generated by abnormal metabolism of APP may self-aggregate within the neurons and the intracellular aggregates eventually impair the viability of the cells. Another possibility is that the aggregated fragments disrupt the lysosomes to liberate lysosomal enzymes into the cytoplasm of neurons. On the other hand, Nishimoto *et al.* (1993) reported that the cytoplasmic domain of the C-terminal sequence of APP contains the Go protein binding sequence (see Figure 1). Overproduction of C-terminal fragments including the Go protein binding domain impairs cell functions that are mediated by Go (e.g. phosholipase C, voltage-dependent calcium channel). If the cytoplasmic domain of APP, which is devoid of the Aβ domain, exerts a neurotoxic effect, then it is suggested that the two domains for neurotoxicity and amyloidogenicity are separable. Another possibility is that Aβ-related fragments generated within the cells are toxic to neurons by forming calcium channels (Arispe *et al.*, 1993) and increase the intracellular calcium concentration (Joseph and Han, 1992).

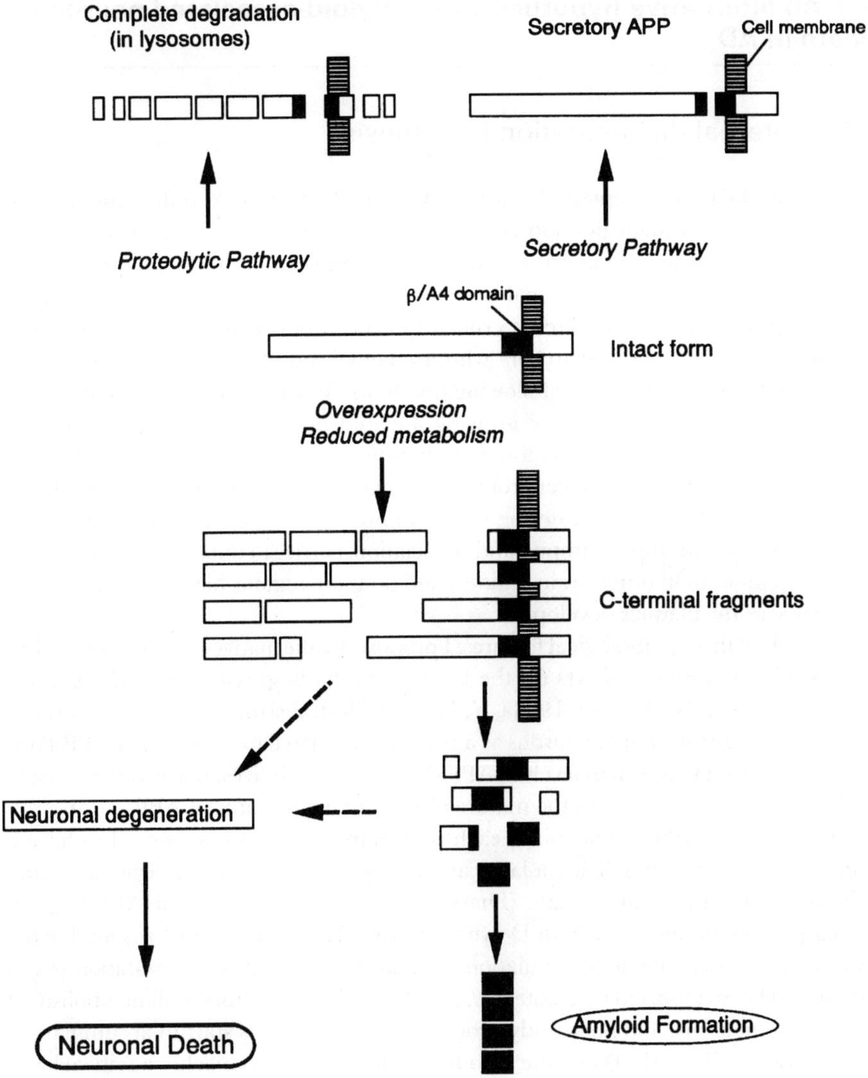

Figure 7 Proposed involvement of abnormal metabolism of APP in neuronal death and amyloid formation. On the basis of the findings with the APP cDNA transfected cell models, a hypothetical mechanism in which the abnormal metabolism of APP induces both neuronal degeneration and generation of Aβ is proposed. See text for details.

8.5 An alternative hypothesis for amyloid-associated neuronal death in AD

8.5.1 Internal disintegration hypothesis

At present, the most accepted idea is that Aβ initially deposits as diffuse amorphous plaques, which eventually develop, over years, into neuritic plaques having condensed amyloid with degenerated neurites. Since the transition from diffuse plaques to neuritic plaques accompanies neuronal degeneration, one can speculate that formation of condensed amyloid structures is crucial to exert a destructive effect on adjacent neurons. However, even researchers who are convinced of such a proposition may have difficulty in answering the following questions: (i) why do neurons remain intact for years despite the presence of growing amyloid masses nearby, and should the amyloid burden be eliminated before manifesting its degenerative effect?; (ii) how do dystrophic neurites form a cluster around the amyloid core, and does the amyloid core first attract and then deteriorate the neuronal processes?; (iii) why do specific brain regions such as the hippocampus and association cortices contain more abundant neuritic plaques than other brain regions, and do these regions have a larger number of loci where the plaques develop?

To explain these pathological features, I propose an alternative hypothesis for APP-associated pathogenesis of AD on the basis of the findings with APP cDNA-transfected cell models (Yoshikawa, 1993a, b). Figure 8 illustrates the alternative hypothesis of internal disintegration of neurons as a result of abnormal metabolism of APP. Post-mitotic neurons express high levels of APP throughout their lives; newly differentiated neurons as well as neurons in the mature CNS contain amounts of APP_{695} transcript (Yoshikawa *et al.*, 1990). Therefore, each post-mitotic neuron is supposed to have a high capability of metabolizing a large amount of APP into nontoxic fragments and thus maintains the healthy state (Phase I). The overproduction of APP (e.g. an increased gene dosage of APP in Down's syndrome), a reduction of lysosomal activity (e.g. an age-dependent accumulation of lipofuscin), or a missense mutation [e.g. a familial AD case reported by Goate *et al.* (1991)] leads to the abnormal metabolism of APP in the neurons, and aberrantly processed APP fragments start to accumulate in the lysosomes (Phase II). When the amount of these fragments reaches a critical level, the neuron starts to degenerate and disintegrate (Phase III). During degeneration, APP fragments are converted into smaller amyloidogenic fragments within the neuron or at the extracellular space. These fragments might form nascent fibrils (e.g. the primordial fibrils; Yoshikawa *et al.*, 1991), which eventually develop into condensed amyloid structures after they have undergone some modifications. After perikarya of degenerated neurons are completely scavenged by glial cells or macrophages, the condensed masses of amyloid and surrounding dystrophic neurites are left behind at the extracellular space. Thus, the dystrophic neurites in neuritic plaques might derive from presynaptic nerve terminals that miss their target neurons.

The neuropathogical findings seen in AD brain could be just a temporary scenario

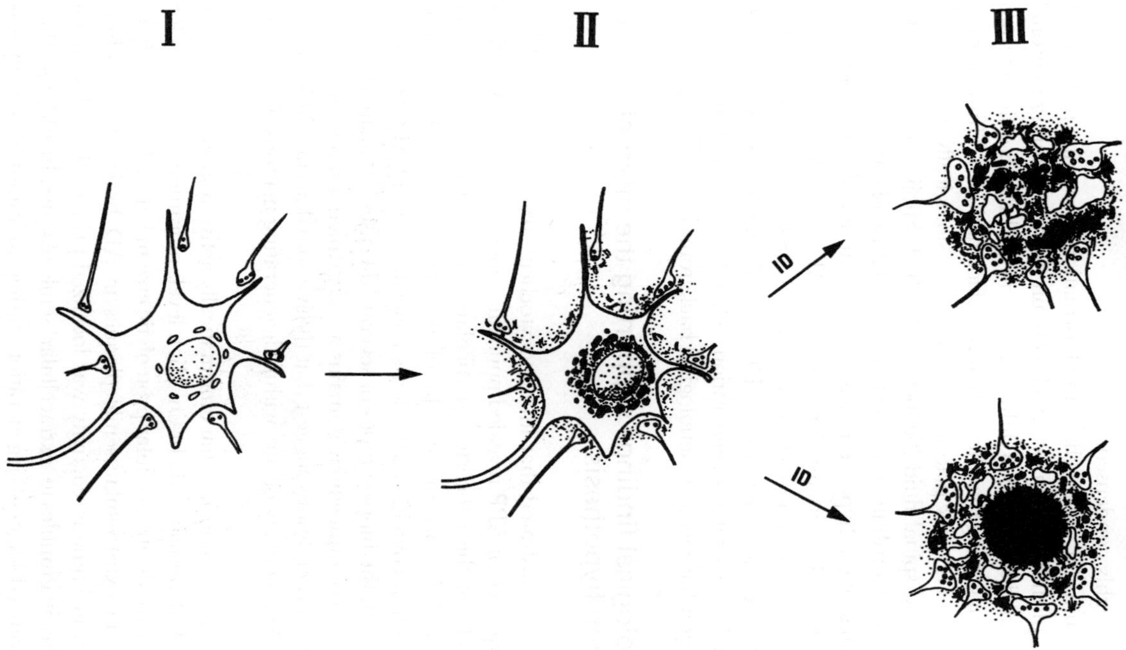

Figure 8 An internal disintegration hypothesis for neuronal death and formation of neuritic plaques in AD. A healthy neuron metabolizes APP completely (Phase I). Overproduction of APP, reduction of APP metabolism or mutation of the APP gene causes abnormal processing, which accumulates APP fragments in the lysosomes (Phase II). When the intracellular accumulation of cytotoxic APP fragments reaches a critical point, neurons are internally disintegrated (ID) (Phase II to Phase III). During this transition period (Phases II–III), APP fragments are processed further into Aβ, which eventually form amyloid fibrils. Finally, the amyloid deposits and dystrophic neurites are left behind as 'extracellular' neuritic plaques following disappearance of the perikaryon.

of continuous changes of neuronal death and repairing processes. Our transient transfection study suggests that formation of amyloid-like fibrils consisting of APP-C100 occurs within a few days (Maruyama *et al.*, 1990). This suggests that the formation of nascent amyloid fibrils *in vivo* is also completed on a time scale of days when amyloidogenic peptides are highly condensed.

This hypothesis accounts for the mechanism of formation of neuritic plaques but not for the generation of diffuse plaques, which contain no degenerating neuronal components. The regional distribution of diffuse plaques in the brain is different from that of neuritic plaques, suggesting that the mechanism of formation of diffuse plaques can be distinct from those of neuritic plaques. Since Aβ is physiologically secreted into the extracellular space and spinal fluid (Seubert *et al.*, 1992; Shoji *et al.*, 1992), the secreted Aβ peptides may contribute to the formation of diffuse plaques.

The internal disintegration hypothesis cannot account for the formation of amyloid angiopathy either. The perivascular amyloid in AD brain, despite the presence of β-pleated sheet structure, is not accompanied by dystrophic neurites. In addition, the absence of neuritic reactions in HCHWA-D suggests that amyloid angiopathy is unrelated to neuronal degeneration. Therefore, amyloid angiopathy, like diffuse plaques, might have its own aetiological background.

8.5.2 Neuropathological findings supporting the internal disintegration hypothesis

Benowitz *et al.* (1989) demonstrated that immunocytochemistry using an antibody against C-terminal sequence of APP revealed punctate concentrations of APP fragments in pyramidal cells of the neocortex, particularly in associative regions, and intense staining in the CA1 pyramidal cells of the hippocampus. By electron microscopy, this distribution coincided with dense concentrations of APP fragments in secondary lysosomes. In the hippocampus of several AD cases, abnormally dense immunostaining in enlarged intracellular domains accompanied a severe atrophy of the CA1 neurons. These observations suggest that the lysosomal accumulation of C-terminal fragments of APP actually occurs within degenerating neurons (Phase II in Figure 8).

Cataldo and Nixon (1990) reported that enzymatically active lysosomal proteases were associated with Aβ deposits in AD brain. High levels of immunoreactivities for cathepsins, which are normally intracellular proteolytic enzymes associated with lysosomes, were localized in extracellular senile plaques in AD brain. At the ultrastructural level, cathepsin immunoreactivity was localized principally to lysosomal dense bodies and lipofuscin granules in 'extracellular' senile plaques. In addition, they found that in more advanced stages of degeneration, cathepsin immunoreactivity was present throughout the cytoplasm, suggesting that cathepsin-containing lysosomes are liberated from degenerating neuronal perikarya into the extracellular space. Thus, these findings strongly support the idea that degenerating neurons in AD brain actually disintegrate to liberate the lysosomes into the extracellular space (Phases II and III in Figure 8).

The presence of active lysosomal enzymes in the extracellular space may be responsible for the further processing of APP into more dense amyloid fibrils. Cathepsins in the lysosomes may play a major role in metabolizing APP into small fragments under physiological conditions, but may cleave different sites of APP in abnormal pH environments such as the cytoplasm outside the lysosomes and the extracellular space. The abnormally processed fragments of APP can be effectively condensed to form amyloid fibrils.

The cathepsin-laden senile plaques were localized in layers III and V of the prefrontal association cortex (Cataldo *et al.*, 1990). A large number of neuropathological studies have demonstrated that neuritic plaques and neurofibrillary tangles were localized in layers III and V of the association cortices in AD. APP mRNA, mainly APP_{695} mRNA, was highly expressed in the association cortices (Tanzi *et al.*, 1988). *In situ* hybridization histochemistry demonstrated that layers III and V of the primate prefrontal cortex contained neurons expressing high levels of APP mRNA, and there was a reduction in the signal intensity of APP mRNA in layer III of the AD brain (Bahmanyar *et al.*, 1987). Taken together, it is conceivable that neuritic plaques are the most numerous in the brain areas that contain the perikarya of neurons expressing high levels of the APP gene. This may account for the preferential distribution of neuritic plaques in the association cortices of AD brain.

The number and size of neurons in the association cortices of the primate brain have increased during evolution to reach the highest figures in man. These neurons are generated at the latest stages of phylogeny. Therefore, it is tempting to speculate that a class of neurons created at the final stage of mammalian evolution greatly expanded the intellectual activities of man, but simultaneously became vulnerable to APP-associated cytotoxicity, which might be closely associated with the pathogenesis of AD.

8.6 Conclusion

On the assumption that the extracellular amyloid deposition is responsible for neuronal degeneration in AD, a large number of studies have been made to understand the mechanisms underlying amyloid formation. Although many researchers would admit that the amyloid deposition is associated with the neuronal degeneration, it is still ambiguous at present whether the extracellular amyloid *per se* is the primary cause of neuronal death seen in AD.

Clarification of the neurotoxicity of Aβ and related peptides is important not only for elucidating the pathogenesis of AD, but also for planning the strategies for prevention and treatment of AD. If the extracellular Aβ deposits kill neurons from outside, then one can expect that AD can be treated by either eliminating the Aβ deposits or preventing the extracellular accumulation of Aβ. However, these strategies might be irrelevant if the extracellular amyloid deposition is simply a by-product of neuronal degeneration owing to the abnormality of intracellular APP metabolism; either increasing the metabolism of APP or decreasing the production of APP within the neurons may be an achievable approach to prevent the progression of AD. Thus, further studies on intraneuronal metabolism of APP and the mechanism of neuronal degeneration as a result of its abnormality should be worthwhile to elucidate the pathogenesis of AD and to establish the therapeutic strategies for this devastating disease.

References

Arispe, N., Rojas, E. & Pollard, H.B. (1993) *Proc. Natl Acad. Sci. USA* **90**, 567–571.

Bahmanyar, S., Higgins, G.A., Goldgaber, D., Lewis, D.A., Morrison, J.H., Wilson, M.C., Shankar, S.K. & Gajdusek, D.C. (1987) *Science* **237**, 77–80.

Benowitz, L.I., Rodriguez, W., Paskevich, P., Mufson, E.J., Schenk, D. & Neve, R.L. (1989) *Exp. Neurol.* **106**, 237–250.

Cataldo, A.M. & Nixon, R.A. (1990) *Proc. Natl Acad. Sci. USA* **87**, 3861–3865.

Cataldo, A.M., Thayer, C.Y., Bird, E.D., Wheelock, T.R. & Nixon, R.A. (1990) *Brain Res.* **513**, 181–192.

Clemens, J.A. & Stephenson, D.T. (1992) *Neurobiol. Aging* **13**, 581–586.

Dyrks, T., Weidemann, A., Multhaup, G., Salbaum, J.M., Lemaire, H-G., Kang, J., Müller-Hill, B., Masters, C.L. & Beyreuther, K. (1988) *EMBO J.* **7**, 949–957.

Fukuchi, K., Kamino, K., Deeb, S.S., Furlong, C.E., Sundstrom, J.A., Smith, A.C. & Martin, G.M. (1992a) *Mol. Brain Res.* **16**, 37–46.

Fukuchi, K., Kamino, K., Deeb, S.S., Smith, A.C., Dang, T. & Martin, G.M. (1992b) *Biochem, Biophys. Res. Commun.* **182**, 165–173.

Games, D., Khan, K.M., Soriano, F.G., Keim, P.S., Davis, D.L., Bryant, K. & Lieberburg, I. (1992) *Neurobiol. Aging* **13**, 569–576.

Glenner, G.G. & Wong, C.W. (1984) *Biochem. Biophys. Res. Commun.* **120**, 885–890.

Goate, A., Chartier-Harlin, M.-C., Mullan, M., Brown, J., Crawford, F., Fidani, L., Giuffra, L., Haynes, A., Irving, N., James, L., Mant, R., Newton, P., Rooke, K., Roques, P., Talbot, C., Pericak-Vance, M., Roses, A., Williamson, R., Rossor, M., Owen, M. & Hardy, J. (1991) *Nature* **349**, 704–706.

Haan, J., Hardy, J.A. & Roos, R.A.C. (1991) *Trends Neurosci.* **14**, 231–234.

Hayashi, Y., Kashiwagi, K. & Yoshikawa, K. (1992) *Biochem. Biophys. Res. Commun.* **187**, 1249–1255.

Hung, A.Y., Koo, E.H., Haass, C. & Selkoe, D.J. (1992) *Proc. Natl Acad. Sci. USA* **89**, 9439–9443.

Joseph, R. & Han, E. (1992) *Biochem. Biophys. Res. Commun.* **184**, 1441–1447.

Kang, J., Lemaire, H.-G., Unterbeck, A., Salbaum, J.M., Masters, C.L., Crzeschik, K.-H., Multhaup, G., Beyreuther, K. & Müller-Hill, B. (1987) *Nature* **325**, 733–736.

Kirschner, D.A., Inouye, H., Duffy, L.K., Sinclair, A., Lind, M. & Selkoe, D.J. (1987) *Proc. Natl Acad. Sci. USA* **84**, 6953–6957.

Kitaguchi, N., Takahashi, Y., Tokushima, Y., Shiojiri, S. & Ito, H. (1988) *Nature* **331**, 530–532.

Koh, J., Yang, L.L. & Cotman, C.W. (1990) *Brain Res.* **533**, 315–320.

Kowall, N.W., Beal, M.F., Busciglio, J., Duffy, L.K. & Yankner, B.A. (1991) *Proc. Natl Acad. Sci. USA* **88**, 7247–7251.

Kowall, N.W., McKee, A.C., Yankner, B.A. & Beal, M.F. (1992) *Neurobiol. Aging* **13**, 537–542.

Levy, E., Carman, M.D., Fernandez-Madrid, I.J., Power, M.D., Lieberburg, I., Van Duinen, S.G., Bots, G.A.M., Luyendijk, W. & Frangione, B. (1990) *Science* **248**, 1124–1126.

Loo, D.T., Copani, A., Pike, C.J., Whittemore, E.R., Walencewicz, A.J. & Cotman, C.W. (1993) *Proc. Natl Acad. Sci. USA* **90**, 7951–7955.

Malouf, A.T. (1992) *Neurobiol. Aging* **13**, 543–551.
Maruyama, K., Terakado, K., Usami, M. & Yoshikawa, K. (1990) *Nature* **347**, 566–569.
Mattson, M.P. & Rydel, R.E. (1992) *Neurobiol. Aging* **13**, 617–621.
May, P.C., Gitter, B.D., Waters, D.C., Simmons, L.K., Becker, G.W., Small, J.S. & Robison, P.M. (1992) *Neurobiol. Aging* **13**, 605–607.
Mullan, M., Crawford, F., Axelman, K., Houlden, H., Lilius, L., Winblad, B. & Lannfelt, L. (1992) *Nature Genet.* **1**, 345–347.
Nishimoto, I., Okamoto, T., Matsuura, Y., Takahashi, S., Okamoto, T., Murayama, Y. & Ogata, E. (1993) *Nature* **362**, 75–79.
Pike, C.J., Walencewicz, A.J., Glabe, C.G. & Cotman, C.W. (1991) *Brain Res.* **563**, 311–314.
Pike, C.J., Burdick, D., Walencewicz, A.J., Glabe, C.G. & Cotman, C.W. (1993) *J. Neurosci.* **13**, 1676–1687.
Podlisny, M.B., Stephenson, D.T., Frosch, M.P., Lieberburg, I., Clemens, J.A. & Selkoe, D.J. (1992) *Neurobiol. Aging* **13**, 561–567.
Ponte, P., Gonzalez-DeWhitt, P., Schilling, J., Miller, J., Hsu, D., Greenberg, B., Davis, K., Wallace, W., Lieberburg, I., Fuller, F. & Cordell, B. (1988) *Nature* **331**, 525–527.
Roher, A.E., Ball, M.J., Bhave, S.V. & Wakade, A.R. (1991) *Biochem. Biophys. Res. Commun.* **174**, 572–579.
Rush, D.K., Aschmies, S. & Merriman, M.C. (1992) *Neurobiol. Aging* **13**, 591–594.
Seubert, P., Vigo-Pelfrey, C., Esch, F., Lee, M., Dovey, H., Davis, D., Sinha, S., Schlossmacher, M., Whaley, J., Swindlehurst, C., McCormack, R., Wolfert, R., Selkoe, D., Lieberburg, I. & Schenk, D. (1992) *Nature* **359**, 325–327.
Shoji, M., Golde, T.E., Ghiso, J., Cheung, T.T., Estus, S., Shaffer, L.M., Cai, X.-D., McKay, D.M., Tintner, R., Frangione, B. & Younkin, S.G. (1992) *Science* **258**, 126–129.
Stein-Behrens, B., Adams, K., Yeh, M. & Sapolsky, R. (1992) *Neurobiol. Aging* **13**, 577–579.
Takadera, T., Sakura, N., Mohri, T. & Hashimoto, T. (1993) *Neurosci. Lett.* **161**, 41–44.
Tanzi, R.E., McClatchey, A.I., Lamperti, E.D., Villa-Komaroff, L., Gusella, J.F. & Neve, R.L. (1988) *Nature* **331**, 528–530.
Waite, J., Cole, G.M., Frautschy, S.A., Connor, D.J. & Thal, L.J. (1992) *Neurobiol. Aging* **13**, 595–599.
Yankner, B.A., Dawes, L.R., Fisher, S., Villa-Komaroff, L., Oster-Granite, M.L. & Neve, R.L. (1989) *Science* **245**, 417–420.
Yankner, B.A., Caceres, A. & Duffy, L.K. (1990a) *Proc. Natl Acad. Sci. USA* **87**, 9020–9023.
Yankner, B.A., Duffy, L.K. & Kirschner, D.A. (1990b) *Science* **250**, 279–282.
Yoshikawa, K., Aizawa, T. & Maruyama, K. (1990) *Biochem. Biophys. Res. Commun.* **171**, 204–209.
Yoshikawa, K., Maruyama, K., Terakado, K., Usami, M. & Aizawa, T. (1991) In *Frontiers of Alzheimer Research* (eds Ishii, T., Allsop. D. & Selkoe, D.), pp. 217–227. Elsevier Science Publishers, Amsterdam.
Yoshikawa, K., Aizawa, T. & Hayashi, Y. (1992) *Nature* **359**, 64–67.
Yoshikawa, K. (1993a) *Nature* (Scientific Correspondence) **361**, 122–123.
Yoshikawa, K. (1993b) *Seikagaku* **65**, 1149–1159 [in Japanese].

CHAPTER 9

BIOLOGICAL FUNCTIONS OF THE AMYLOID β-PROTEIN PRECURSOR

STEVEN L. WAGNER

SIBIA Inc., 505 Coast Boulevard South, La Jolla, California 92037, USA

Table of Contents

9.1 Introduction

The intense study of the amyloid β-protein precursor (APP) molecule during the last decade is a result of several key observations. The first was the isolation and characterization of amyloid filaments from both congophilic meningeal vessels (Glenner and Wong, 1984a, b) and neuritic plaque cores (Masters *et al.*, 1985) in post-mortem brain tissue from individuals with Alzheimer's disease (AD) and Down's syndrome. Both of these neuropathological lesions contain as their primary constituent a 4 kDa peptide which was termed the amyloid β-protein (Aβ). Amino-acid sequencing of this solubilized peptide from meningovascular deposits provided essential information which permitted the cloning of various Aβ cDNAs, and ultimately resulted in the identification and localization of the APP gene (Goldgaber *et al.*, 1987; Kang *et al.*, 1987; Robakis *et al.*, 1987; Tanzi *et al.*, 1987).

Another noteworthy observation was the detection of a missense mutation in the APP gene, which cosegregated in a family with a very early-onset form of AD (Goate *et al.*, 1991). The absence of this mutation in unaffected individuals strongly suggested that this mutation was sufficient to be causative for AD and that APP metabolism played a key role in this disease.

This chapter will selectively focus on only a few of the many studies which have been published concerning the possible biological functions and localization of APP, with an emphasis on the secreted isoforms. Some of these studies have led to an increased understanding of certain key physiological processes of the brain

Pathobiology of Alzheimer's Disease
ISBN 0-12-286965-5

in which APP has been shown to play a major role. Such studies have provided considerable insight into the pathogenesis of AD as well as other related cerebral amyloidotic disorders.

9.2 Characterization and distribution of APP

9.2.1 Initial cDNA cloning and preliminary characterization of APP

The first reported isolation of a full-length complementary DNA (cDNA) clone encoding Aβ revealed a 695-residue precursor protein (calculated M_r 78 644) whose predicted amino-acid sequence contained an amino-terminal signal sequence, three extracellular domains, both N- and O-glycosylation sites, a single transmembrane domain and a short (<50 amino acids) cytoplasmic domain, therefore resembling a transmembrane cell surface receptor (Kang *et al.*, 1987). The Aβ peptide was determined to be a proteolytic breakdown product of this large precursor protein (Kang *et al.*, 1987). The location of the Aβ peptide within the precursor protein was predicted to be partially amino-terminal to and partially embedded within the transmembrane domain (see Figure 1).

Since this initial isolation, there have been a number of different Aβ precursor proteins identified by cDNA cloning. Virtually all of these (at least those discussed in the present chapter) appear to be the product of a single gene resulting from alternative splicing of the primary APP mRNA transcript.

In 1988, three groups simultaneously reported the isolation of Aβ-specific cDNA clones, which were shown to contain either one or two additional domains (compared to the original APP_{695} molecule), one of which showed extensive similarity to the Kunitz family of serine protease inhibitors (Kitaguchi *et al.*, 1988; Ponte *et al.*, 1988; Tanzi *et al.*, 1988). The other domain showed homology to the MRC OX2 antigen of thymus-derived lymphoid cells (Clark *et al.*, 1985). The inclusion of the serine protease inhibitor domain (abbreviated KPI for Kunitz protease inhibitor) results in a 56 amino-acid insertion within a highly acidic region of the putative extracellular domain of the precursor protein. APP transcripts which contain nucleotides encoding this insert are denoted as APP_{751}. The full-length cDNA isolated by Kitaguchi and coworkers harboured, in addition to the KPI domain, a further contiguous sequence of 19 amino acids directly carboxy-terminal to the serine protease inhibitor domain resulting in a 75 amino-acid extension of the original APP_{695}. This isoform is therefore denoted as APP_{770}.

9.2.2 Localization of APP expression

9.2.2.1 APP mRNA distribution studies

The first Aβ cDNA identified (APP_{695}) was isolated from a cDNA library constructed from 5-month-old fetal brain tissue (Kang *et al.*, 1987). Subsequent full-length clones

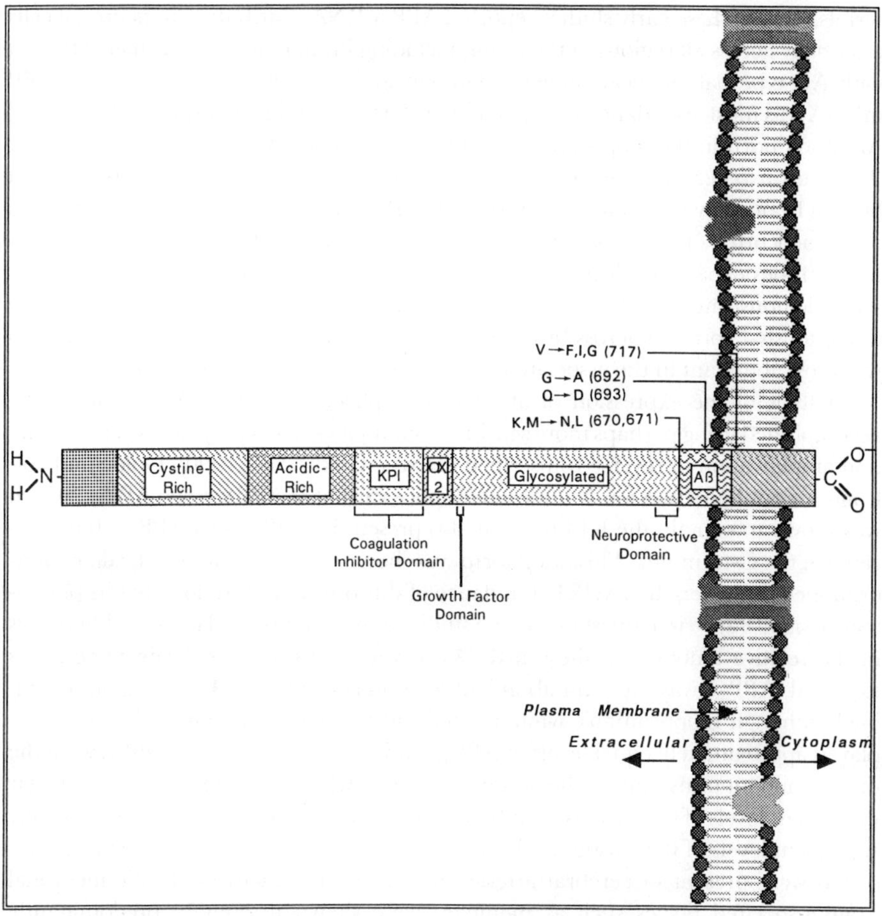

Figure 1 Putative structural and functional domains of the APP molecule. Arrows refer to missense mutations known to be tightly linked to early-onset familial Alzheimer's disease and/or hereditary cerebral haemorrhage with amyloidosis. OX2 is the region of APP with structural homology to the MRC OX2 antigen of thymocytes.

containing DNA encoding the serine protease inhibitor domain were isolated from a number of different cDNA libraries ranging from fibroblast cells, lymphocytes, post-mortem aged brain tissue (normal and AD), promyelocytic leukaemia cells and glioblastoma cells. These early findings implied that APP transcripts, which harboured nucleotides encoding the protease inhibitory domain (APP_{751} and APP_{770}), could be found in a variety of tissues whereas the transcript lacking nucleotides encoding this domain (APP_{695}) could only be demonstrated in tissues of neural origin.

Numerous studies have been published concerning the distribution of these various APP mRNA transcripts in mammalian tissues. Initial mRNA distribution studies of APP utilized cDNA probes, which could not distinguish among the various APP transcripts. Two of these early studies reported APP mRNA distribution in neuronal cells essentially across all regions of the brain, including brain regions not usually affected with AD-type pathological changes (Bahmanyar *et al.*, 1987; Goedert, 1987). APP mRNA was most abundantly expressed in the large pyramidal neurons in the fields of Ammon's horn in the hippocampus and in layers III and V of the prefrontal cortex (Bahmanyar *et al.*, 1987). Another study revealed a regional pattern of expression in human brain showing highest levels of APP mRNA expression in associative regions of the neocortex (Tanzi *et al.*, 1987). Subsequent studies by these same investigators using probes capable of distinguishing the alternate splice variants of the APP transcript demonstrated that the KPI-containing transcripts (predominantly APP_{751}) were ubiquitously expressed across brain regions and the KPI-lacking transcript (APP_{695}) was most abundant in the association cortex (Neve *et al.*, 1988; Tanzi *et al.*, 1988).

Analysis of the expression of alternatively spliced APP mRNAs has also been accomplished using perhaps more sensitive techniques such as polymerase chain reaction amplification of reverse transcribed cDNAs (Golde *et al.*, 1990). Golde and colleagues identified a previously undetected splice variant of APP termed APP_{714}, which was shown to contain the KPI domain also present in APP_{751} and APP_{770}, but truncated at the C-terminus. This novel form was shown to be present in all brain regions examined, however, the level of abundance of this transcript is so low that its physiological significance is in question. These studies also demonstrated marked differences in the relative amounts of these mRNAs in various tissues and brain regions. As expected, APP_{695} was the most abundant form in essentially all brain regions examined, although hippocampus, white matter and the nucleus basalis of Meynert displayed approximately equal levels of APP_{695} and APP_{751}. APP_{751} and APP_{770} were the most abundant transcripts in the peripheral tissues, where the kidney contained by far the highest levels of the various peripheral tissues that were examined. Approximately equivalent levels of the various APP transcripts (excluding of course the APP_{714} transcript) were present in cerebral arteries, arterioles and microvessels. Nonneuronal brain-associated tissues such as meninges were shown to contain predominantly APP_{770} transcripts.

One of the more intriguing studies involved quantitative analyses of *in situ* hybridization patterns of APP mRNA in AD neocortex. APP-specific hybridization signal intensities were shown to negatively correlate with neurofibrillary tangles (NFTs) (Lewis *et al.*, 1988). High numbers of NFTs in the superior frontal cortex of AD brains were associated with diminished numbers of positively hybridizing neurons. In control brains, neurons normally prone to NFT formation in AD (i.e. large pyramidal neurons of layers III and V in prefrontal cortex) were generally found to contain high levels of APP mRNA (Lewis *et al.*, 1988). The virtual absence in all AD cases examined in these studies of neurons containing both NFTs and APP mRNA suggested the possibility that APP down-regulation accompanied NFT formation. This observation could also have been explained by a general decrease in viability of neurons, which normally express high levels of APP mRNA; nevertheless, this study interestingly demonstrates an inverse correlation of APP gene expression with neurofibrillary degeneration.

Experiments employing antisense complementary RNA probes capable of distinguishing between the various APP mRNA transcripts support the hypothesis that the majority of APP-specific mRNA transcripts in brain arise from neurons. *In situ* hybridization analysis of a relatively small number of AD and control cases suggested a significant increase in AD brain (cholinergic neurons of the locus ceruleus and nucleus basalis of Meynert) of KPI-lacking mRNA transcripts with essentially no change in the level of KPI-containing transcripts (Palmert *et al.*, 1988). This same study, however, failed to reveal altered APP mRNA levels between AD and control brains in neuronal populations from other areas such as basis pontis, subiculum and cerebral cortex. A similar *in situ* hybridization study, which focused on serial hippocampal sections from AD and control brains, demonstrated that individual hippocampal pyramidal neurons contain both APP_{751} and APP_{695} mRNA transcripts (Johnson *et al.*, 1990). These investigators also observed an AD-specific increase in APP_{751} mRNA in hippocampal neurons (in disagreement with Palmert *et al.*), which correlated to neuritic plaque density.

Specific APP mRNA transcripts have also been directly measured in several brain regions of subjects with AD and controls of various ages using S-1 nuclease protection analyses (Koo *et al.*, 1990). Relative levels of APP_{695} and $APP_{751/770}$ mRNAs were shown to undergo reciprocal changes during development and ageing. In fetal human brain, APP_{695} mRNA transcripts predominated while in aged human brain $APP_{751/770}$ mRNA transcripts were more abundant. Young adults were shown to contain approximately equivalent levels of APP_{695} and $APP_{751/770}$ mRNA. These ratios appeared to remain constant throughout different brain regions of a given individual. In addition, the latter study did not detect any specific changes in APP mRNA transcripts between AD and controls.

In rats, examination of the tissue distribution and cellular localization of mRNA transcripts for the APP_{695} homologue by *in situ* hybridization demonstrated high levels of expression throughout the cerebral cortex and cerebellum with a neuron-specific localization (Shivers *et al.*, 1988). White matter oligodendroglia failed to show any hybridization signal. Immunocytochemical studies of APP localization in rat brain demonstrated a punctate staining pattern on the plasma membrane of neurons suggesting that the protein existed in aggregates, typical of an integral membrane protein. This suggests that, in rat brain, APP functions as a mediator of cell–cell or cell–matrix stabilization (Shivers *et al.*, 1988). Localization of APP antigen in various rat brain tissue sections immunohistochemically has demonstrated that astrocytes and neurons contain APP (Card *et al.*, 1988).

9.2.2.2 Immunological studies of APP distribution

In vitro translation studies with human, dog and rat brain polysomes followed by immunoprecipitation with domain-specific antipeptide antisera capable of differentiating between KPI-containing and KPI-lacking APP isoforms revealed significant differences in isoform expression patterns between these species (Anderson *et al.*, 1989). Human brain was found to contain the highest ratios of APP_{751}/APP_{695} and rat brain the lowest. Dog brain was found to contain intermediate ratios of APP_{751}/APP_{695}. In human brain, APP_{751} expression was highest in areas of the brain prone to neuritic plaque formation (i.e. cerebral cortex) and lowest in brain regions lacking these lesions (i.e. thalamus, cerebellum). As a result of the apparent lack of APP_{751} expression in rat brain (a species which does not develop neuritic plaques), this particular isoform was proposed to play an essential role in amyloid plaque formation.

Numerous studies have been published concerning immunolocalization in mammalian tissues of the various APP protein isoforms with a variety of antibodies. One of the earlier studies utilized antipeptide antibodies directed against the predicted carboxyl-terminus of APP (Selkoe *et al.*, 1988). These antibodies were shown to immunolabel 110–135 kDa membrane-associated proteins in certain neural and nonneural tissues. Subsequent studies, which employed monoclonal and polyclonal antibodies directed against the amino-termini of the various APP isoforms, revealed that these antibodies immunolabelled senile plaque coronas and extraneuronal tangles in AD brain tissue (Van Nostrand *et al.*, 1989; Arai *et al.*, 1991; Cummings *et al.*, 1992). Other structures reported to be labelled by these anti-amino-terminal APP antibodies were diverse neurons and their processes localized throughout the cerebral cortex and medulla, cerebellum (Purkinje cells), neurohypophysis and spinal cord (Arai *et al.*, 1991). In this same study, examination of several peripheral tissues revealed APP immunoreactivity in ganglia and proximal nerves of the intestinal tract, adenohypophysis, dorsal root and trigeminal ganglia (both neurons and satellite glial cells), cardiac muscle and megakaryocytes. Other peripheral tissues examined such as skin, liver, spleen, kidney, thyroid, lung, striated muscle, tongue, choroid plexus, ovary, aorta and urinary bladder failed to demonstrate any APP-positive immunoreactivity. Consistent with these findings was a report (Kim *et al.*, 1992) describing immunoreactive APP within the cytoplasm of human dorsal root ganglion neurons.

Semiquantitative immunoblot analysis using a monoclonal antibody raised against native human APP, which recognizes all APP isoforms (APP_{695}, APP_{751} and APP_{770}), failed to detect APP in several peripheral tissues from nonhuman primates (skeletal muscle, lung, liver, spleen, pancreas, heart and spinal cord) (Wagner *et al.*, 1993). In this same study, APP immunoreactivity was shown to be abundant in brain with relatively low amounts in kidney and testis. Within the primate brain, APP immunoreactivity was highest in the cerebral cortical areas, and lowest in the cerebellum and brainstem.

In a detailed immunohistochemical study of APP in AD and aged control brains using the same monoclonal antibody against native human APP, a number of interesting observations were cited (Cummings *et al.*, 1992). Aged control brain tissue revealed a very faint homogeneous cytoplasmic neuronal staining quite distinct from the granular staining pattern obtained with AD tissue. In close proximity to the APP-immunopositive plaques in AD brain were APP-immunopositive dystrophic neurites. These immunopositive neurites (most prominent in the pyramidal cells of hippocampal fields CA3-CA1 and entorhinal cortex) often contained varicosities along their lengths. This type of staining pattern was not seen in the aged control brains and suggested a neuronal source for the APP-immunoreactivity found in senile plaques. In addition to senile plaques, amino-terminal APP antibodies have been shown to immunolabel prion plaques in Gerstmann–Straüssler–Scheinker disease (Bugiani *et al.*, 1993; Farlow *et al.*, 1994) and congophilic vessels in the Dutch form of hereditary cerebral haemorrhage with amyloidosis (Rozemuller *et al.*, 1993).

In the most recent immunohistochemical study using AD brain tissue, antibodies directed against amino-, mid- and carboxyl-terminal domains of APP were found to label NFTs (both intraneuronal and extracellular) (Perry *et al.*, 1994), suggesting that APP present in NFTs potentially provided a source for the Aβ deposition detected on NFTs.

Antibodies specific for the KPI-containing APP isoforms have been utilized to compare expression patterns in various tissues. Brain and cerebrospinal fluid (CSF) have been shown to contain soluble derivatives of APP isoforms, which both contain and lack the KPI domain (Palmert *et al.*, 1989; Van Nostrand *et al.*, 1991). The regional distribution of the relative abundance of KPI-containing in comparison to KPI-lacking APP isoforms in the human brain is not currently known, but may depend on certain factors such as age and disease status. Several studies have shown that CSF contains primarily APP_{695} (Palmert *et al.*, 1989; Henriksson *et al.*, 1991; Prior *et al.*, 1991; Van Nostrand *et al.*, 1992a). Human platelets have also been shown to contain the KPI-containing APP isoforms within their α-granules (Bush *et al.*, 1990; Smith *et al.*, 1990; Van Nostrand *et al.*, 1990a). KPI-containing isoforms of APP have also been immunolocalized to proacrosomal granules and acrosomal vesicles of spermatid flagella (Shoji *et al.*, 1990).

Collectively, these studies suggest that APP expression in mammalian tissue occurs to a limited extent outside the nervous system, which involves predominantly the KPI-containing isoforms, and that brain contains by far the highest levels of APP gene expression products harbouring both KPI-containing and KPI-lacking APP isoforms. At present, there is no consensus regarding specific alterations of APP expression in AD.

9.3 Potential physiological roles of APP

9.3.1 Protease inhibitory functions of APP

9.3.1.1 Identity of KPI-containing APP and protease nexin-II

The initial cloning of KPI-containing APP cDNAs led to the identification of secreted forms of KPI-containing APPs as protease nexin II (PN-II). Although tissue and cellular distribution studies indicate that APP is a membrane glycoprotein, soluble/secreted derivatives of APP can be detected in a variety of extracellular fluids using amino-terminal antibodies. In fact, PN-II was originally discovered in culture medium of human fibroblasts based on its ability to form sodium dodecyl sulphate (SDS)-stable inhibitory complexes with ^{125}I-labelled epidermal growth factor binding protein, a serine protease found in mouse submaxillary gland believed to be involved in the proteolytic processing of epidermal growth factor (Knauer and Cunningham, 1982). PN-II was later purified to homogeneity from conditioned medium of large-scale cultures of human fibroblasts and was shown to have an apparent M_r of 106 000 and an isoelectric point of 7.2 (Van Nostrand and Cunningham, 1987). The purified PN-II protein was also capable of forming SDS-stable inhibitory complexes with two other serine proteases; namely, nerve growth factor-γ, another mouse submaxillary gland serine protease whose apparent role is in the proteolytic processing of nerve growth factor, and the pancreatic enzyme trypsin. PN-II was shown to retain full activity even after treatment with 0.05% SDS or after prolonged incubation at pH 1.5. A monoclonal antibody prepared against the purified PN-II (mAbp2-1) recognized proteins of apparently identical molecular weights in immunoblots of culture medium from human neuroblastoma cells, human glioblastoma cells and human fibroblasts. Furthermore, this anti-PN-II monoclonal antibody was shown to immunostain Alzheimer's disease neuritic plaques as well as pyramidal neurons in the hippocampus (Van Nostrand *et al.*, 1989). Identification of PN-II as $APP_{751/770}$ was confirmed based on alignment of the amino-terminal sequence of PN-II (Van Nostrand and Cunningham, 1987) with the deduced amino-acid sequence of APP (Oltersdorf *et al.*, 1989; Van Nostrand *et al.*, 1989).

9.3.1.2 Inhibition of blood coagulation factors by $APP_{751/770}$ and the KPI domain

It has been demonstrated that PN-II and/or KPI-containing isoforms of APP are potent reversible inhibitors of chymotrypsin ($K_i = 6\times10^{-10}$ M) (Van Nostrand *et al.*, 1989). Subsequent studies with monoclonal antibody-purified $APP_{751/770}$/PN-II displayed protease inhibitory properties of this protein towards a number of serine proteases (Van Nostrand *et al.*, 1990b). Most notable was the inhibition of blood coagulation Factor XIa ($K_i = 2.9\times10^{-10}$ M), which could be augmented considerably by low concentrations of heparin ($K_i = 5.5\times10^{-11}$ M). This latter inhibitory activity is of potential physiological significance because $APP_{751/770}$/PN-II is localized to the platelet α-granule and is secreted following platelet activation (Van Nostrand *et al.*, 1990a). These observations suggest a potential role of $APP_{751/770}$ in blood coagulation.

The previously described protease inhibitory activities of $APP_{751/770}$/PN-II (including Factor XIa) were shown to be fully accommodated by the KPI domain. High-level recombinant expression and purification of the 57 amino-acid KPI domain, including an additional four amino acids of APP, which directly precede it, produced an inhibitor of equivalent potency and specificity to that of $APP_{751/770}$/PN-II (Wagner *et al.*, 1992). The only difference observed between the KPI peptide and secreted $APP_{751/770}$/PN-II was the inability of heparin to augment the inactivation of Factor XIa by the recombinant KPI peptide. Incubation of heparin (10 U/ml) alters the protease inhibition constant of secreted $APP_{751/770}$/PN-II for Factor XIa from approximately 4.0×10^{-10} M to approximately 5.5×10^{-11} M suggesting that heparin binds to regions on secreted $APP_{751/770}$/PN-II outside of the KPI domain (Wagner *et al.*, 1992).

Factor XIa is the initial protease of the intrinsic coagulation cascade, however, abnormalities involving Factor XIa do not result in spontaneous haemorrhaging (Bolton-Maggs *et al.*, 1988). This is a result of the many compensatory mechanisms involved with blood clotting. If $APP_{751/770}$/PN-II is truly involved in controlling blood coagulation it would be expected to interact with additional proteases within the clotting cascade.

Secreted $APP_{751/770}$/PN-I and the recombinant KPI peptide were subsequently shown to be potent inhibitors of coagulation Factor IXa (Schmaier *et al.*, 1993). The equilibrium inhibition constants (K_i) for inhibition of Factor IXa by KPI and $APP_{751/770}$/PN-II were 1.9×10^{-7} M and 7.9×10^{-11} M, respectively. Although $APP_{751/770}$/PN-II inhibits comparably both Factors IXa and XIa ($K_i = 7.9\times10^{-11}$ M and 4.0×10^{-10} M, respectively), the recombinant KPI peptide is a much better inhibitor of Factor XIa than of Factor IXa ($K_i = 4.5\times10^{-10}$ M and 1.9×10^{-7} M, respectively). The large difference between $APP_{751/770}$/PN-II and KPI with respect to inhibition of Factor IXa suggests that portions of the $APP_{751/770}$/PN-II molecule other than the KPI domain are involved in the inhibition of Factor IXa.

Factor IXa is a component of one of the two enzyme complexes capable of activating coagulation Factor Xa, which is ultimately responsible for converting prothrombin to thrombin. *In vitro* clotting assays revealed that both $APP_{751/770}$/PN-II and the recombinant KPI peptide were capable of prolonging clotting time approximately twofold over pooled normal plasma. The plasma-borne serine protease inhibitor antithrombin III (AT-III) [believed to be the physiological inhibitor of Factor IXa (Rosenberg *et al.*, 1975)] has a second-order association rate constant approximately 70-fold slower than does $APP_{751/770}$/PN-II in the absence of heparin and 95-fold slower in the presence of heparin (Schmaier *et al.*, 1993).

The potent inhibition of Factors XIa and IXa by $APP_{751/770}$/PN-II strongly suggests that this protein influences the activation and maintenance of the coagulation cascade (see Figure 2). The rich investment of $APP_{751/770}$/PN-II and the relative lack of thrombomodulin, a potent endothelial anticoagulant protein (Ishii *et al.*, 1986) in brain, suggest that $APP_{751/770}$/PN-II may function as a critical intracerebral anticoagulant (Van Nostrand *et al.*, 1992b). The physiological significance of Factor IXa in the coagulation cascade is based on the spontaneous bleeding disorders (including intracerebral haemorrhages) of individuals with deficiencies in this protein (Biggs, 1974). Interestingly, tissue factor pathway inhibitor (TFPI), another Kunitz-type protease inhibitor, has been demonstrated to modulate the activation of coagulation Factor Xa by inhibiting Factor VIIa-tissue factor complex (Rao and Rapaport, 1987). Thus, $APP_{751/770}$/PN-II would be the second Kunitz-type protease inhibitor implicated in the modulation of haemostasis.

9.3.1.3 Identification of an additional domain of APP that displays protease inhibitory activity

In addition to the KPI domain, another domain within the APP molecule has been shown to contain protease inhibitory activity. Culture medium from EJ-1 bladder carcinoma cells contains a 100 kDa inhibitor of matrix-degrading metalloproteinases

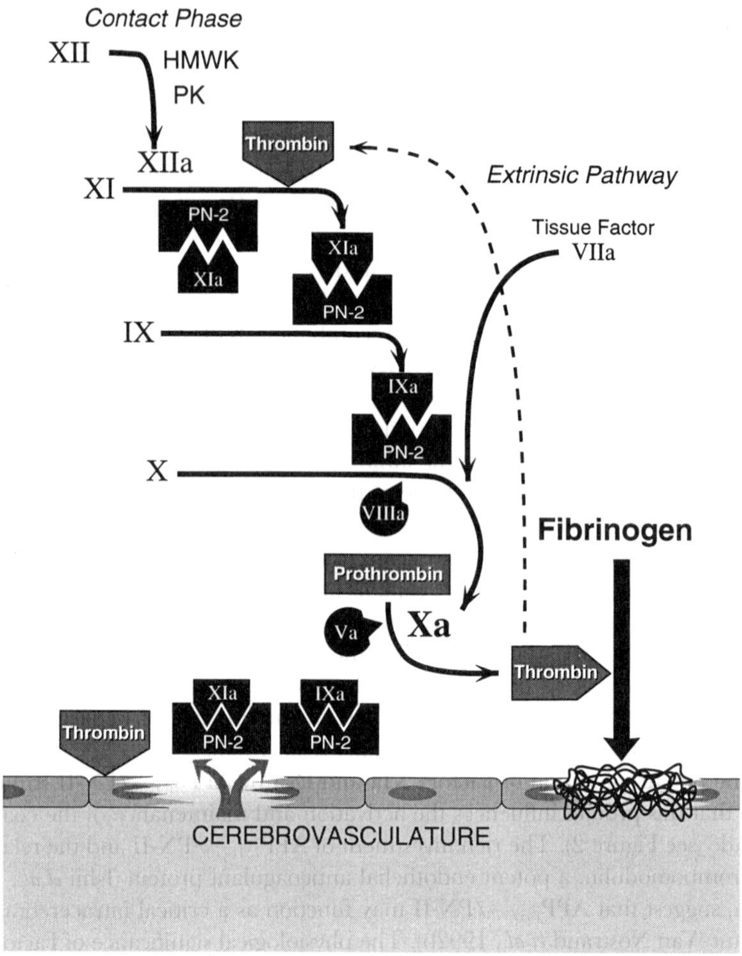

Figure 2 Proposed role of protease nexin-II (PN-II/$APP_{751/770}$) in controlling the coagulation cascade in the brain. HMWK, high molecular weight kininogen; PK, prekallikrein.

(MMPs) (Miyazaki *et al.*, 1993). The amino-terminal sequence of the purified 100 kDa MMP inhibitor was found to be identical to that of APP without the signal peptide. The MMP inhibitory domain was localized within the APP molecule by reverse zymography of endoproteinase Lys-C-digested APP. A 28 kDa endoproteinase Lys-C fragment of APP was shown to retain gelatinase inhibitory activity according to reverse zymography. Amino-terminal sequence analysis of this 28 kDa fragment (Val-Glu-Ser-Leu-Glu-Gln-Glu-Ala) corresponded to amino acids 439–446 of APP (according to the APP_{770} numbering system). This domain is located within the C-terminal glycosylated region of all secreted isoforms of APP including APP_{695}. One member of the MMP family (gelatinase A) was shown to be capable of hydrolysing a synthetic peptide corresponding to amino acids 10–20 of Aβ. The hydrolysis occurred precisely at the so-called α-secretase site (Lys 16–Leu 17 of Aβ) (Esch *et al.*, 1990; Sisodia *et al.*, 1990), implicating gelatinase A as a candidate for the as yet uncharacterized α-secretase enzyme. In addition, the 100 kDa secretory form of $APP_{751/770}$ purified from the medium of the bladder carcinoma cells was found capable of competitively inhibiting the gelatinase A-mediated hydrolysis of this peptide (K_i = 40 nM). Thus, these studies have suggested modulation of MMPs as a potential physiological role for both KPI-containing and KPI-lacking forms of APP. Confirmation of the ability of KPI-lacking forms of APP to inhibit gelatinase A could be provided through studies of the inhibitory properties of APP_{695} or the 28 kDa Lys-C fragment of APP.

9.3.2 Growth-promoting functions of APP

APP is a large molecule with multiple domains, which display a high degree of evolutionary conservation (Rosen *et al.*, 1989; Okado and Okamoto, 1992; Daigle and Li, 1993). Thus, it is not surprising that multiple biological functions have been attributed to it. A fundamental biological role proposed for APP concerned its effects on the growth rates of cultured cells. Suppression of APP synthesis in cultured cells through transfection with a vector driving the expression of an anti-sense APP RNA resulted in a dramatically reduced growth rate for a human lung fibroblast cell line (Saitoh *et al.*, 1989). These fibroblasts (A-1 lung fibroblasts) returned to a normal growth rate after addition of conditioned medium from the parental lung fibroblasts, but not if the medium had been passed over an anti-APP affinity column. Purified APP from brain tissue was also capable of restoring the growth of these A-1 fibroblasts in a dose-dependent manner with optimal concentrations occurring in the picomolar range. Normal growth rates could be achieved with either $APP_{751/770}$ or APP_{695} added exogenously.

Subsequent studies identified regions of the APP molecule responsible for these growth-promoting activities (Roch *et al.*, 1992). Fragments of APP_{695} spanning different regions of the molecule were produced in an *Escherichia coli* expression system. The largest of these fragments (571 amino acids) retained full activity. Two other fragments (316 and 150 amino acids), which overlapped by only 40 amino acids in the ectodomain near the KPI insertion site, were also functionally active. In addition, a synthetic 40 amino-acid peptide corresponding to the overlapping region of the 316 and 150 amino-acid fragments was also shown to be capable of promoting growth in the A-1 lung fibroblasts. This 40 amino-acid peptide corresponded to amino acids 296–335 of APP_{695}. A battery of synthetic and recombinant peptides within this region of APP_{695} was tested for growth-promoting activities in cultured A-1 fibroblasts (Ninomiya *et al.*, 1993). A pentapeptide (RERMS) localized within this functional domain (amino acids 328–332 of APP_{695}) was found to harbour the majority of this activity. The reverse sequence of this peptide was found to have no activity. Inactive peptides, which only partially overlapped the active sequence RERMS, were shown to be capable of antagonizing the growth-promoting activity of RERMS, suggesting that an interaction of APP with specific cell surface molecules was responsible for this activity.

A related property of APP is based on its ability to stimulate adhesion of phaeochromocytoma cells (PC12 cells) to cell substratum (Schubert *et al.*, 1989). Highly purified soluble derivatives of APP_{695} and APP_{751}, at subpicomolar concentrations, were approximately tenfold better than laminin at stimulating PC12 cell adhesion in short-term adhesion assays. These studies suggest that APPs fall into the class of adhesion molecules that mediate the binding of cells to the extracellular matrix. In addition to stimulating the adhesion of PC12 cells to substratum, both soluble and membrane-bound forms of APP (10–100 ng/ml) were capable of increasing neurite length and branching of PC12 cells (Milward *et al.*, 1992). In these same studies, higher concentrations of either membrane-associated or soluble APP (500 ng/ml) were cytotoxic to these cells. Subsequent studies suggested that the effects of APP on neurite outgrowth in cultures of chick sympathetic and mouse hippocampal neurons required an interaction with substratum-bound heparin sulphate proteoglycans (HSPGs) (Small *et al.*, 1994). These investigators also identified a potential heparin-binding region on APP close to the amino terminus. APP has also been shown to be the same as LBP110, a 110 kDa membrane-associated laminin-binding protein (Kibbey *et al.*, 1993). APP binds specifically to a pentapeptide of the laminin A chain and is up-regulated as migrating neural crest cells come into contact with laminin. This further demonstrates the involvement of APP with neurite outgrowth and neuronal cell migration (Pomeranz *et al.*, 1993). Taken together, these growth-promoting and cell-adhesive properties of APP suggest that it may play a role in the developmental and synaptic plasticity of the nervous system that occurs under a variety of physiological and pathophysiological conditions.

9.3.3 Neuroprotective functions of APP

The early observation that nerve growth factor (NGF) caused an increase in APP mRNA expression in the septal nuclei of the hamster suggested a role for APP in synaptogenesis (Mobley *et al.*, 1988). Subsequent studies showed that expression of APP is dramatically increased following an intracerebroventricular injection of the neurotoxin kainic acid (Siman *et al.*, 1989). The neuronal damage was accompanied by a long-lasting induction of APP expression mediated by reactive astrocytes. APP is increased in the rat cerebral cortex following *N*-methyl-D-aspartate (NMDA)-induced unilateral lesions of the nucleus basalis of Meynert (Wallace *et al.*, 1993). In addition, subcortical lesions of the cortically projecting noradrenergic and serotonergic systems similarly induced cortical APP expression. CNS insults such as head trauma and cerebral ischaemia alter APP expression (Abe *et al.*, 1991; Stephenson *et al.*, 1992; Gentleman *et al.*, 1993), however, general central nervous system perturbations (i.e. adrenalectomy or streptozotocin treatment), which do not involve neurotransmitter deficits, fail to induce APP expression (Wallace *et al.*, 1993). This suggests a role for APP in interneuronal communications.

Some of the more compelling evidence thus far in support of a neuroprotective role for APP comes from experiments concerning the effects of secreted forms of APP (APP_{695} and APP_{751}) on neuronal survival in cultured cell systems (Mattson *et al.*, 1993;

Mattson, 1994). In these studies, rat hippocampal and septal neurons, as well as human cortical neurons, were exposed to hypoglycaemic conditions for prolonged periods of time. Neurons are normally unable to survive these hypoglycaemic conditions, however, when the neurons were treated with nanomolar concentrations of secreted APP_{695} or APP_{751}, a highly significant percentage of the neurons survived. The extent of neuronal survival correlated with the exogenously administered APP concentrations in the range of 10 pM to 100 nM. The neuroprotective activities exhibited by the secreted APP_{695} and APP_{751} isoforms appear to reside in a region just amino-terminal to the Aβ region (see Figure 1) based on studies using a panel of peptide-specific antibodies. A specific monoclonal antibody (epitope amino acids 444–592 of APP_{695}) prevented the neuroprotective effects of secreted APP_{695} and APP_{751}. Antibodies to other regions of APP were not capable of inhibiting the neuroprotective activities.

The ability of secreted APP_{695} and APP_{751} isoforms to prevent the hypoglycaemia-induced neuronal cell death appeared to be related to their effects on intracellular calcium concentrations $[Ca^{2+}]_i$. Concentrations of secreted APP_{695} or APP_{751} as low as 1 pM significantly reduced $[Ca^{2+}]_i$ in resting hippocampal pyramidal neurons. Nanomolar concentrations of secreted APP_{695} or APP_{751} prevented the threefold increase in $[Ca^{2+}]_i$ that accompanied hypoglycaemic insults to hippocampal pyramidal neurons. In addition, nanomolar concentrations of secreted APP_{695} or APP_{751} prevented neuronal cell death in hippocampal cultures normally elicited by high concentrations (100–200 μM) of glutamate.

These neuroprotective activities of secreted APP isoforms, which apparently result from their ability to control intracellular calcium ion concentrations, are analogous to effects elicited by other neuronal growth factors such as nerve growth factor and basic fibroblast growth factor. Taken together, the studies demonstrating the ability of APP to support the growth of cultured cells, to promote adhesion and neurite extension of cultured neurons, and to protect them from environmental insults, shed new light on the potential biological role(s) of this protein.

9.4 Conclusions

APP has evolved from literally being the precursor of a pathological peptide (Aβ) into a molecule with potentially diverse biological functions. The initial cloning of a cDNA encoding APP revealed primary structural homology to a single membrane-spanning cell surface receptor (Kang *et al.*, 1987). Although to date there has been no ligand found to support this notion, a recent report has shown that a region of the APP cytoplasmic sequence forms a specific complex with G_o, a major GTP-binding protein in brain (Nishimoto *et al.*, 1993). Future studies may indeed demonstrate that APP is involved in the transportation of some very critical molecule(s).

The identification of APP isoforms containing a Kunitz-type protease inhibitor domain (APP_{751} and APP_{770}) led to the discoveries that $APP_{751/770}$ can modulate two key pro-coagulant enzymes better than any physiological protease inhibitors known, thus providing the first unequivocal biological function attributed to this molecule. Potential physiological role(s) for the APP_{695} isoform, although extremely exciting, appear to be at this stage somewhat more speculative; however, the rich investment of this particular isoform in human brain will no doubt lead to findings which have a major impact on neuroscience. Finally, the intimate association of this molecule with the hallmark pathological lesions of AD will continue to demand intense scientific attention.

References

Abe, K., Tanzi, R.E. & Kogure, K. (1991) *Neurosci. Lett.* **125**, 172–174.

Anderson, J.P., Refolo, L.M., Wallace, W., Mehta, P., Krishnamurthi, M., Gotlib, J., Bierer, L., Haroutunian, V., Perl, D. & Robakis, N.K. (1989) *EMBO J.* **8**, 3627–3632.

Arai, H., Lee, V.M.Y., Messinger, M.L., Greenberg, B.D., Lowery, D.E. & Trojanowski, J.Q. (1991) *Ann. Neurol.* **30**, 686–693.

Bahmanyar, S., Higgins, G.A., Goldgaber, D., Lewis, D.A., Morrison, J.H., Wilson, M.C., Shankar, S.K. & Gajdusek, D.C. (1987) *Science* **237**, 77–80.

Biggs, R. (1974) *Br. J. Haematol.* **35**, 487–504.

Bolton-Maggs, P.H.B., Young-Wan-Yin, B., McCraw, A.H., Slack, J. & Kernoff, P.B.A. (1988) *Br. J. Haematol.* **69**, 521–528.

Bugiani, O., Giaccone, G., Verga, L., Pollo, B., Frangione, B., Farlow, M.R., Tagliavini, F. & Ghetti, B. (1993) *J. Neuropathol. Exp. Neurol.* **52**, 64–70.

Bush, A.I., Martins, R.N., Rumble, B., Moir, R., Fuller, S., Milward, E., Currie, J., Ames, D., Wedemann, A., Fischer, P., Multhaup, G., Beyreuther, K. & Masters, C.L. (1990) *J. Biol. Chem.* **265**, 15977–15983.

Card, J.P., Meade, R.P. & Davis, L.G. (1988) *Neuron* **1**, 835–846.

Clark, M.J., Gagnon, J., Williams, A.F. & Barclay, A.N. (1985) *EMBO J.* **4**, 113–118.

Cummings, B.J., Su, J.W., Van Nostrand, W.E., Wagner, S.L., Cunningham, D.D. & Cotman, C.W. (1992) *Neuroscience* **4**, 763–777.

Daigle, I. & Li, C. (1993) *Proc. Natl Acad. Sci. USA* **90**, 12045–12049.

Esch, F.A., Keim, P.S., Beattie, E.C., Blacher, R.W., Culwell, A.R., Oltersdorf, T., McClure, D. & Ward, P.J. (1990) *Science* **248**, 1122–1124.

Farlow, M.R., Ghetti, B., Dlouhy, S., Giaccone, G., Bugiani, O., Tagliavini, F. & Wagner, S.L. (1994) *Neurology* **44**, 1508–1510.

Gentleman, S.M., Nash, M.J., Sweeting, C.J., Graham, D.I. & Roberts, G.W. (1993) *Neurosci. Lett.* **160**, 139–144.

Glenner, G.G. & Wong, C.W. (1984a) *Biochem. Biophys. Res. Commun.* **120**, 885–890.

Glenner, G.G. & Wong, C.W. (1984b) *Biochem. Biophys. Res. Commun.* **122**, 1131–1135.

Goate, A., Chartier-Harlin, M.C., Mullan, M., Brown, J., Crawford, F., Fidani, L., Giuffra, L., Haynes, A., Irving, N., James, L., Mant, R., Newton, P., Rooke, K., Roques, P., Talbot, C., Pericak-Vance, M., Roses, A., Williamson, R., Rossor, M., Owen, M. & Hardy, J. (1991) *Nature* **349**, 704–706.

Goedert, M. (1987) *EMBO J.* **6**, 3627–3632.

Golde, T.E., Estus, S., Usiak, M., Younkin, L.H. & Younkin, S.G. (1990) *Neuron* **4**, 253–267.

Goldgaber, D., Lerman, M.J., McBride, O.W., Saffiotti, V. & Gajdusek, D.C. (1987) *Science* **235**, 877–880.

Henriksson, T., Barbour, R.M., Braa, S., Ward, P., Fritz, L.C., Johnson-Wood, K., Chung, H.D., Burke, W., Reinikainen, K.J., Reikkinen, P. & Schenk, D.B. (1991) *J. Neurochem.* **56**, 1037–1042.

Ishii, H., Salem, H.H., Bell, C.E., Laposata, E.A. & Majerus, P.W. (1986) *Blood* **67**, 362–365.
Johnson, S.A., McNeill, T., Cordell, B. & Finch, C.E. (1990) *Science* **248**, 854–857.
Kang, J., Lemaire, H.G., Unterbeck, A., Salbaum, J.M., Masters, C.L., Grzeschik, K.H., Multhaup, G., Beyreuther, K. & Muller-Hill, B. (1987) *Nature* **325**, 733–736.
Kibbey, M.C., Jucjer, M.J., Weeks, B.S., Neve, R.L., Van Nostrand, W.E. & Kleinman, H.K. (1993) *Proc. Natl Acad. Sci. USA* **90**, 10150–10153.
Kim, T., Choi, B.H., Choe, W., Kim, R.C., Van Nostrand, W.E., Wagner, S.L. & Cunningham, D.D. (1992) *Mol. Chem. Neuropathol.* **16**, 225–239.
Kitaguchi, N., Takahashi, Y., Tokushima, Y., Shiojiri, S. & Ito, H. (1988) *Nature* **331**, 530–532.
Knauer, D.J. & Cunningham, D.D. (1982) *Proc. Natl Acad. Sci. USA* **79**, 2310–2314.
Koo, E., Sisodia, S.S., Cork, L.C., Unterbeck, A., Bayney, R.M. & Price, D.L. (1990) *Neuron* **2**, 97–104.
Lewis, D.A., Higgins, G.A., Young, W.G., Goldgaber, D., Gajdusek, D.C., Wilson, M.C. & Morrison, J.H. (1988) *Proc. Natl Acad. Sci. USA* **85**, 1691–1695.
Masters, C.L., Simms, G., Weinmann, N.A., Multhaup, G., McDonald, B.L. & Beyreuther, K. (1985) *Proc. Natl Acad. Sci. USA* **82**, 4245–4249.
Mattson, M.P. (1994) *J. Neurobiol.* **25**, 439–450.
Mattson, M.P., Cheng, B., Culwell, A.R., Escg, F.S., Lieberburg, I. & Rydel, R.E. (1993) *Neuron* **10**, 243–254.
Milward, E.A., Papadopoulos, R., Fuller, S.J., Moir, R.D., Small, D., Beyreuther, K. & Masters, C.L. (1992) *Neuron* **9**, 129–137.
Miyazaki, K., Hasegawa, M., Funahashi, K. & Umeda, M. (1993) *Nature* **362**, 839–841.
Mobley, W.C., Neve, R.L., Pruisner, S.B. & McKinley, M.P. (1988) *Proc. Natl Acad. Sci. USA* **85**, 9811–9815.
Neve, R.L., Finch, C.E. & Dawes, L.R. (1988) *Neuron* **1**, 669–677.
Ninomiya, H., Roch, J.M., Sundsmo, M.P., Otero, D.A. & Saitoh, T. (1993) *J. Cell Biol.* **121**, 879–886.
Nishimoto, I., Okamoto, T., Matsuura, Y., Takahashi, S., Okamoto, T., Murayama, Y. & Ogata, E. (1993) *Nature* **362**, 75–79.
Okado, H. & Okamoto, H. (1992) *Biochem. Biophys. Res. Commun.* **189**, 1561–1568.
Oltersdorf, T., Fritz, L.C., Schenk, D.B., Lieberburg, I., Johnson-Wood, K.L., Beattie, E.C., Ward, P.J., Blacher, R.W., Dovey, H.F. & Sinha, S. (1989) *Nature* **341**, 144–147.
Palmert, M.R., Golde, T.E., Cohen, M.L., Kovacs, D.M., Tanzi, R.E., Gusella, J.F., Usiak, M.F., Younkin, L.H. & Younkin, S.G. (1988) *Science* **241**, 1080–1084.
Palmert, M.R., Podlisny, M.B., Witker, D.S., Oltersdorf, T., Younkin, L.H., Selkoe, D.J. & Younkin, S.G. (1989) *Proc. Natl Acad. Sci. USA* **86**, 6338–6342.
Perry, G., Richey, P.L., Siedlak, S.L., Smith, M.A., Mulvihill, P., DeWitt, D.A., Barnett, J., Greenberg, B.D. & Kalaria, R. (1994) *Am. J. Pathol.* **143**, 1586–1593.
Pomeranz, H.D., Rothman, T.P., Chalazonitis, A., Tennyson, V.M. & Gershon, M.D. (1993) *Dev. Biol.* **156**, 341–361.
Ponte, P., Gonzalez-DeWhitt, P., Schilling, J., Miller, J., Hsu, D., Greenberg, B., Davis, K., Wallace, W., Lieberburg, I., Fuller, F. & Cordell, B. (1988) *Nature* **331**, 525–528.
Prior, R., Monning, U., Schreiter-Gasser, U., Wedemann, A., Blennow, K., Gottfries, C.G., Masters, C.L. & Beyreuther, K. (1991) *Neurosci. Lett.* **124**, 69–73.
Rao, L.V.M. & Rapaport, S.I. (1987) *Blood* **69**, 645–651.
Robakis, N.K., Ramakrishna, N., Wolfe, G. & Wisniewski, H.M. (1987) *Proc. Natl Acad. Sci. USA* **84**, 4190–4194.
Roch, J.M., Shapiro, P., Sundsmo, M.P., Otero, D.A.C., Refolo, L.M., Robakis, N.K. & Saitoh, T. (1992) *J. Biol. Chem.* **267**, 2214–2221.
Rosen, D.R., Martin-Morris, L., Luo, L. & White, K. (1989) *Proc. Natl Acad. Sci. USA* **86**, 2478–2482.
Rosenberg, J.S., McKenna, P.W. & Rosenberg, R.D. (1975) *J. Biol. Chem.* **250**, 8883–8888.

Rozemuller, A.J.M., Roos, R.A.C., Bots, G.T.A.M., Kamphorst, W., Eikelenboom, P. & Van Nostrand, W.E. (1993) *Am. J. Pathol.* **142**, 1449–1457.

Saitoh, T., Sundsmo, M., Roch, J.M., Kimura, N., Cole, G., Schubert, D., Oltersdorf, T. & Schenk, D.B. (1989) *Cell* **58**, 615–622.

Schmaier, A.H., Dahl, L.D., Rozemuller, A.J.M., Roos, R.A.C., Wagner, S.L., Chung, R. & Van Nostrand, W.E. (1993) *J. Clin. Invest.* **92**, 2540–2545.

Schubert, D., Jin, L.W., Saitoh, T. & Cole, G. (1989) *Neuron* **3**, 689–694.

Selkoe, D.J., Podlisny, M.B., Joachim, C.L., Vickers, E.A., Lee, G., Fritz, L.C. & Oltersdorf, T. (1988) *Proc. Natl Acad. Sci. USA* **85**, 7341–7345.

Shivers, B.D., Hilbich, C., Multhaup, G., Salbaum, M., Beyreuther, K. & Seeburg, P.H. (1988) *EMBO J.* **7**, 1365–1370.

Shoji, M., Kawarabayashi, T., Harigaya, Y., Yamaguchi, H., Hirai, S., Kamimura, T. & Sugiyama, T. (1990) *Am. J. Pathol.* **137**, 1027–1032.

Siman, R., Card, J.P., Nelson, R.B. & Davis, L.G. (1989) *Neuron* **3**, 275–285.

Sisodia, S.S., Koo, E.H., Beyreuther, K., Unterbeck, A. & Price, D.L. (1990) *Science* **248**, 492–495.

Small, D.H., Nurcombe, V., Reed, G., Clarris, H., Moir, R., Beyreuther, K. & Masters, C.L. (1994) *J. Neurosci.* **14**, 2117–2127.

Smith, R.P., Higuchi, D.A. & Broze, G.J. (1990) *Science* **248**, 1126–1128.

Stephenson, D.T., Rash, K. & Clemens, J.A. (1992) *Brain Res.* **593**, 128–135.

Tanzi, R.E., Gusella, J.F., Watkins, P.C., Bruns, G.A.P., St George-Hyslop, P., Van Keuren, M.L., Patterson, D., Pajan, S., Kurnit, D.M. & Neve, R.L. (1987) *Science* **235**, 880–884.

Tanzi, R.E., McClatchey, A.I., Lamberti, E.D., Villa-Komaroff, L., Gusella, J.F. & Neve, R.L. (1988) *Nature* **331**, 528–530.

Van Nostrand, W.E. & Cunningham, D.D. (1987) *J. Biol. Chem.* **262**, 8508–8514.

Van Nostrand, W.E., Wagner, S.L., Suzuki, M., Choi, B.H., Farrow, J.S., Geddes, J.W., Cotman, C.W. & Cunningham, D.D. (1989) *Nature* **341**, 546–549.

Van Nostrand, W.E., Schmaier, A.H., Farrow, J.S. & Cunningham, D.D. (1990a) *Science* **248**, 745–748.

Van Nostrand, W.E., Wagner, S.L., Farrow, J.S. & Cunningham, D.D. (1990b) *J. Biol. Chem.* **265**, 9591–9594.

Van Nostrand, W.E., Farrow, J.S., Wagner, S.L., Bhasin, R., Goldgaber, D., Cotman, C.W. & Cunningham, D.D. (1991) *Proc. Natl Acad. Sci. USA* **88**, 10302–10306.

Van Nostrand, W.E., Wagner, S.L., Shankle, W.R., Farrow, J.S., Dick, M., Rozemuller, J.M., Kuiper, M.A., Wolters, E.C., Zimmerman, J., Cotman, C.W. & Cunningham, D.D. (1992a) *Proc. Natl Acad. Sci. USA* **89**, 2551–2555.

Van Nostrand, W.E., Schmaier, A.H. & Wagner, S.L. (1992b) *Ann. N. York Acad. Sci.* **674**, 243–252.

Wagner, S.L., Siegel, R.S., Vedvick, T.S., Raschke, W.C. & Van Nostrand, W.E. (1992) *Biochem. Biophys. Res. Commun.* **186**, 1138–1145.

Wagner, S.L., Van Nostrand, W.E., Lau, A.L., Farrow, J.S., Suzuki, M., Bartus, R., Schuppek, R., Nguyen, A., Cotman, C.W. & Cunningham, D.D. (1993) *Brain Res.* **626**, 90–98.

Wallace, W., Ahlers, S.T., Gotlib, J., Bragin, V., Sugar, J., Gluck, R., Shea, P.A., Davis, K.L. & Haroutunian, V. (1993) *Proc. Natl Acad. Sci. USA* **90**, 8712–8716.

CHAPTER 10

CELLULAR AND ANIMAL MODELS OF AMYLOID β-PROTEIN AMYLOIDOSIS

Sangram S. Sisodia*, David R. Borchelt* and Donald L. Price*†

*Department of *Pathology, †Neurology, and †Neuroscience and the *Neuropathology Laboratory, The Johns Hopkins University School of Medicine, Baltimore, Maryland 21205–2196, USA*

Table of Contents

10.1 Introduction

Alzheimer's disease (AD), the most common dementing disorder of late life, is a major cause of disability and death in the elderly (McKhann *et al.*, 1984; Khachaturian, 1985; Evans *et al.*, 1989). The disease is manifested by the appearance of abnormalities in the brain, particularly involving the hippocampus, amygdala, thalamus and neocortex. Lesions in these regions are associated with dysfunction/death of neurons and deafferentation of targets (Whitehouse *et al.*, 1982; Hyman *et al.*, 1984; Arnold *et al.*, 1991; Braak and Braak, 1991a, b; Terry *et al.*, 1991). The principal pathological hallmarks of AD are deposits of the amyloid β-protein (Aβ) in extracellular parenchyma and cerebral vessels, and neurofibrillary tangles (NFT). This review focuses on the biology of the amyloid β-protein precursor (APP) as studied *in vitro*, with an emphasis on mechanisms relevant to the generation and deposition of Aβ, and studies of animal models, particularly the aged nonhuman primate.

Pathobiology of Alzheimer's Disease
ISBN 0-12-286965-5

10.2 Biology of APP and Aβ

10.2.1 *In vitro* studies of APP and Aβ

A principal hallmark of AD is the presence of deposits of Aβ in parenchyma, particularly in the hippocampus and neocortex. Extracellular Aβ deposits are generally associated with neurites in senile plaques (Glenner and Wong, 1984; Masters *et al.*, 1985a, b; Selkoe *et al.*, 1987; Probst *et al.*, 1991) but are also present in pre-amyloid deposits and within the walls of leptomeningeal and cerebral vessels (Selkoe *et al.*, 1987; Martin *et al.*, 1991).

Aβ, a 39–43 amino-acid peptide, is comprised of 11–15 amino acids of the transmembrane domain and 28 amino acids of the extracellular domain of the larger APP (Kang *et al.*, 1987; Kitaguchi *et al.*, 1988; Ponte *et al.*, 1988; Tanzi *et al.*, 1988). The APP gene, encompassing ~400 kilobases of DNA (Lamb *et al.*, 1993), gives rise to alternatively spliced APP mRNAs that encode Aβ-containing proteins of 695, 714, 751, and 770 amino acids (Kang *et al.*, 1987; Kitaguchi *et al.*, 1988; Ponte *et al.*, 1988; Tanzi *et al.*, 1988). APP is matured through the constitutive secretory pathway and is modified by the addition of both N- and O-linked carbohydrates, phosphate, and sulfate moieties (Oltersdorf *et al.*, 1989; Weidemann *et al.*, 1989; Hung and Selkoe, 1994). Varying levels of newly synthesized APP molecules appear at the cell surface (Weidemann *et al.*, 1989; Haass *et al.*, 1992a; Sisodia, 1992); some of these molecules are cleaved endoproteolytically by APP 'α-secretase' within the Aβ sequence (Esch *et al.*, 1990; Sisodia *et al.*, 1990; Anderson *et al.*, 1991; Wang *et al.*, 1991) to release the ectodomain of APP, including residues 1–16 of Aβ, into the medium. Thus, APP cleavage within the Aβ domain precludes the formation of Aβ.

The presence of secreted APP isoforms that contain Aβ epitopes in cerebrospinal fluid suggests that similar processing events occur *in vivo* (Palmert *et al.*, 1989a, b, 1990; Weidemann *et al.*, 1989). A fraction of cell-surface APP is also internalized and degraded via endosomal-lysosomal pathways (Cole *et al.*, 1989; Golde *et al.*, 1992; Haass *et al.*, 1992a). Processing via the endosomal–lysosomal pathway results in the production of fragments that contain the entire Aβ region and APP C-terminus and are, hence, potentially amyloidogenic (Golde *et al.*, 1992; Haass *et al.*, 1992a). Recent reports indicate that peptides similar to Aβ (β1–40) and truncated forms of Aβ (β17–40) are secreted constitutively by primary and tissue culture cells (Haass *et al.*, 1992b; Seubert *et al.*, 1992; Shoji *et al.*, 1992; Busciglio *et al.*, 1993) and are present in cerebrospinal fluid (Shoji *et al.*, 1992; Seubert *et al.*, 1993). The cellular and molecular mechanisms of Aβ generation are not clearly understood, but several recent studies have shed some light on the process. For example, despite excitement created by the discovery of potential amyloidogenic fragments generated in endosomal–lysosomal pathways, several lines of evidence now suggest that the lysosomal degradation of APP is unlikely to contribute to the production of Aβ. Kinetic studies show that: Aβ is released in parallel to APP_s (Busciglio *et al.*, 1993; Haass *et al.*, 1993); Aβ is not detected in purified lysosomes (Haass *et al.*, 1992 b); Aβ production is not inhibited by leupeptin, an inhibitor of lysosomal protease function (Haass *et al.*, 1992b); and, finally, Aβ is released by cultured fibroblasts from patients with I-cell disease, in which proteases fail to target to lysosomes (Haass *et al.*, 1993). However, agents that interfere with pH gradients (i.e. ammonium chloride and chloroquine) inhibit the production of Aβ (Haass *et al.*, 1992b; Shoji *et al.*, 1992), suggesting that Aβ may be generated in acidic compartments (i.e. endosomes or late Golgi). Furthermore, diminished levels of soluble Aβ are released from cells that express APP deleted of the entire cytoplasmic tail (APPΔC) (Citron *et al.*, 1992; Haass *et al.*, 1993). These results suggest indirectly that the reinternalization of APP from the cell surface may favour the generation of Aβ. Studies showing Aβ generation from surface-labelled APP support this model (Koo, 1993; Koo and Squazzo, 1994). Although these studies suggest that Aβ may be produced and released *in vitro* and *in vivo*, the relationship of these Aβ-related fragments to deposits of Aβ isolated from AD (which is principally 42–43 amino acids) is not clear (see below).

10.2.2 APP mutations

Autosomal dominant linkage to missense mutations of APP has been demonstrated in a relatively small subset of patients with early-onset familial AD (Goate *et al.*, 1991; Mullan *et al.*, 1992). In ~12 early-onset pedigrees, missense mutations generated amino-acid substitutions at residue 717 (of APP_{770}) within the transmembrane domain of APP (Chartier-Harlin *et al.*, 1991; Goate *et al.*, 1991; Naruse *et al.*, 1991). In one newly described family with a mutation at codon 692 of APP that resulted in a Gly–Ala substitution, biopsies from affected individuals disclosed diffuse deposits of Aβ, congophilic angiopathy, and scattered senile plaques but no NFT (Hendricks *et al.*, 1992). Finally, in two large, related families from Sweden (Mullan *et al.*, 1992), a double mutation at codons 670 and 671 that resulted in the substitution of Lys–Met to Asn-Leu was linked to early-onset AD.

Cellular transfection approaches have provided considerable insight regarding the mechanism whereby mutations in APP affect the processing of the precursor protein and contribute to Aβ production. For example, tissue culture cells expressing APP harbouring the 'Swedish' substitutions secrete higher levels of Aβ-containing peptides as compared to cells expressing wild-type constructs (Citron *et al.*, 1992; Cai *et al.*, 1993). Additional studies of cells that express APP harbouring the Ala-Gly substitution at amino acid 692 (Hendricks *et al.*, 1992) reveal that α-secretase processing of this mutant polypeptide is inefficient and that secreted Aβ species exhibit considerable microheterogeneity including the appearance of more hydrophic species (Haass *et al.*, 1994). Finally, recent studies have demonstrated that cells expressing APP harbouring '717' substitutions do not appear to secrete higher levels of Aβ but, rather, secrete a higher fraction of longer Aβ peptides (i.e. extending to Aβ residue 42) relative to cells that express wild-type APP (Suzuki *et al.*, 1994). Furthermore, because cells that express APP harbouring the Swedish mutation secrete higher levels of Aβ, the level of secreted Aβ1–42 peptides increases accordingly (Suzuki *et al.*, 1994). The finding that APP harbouring 717 mutations is processed to generate higher levels of longer Aβ peptides is of considerable interest, because recent physicochemical studies have indicated that amyloid formation is a nucleation-dependent phenomenon and that the C-terminus may be a critical determinant of the rate of amyloid formation (Jarrett and Lansbury, 1993; Jarrett *et al.*, 1993). These studies argue that Aβ1–42 and/or Aβ1–43, rather than Aβ1–40, may be the pathogenic proteins in AD. It is possible that the formation of Aβ1–40 can be 'seeded' by trace amounts of Aβ1–42 fibrils (Jarrett and Lansbury; Jarrett *et al.*, 1993). In any event, the observation that APP mutations in early-onset familial AD invariably flank the Aβ sequence suggests that altered processing of APP is central to the formation of amyloid in these individuals (see below). In support of the latter physicochemical studies, elegant immunocytochemical studies of human AD cortex performed with antibodies uniquely specific for Aβ1–40 or Aβ1–42 revealed that Aβ-deposition most likely begins with Aβ1–42(43) and not with Aβ1–40 (Iwatsubo *et al.*, 1994).

10.3 Animal models of AD

10.3.1 Aged nonhuman primates

Aged nonhuman primates develop age-associated impairments in performance on cognitive and memory tasks early in the third decade of life (Presty *et al.*, 1987; Rapp and Amaral, 1989, 1991; Bachevalier *et al.*, 1991). These impairments in performance on specific behavioural tasks are thought to be associated, in individual animals, with the deposition of Aβ (Wisniewski and Terry, 1973; Struble *et al.*, 1985; Selkoe *et al.*, 1987) and the formation of neurites (Struble *et al.*, 1982; Kitt *et al.*, 1984; Cork *et al.*, 1990; Martin *et al.*, 1991). In many old animals, Aβ appears as diffuse deposits, in the cores of senile plaques, and around blood vessels (Wisniewski and Terry, 1973; Struble *et al.*, 1982, 1985; Selkoe *et al.*, 1987; Walker *et al.*, 1987, 1988, 1990; Abraham *et al.*, 1989; Cork *et al.*, 1990). The earliest lesions in the parenchyma are the presence of slightly enlarged neurites (i.e. distal axons, nerve terminals, and dendrites) and pre-amyloid deposits (Struble *et al.*, 1982, 1985; Selkoe *et al.*, 1987; Walker *et al.*, 1988; Cork *et al.*, 1990; Martin *et al.*, 1991). Neurites often contain membranous elements, mitochondria (some degenerating), lysosomes, APP, phosphorylated neurofilaments, synaptophysin and transmitter markers (Martin *et al.*, 1991). In individual plaques, APP- and synaptophysin-immunoreactive neurites are often surrounded by a halo of distorted neuropil and Aβ immunoreactivity (Martin *et al.*, 1991). The presence of APP-like immunoreactivity in neuronal perikarya, axons and neurites within Aβ-containing plaques suggests that neurons can serve as one source for some Aβ deposits. In addition, the proximity of Aβ to reactive astrocytes and microglia, and to vascular cells suggests that these populations of nonneuronal cells may participate in the formation of Aβ (Abraham *et al.*, 1988, 1989; Wisniewski *et al.*, 1989; Wegiel and Wisniewski, 1990; Koo *et al.*, 1991; Wisniewski and Wegiel, 1991; Strittmatter *et al.*,

1993). Neurites appear to represent disconnected axons and swollen dendrites. Research focusing on Aβ amyloidogenesis has focused upon understanding the interactions of cells (i.e. neurons, astroglia, astrocytes and vascular cells), colocalized proteins [i.e. α_1-antichymotrypsin (Abraham *et al.*, 1988, 1989; Pasternack *et al.*, 1989; Koo *et al.*, 1991), apolipoprotein E (Strittmatter *et al.*, 1993)], SP40,40 (apolipoprotein J) (Ghiso *et al.*, 1993), and constituents of the complement cascade (McGeer *et al.*, 1989; Johnson *et al.*, 1992).

10.3.2 Transgenic mice

Transgenic approaches can directly test whether the expression of exogenous wild-type or mutant APP, or the expression of Aβ-containing fragments, is involved in the pathogenesis of AD-type abnormalities. Over the past few years, several groups have attempted to produce mice with Aβ deposits using cDNA and yeast artificial chromosome (YAC)-based transgenic technologies. Some of these efforts are reviewed below. Unfortunately, although a variety of transgenic lines have been developed, with one possible exception (Quon *et al.*, 1991), these animals have not developed the brain abnormalities characteristic of AD. Initial transgenic studies (Wirak *et al.*, 1991) attempted to overexpress the human Aβ peptide or the C-terminal 100 amino acids of APP (Kawabata *et al.*, 1991). In the first of these studies (Wirak *et al.*, 1991), clusters of Aβ-immunoreactive structures were visualized in the hippocampus, but subsequent analyses (Jucker *et al.*, 1992) revealed that these lesions were age-related intracytoplasmic inclusions that occur independently of the transgene. In the second study (Kawabata *et al.*, 1991), animals were reported to show neuritic plaques, NFT, amyloid deposition and neuronal degeneration. However, failure to replicate the initial neuropathological abnormalities resulted in retraction of this paper (Kawabata *et al.*, 1992). More recently, animals with a transgene encoding the C-terminal 100 amino acids of APP, driven by the brain dystrophin promotor, were reported to develop Aβ-immunoreactive deposits in cell bodies and neuropil as well as abnormal neurites and neuronal degeneration (Kammesheidt *et al.*, 1992). However, failure to document transgene-derived polypeptide expression in these animals limits interpretation of the reported pathology.

On the basis of *in situ* hybridization studies that indicated an increase in levels of transcripts encoding APP_{751} in hippocampal neurons in AD (Johnson *et al.*, 1990), several investigators have attempted to produce transgenic mice that overexpress human APP_{751} in brain. The initial report describing transgenic mice expressing APP_{751} (Quon *et al.*, 1991) showed poorly resolved extracellular Aβ deposits and A68-immunoreactive processes in cortex. However, the authors provided no information pertaining to the absolute levels of transgene-derived mRNA or polypeptide products relative to endogenous species. Moreover, to date, these brain abnormalities have not been reproduced by other laboratories using transgenic approaches with similar constructs. More recent studies from this group (Higgins *et al.*, 1994) report that the murine deposits closely resemble the pre-amyloid deposits in the brains of young adults with Down's syndrome or patients with early-stage AD. The authors utilized a human Aβ-specific antibody and several other Aβ antibodies to visualize the diffuse deposits. In the absence of peptide competition studies or other biochemical controls for specificity of the immunological reagents, and without documentation of transgene expression levels, it has been difficult to interpret the significance of the study.

The finding that several missense mutations in APP are genetically linked to pedigrees with early-onset AD has led investigators to assess the phenotype of transgenic mice that overexpress mutant APP. In one such study, transgene encoding myc epitope-tagged human APP_{695} ($HuAPP_{695myc}$) or $HuAPP_{695myc}$ harbouring the APP_{717} mutation were placed under the transcriptional control of the hamster prion gene (PrP) promoter (Scott *et al.*, 1992), and several lines of transgenic mice were generated with these constructs. No developmental or pathological abnormalities were evident in 'wild-type' animals despite abundant $HuAPP_{695myc}$ expression in all neurons of the central nervous system; the level of total APP was elevated ~2.5-fold in wild-type lines. Remarkably, mice that expressed mutant APP at levels similar to or slightly higher than wild-type expressers showed markedly reduced life spans (50–100 days). Death was not preceded by wasting, could not be attributed to a specific injury, and was not accompanied by Aβ plaques or NFT (Borchelt *et al.*, 1994; Hsiao *et al.*, 1994). The biochemical or physiological bases for the premature death phenotype are presently uncertain.

The overexpression of APP, as occurs in individuals with Downs's syndrome, is thought to lead to the premature deposition of Aβ in the brain (Glenner and Wong, 1984; Masters *et al.*, 1985a, b; Giaccone *et al.*, 1989; Rumble *et al.*, 1989; Cork *et al.*, 1990). To mimic the trisomic APP dosage imbalance observed in individuals with Down's syndrome, a YAC containing ~650 kilobases of human genomic DNA, including the APP gene, was transfected into embryonic stem (ES) cells (Lamb *et al.*, 1993). ES cells that contain stably integrated YAC DNA were microinjected into mouse blastocysts and chimeric mice were generated. After breeding, it was established that human APP sequences were transmitted to the mouse germline. Furthermore, human APP mRNA is actively transcribed in mouse tissue and the splicing pattern of human APP transcripts in transgenic mouse tissue mirrored the endogenous pattern of alternatively spliced mRNA. Using antibodies specific for human APP, Western blot analysis of transgenic mouse brain extracts revealed that human APP contributed ~40% of total APP levels. No AD-type pathology was demonstrable in young (<2 months) animals. Furthermore, no pathology has been observed in animals as old as 14 months (Kitt, C.A., Lamb, B.T., Price, D.L. and Gearhart, J.D., unpublished observations). Ongoing additional breeding approaches are intended to increase human APP gene copy number and, hence, APP levels. The YAC-ES strategy can be used to introduce modified human APP YAC that encode familial AD mutations into the mouse germline and to determine whether the presence of these mutations predisposes to Aβ deposition and, possibly, other brain abnormalities that occur in individuals with AD. Moreover, these mice produced by transgenic strategies will allow analyses of the sequential biochemical, cellular and molecular pathologies characteristic of early-onset familial AD.

Acknowledgements

The authors thank Drs Cheryl A. Kitt, John D. Gearhart, Bruce T. Lamb, Philip Wong, Lary C. Walker, Lee J. Martin, Gopal Thinakaran and Linda C. Cork for helpful discussions.

This work was supported by grants from the US Public Health Service (AG 05146, NS 20471) as well as the American Health Assistance Foundation and the Metropolitan Life Foundation. Dr Price is the recipient of a Leadership and Excellence in Alzheimer's Disease (LEAD) award (AG 07914) and a Javits Neuroscience Investigator Award (NS 10580).

References

Abraham, C.R., Selkoe, D.J. & Potter, H. (1988) *Cell* **52**, 487–501.

Abraham, C.R., Selkoe, D.J., Potter, H., Price, D.L. & Cork, L.C. (1989) *Neuroscience* **32**, 715–720.

Anderson, J.P., Esch, F.S., Keim, P.S., Sambamurti, K., Lieberburg, I. & Robakis, N.K. (1991) *Neurosci. Lett.* **128**, 126–128.

Arnold, S.E., Hyman, B.T., Flory, J., Damasio, A.R. & Van Hoesen, G.W. (1991) *Cereb. Cortex* **1**, 103–116.

Bachevalier, J., Landis, L.S., Walker, L.C., Brickson, M., Mishkin, M., Price, D.L. & Cork, L.C. (1991) *Neurobiol. Aging* **12**, 99–111.

Borchelt, D.R., Shen, J., Johannsdottir, R., Kitt, C.A., Mojekwu, J.C., Thinakaran, G., Sisodia, S.S., Price, D.L., Carlson, G. & Hsiao, K.K. (1994) *Soc. Neurosci.* Abstract.

Braak, H. & Braak, E. (1991a) *Acta Neuropathol.* **82**, 239–259.

Braak, H. & Braak, E. (1991b) *Acta Neuropathol.* **81**, 261–268.

Busciglio, J., Gabuzda, D.H., Matsudaira, P. & Yankner, B.A. (1993) *Proc. Natl Acad. Sci. USA* **90**, 2092–2096.

Cai, X.-D., Golde, T.E. & Younkin, S.G. (1993) *Science* **259**, 514–516.

Chartier-Harlin, M.-C., Crawford, F., Houlden, H., Warren, A., Hughes, D., Fidani, L., Goate, A., Rossor, M., Roques, P., Hardy, J. & Mullan, M. (1991) *Nature* **353**, 844–846.

Citron, M., Oltersdorf, T., Haass, C., McConlogue, L., Hung, A.Y., Seubert, P., Vigo-Pelfrey, C., Lieberburg, I. & Selkoe, D.J. (1992) *Nature* **360**, 672–674.

Cole, G.M., Huynh, T.V. & Saitoh, T. (1989) *Neurochem. Res.* **14**, 933–939.

Cork, L.C., Masters, C., Beyreuther, K. & Price, D.L. (1990) *Am. J. Pathol.* **137**, 1383–1392.

Esch, F.S., Keim, P.S., Beattie, E.C., Blacher, R.W., Culwell, A.R., Oltersdorf, T., McClure, D. & Ward, P.J. (1990) *Science* **248**, 1122–1124.

Evans, D.A., Funkenstein, H.H., Albert, M.S., Scherr, P.A., Cook, N.R., Chown, M.J., Hebert, L.E., Hennekens, C.H. & Taylor, J.O. (1989) *JAMA* **262**, 2551–2556.

Ghiso, J., Matsubara, E., Koudinov, A., Choi-Miura, N.H., Tomita, M., Wisniewski, T. & Frangione, B. (1993) *Biochem. J.* **293**, 27–30.

Giaccone, G., Tagliavini, F., Linoli, G., Bouras, C., Frigerio, L., Frangione, B. & Bugiani, O. (1989) *Neurosci. Lett.* **97**, 232–238.

Glenner, G.G. & Wong, C.W. (1984) *Biochem. Biophys. Res. Commun.* **120**, 885–890.

Goate, A., Chartier-Harlin, M.-C., Mullan, M., Brown, J., Crawford, F., Fidani, L., Giuffra, L., Haynes, A., Irving, N., James, L., Mant, R., Newton, P., Rooke, K., Roques, P., Talbot, C., Pericak-Vance, M., Roses, A., Williamson, R., Rossor, M., Owen, M. & Hardy, J. (1991) *Nature* **349**, 704–706.

Golde, T.E., Estus, S., Younkin, L.H., Selkoe, D.J. & Younkin, S.G. (1992) *Science* **255**, 728–730.
Haass, C., Koo, E.H., Mellon, A., Hung, A.Y. & Selkoe, D.J. (1992a) *Nature* **357**, 500–503.
Haass, C., Schlossmacher, M.G., Hung, A.Y., Vigo-Pelfrey, C., Mellon, A., Ostaszewski, B.L., Lieberburg, I., Koo, E.H., Schenk, D., Teplow, D.B. & Selkoe, D.J. (1992b) *Nature* **359**, 322–325.
Haass, C., Hung, A.Y., Schlossmacher, M.G., Teplow, D.B. & Selkoe, D.J. (1993) *J. Biol. Chem.* **268**, 3021–3024.
Haass, C., Hung, A.Y., Selkoe, D.J. & Teplow, D.B. (1994) *J. Biol. Chem.* **269**, 17741–17748.
Hendricks, L., van Duijn, C.M., Cras, P., Cruts, M., Van Hul, W., van Harskamp, F., Warren, A., McInnis, M.G., Antonarakis, S.E., Martin, J.-J., Hofman, A. & Van Broeckhoven, C. (1992) *Nature Genetics* **1**, 218–221.
Higgins, L.S., Holtzman, D.M., Rabin, J., Mobley, W.C. & Cordell, B. (1994) *Ann. Neurol.* **35**, 598–607.
Hsiao, K.K., Olson, K., Borchelt, D.R., Johannsdottir, R., Shen, Z., Yunis, W.S., Xu, X., Kitt, C.A., Diedrich, J., Sisodia, S.S., Carlson, G., Price, D., Clark, H.B. & Iadecola, C. (1994) Fourth International Conference on Alzheimer's Disease and Related Disorders, Minneapolis, Minn. July 29–Aug. 3, 1994.
Hung, A.Y. & Selkoe, D.J. (1994) *EMBO J.* **13**, 534–542.
Hyman, B.T., Van Hoesen, G.W., Damasio, A.R. & Barnes, C.L. (1984) *Science* **225**, 1168–1170.
Iwatsubo, T., Odaka, A., Suzuki, N., Mizusawa, H., Nukina, N. & Ihara, Y. (1994) *Neuron* **13**, 45–53.
Jarrett, J.T. & Lansbury, P.T., Jr (1993). *Cell* **73**, 1055–1058.
Jarrett, J.T., Berger, E.P. & Lansbury, P.T., Jr (1993) *Biochemistry* **32**, 4693–4697.
Johnson, S.A., McNeill, T., Cordell, B. & Finch, C.E. (1990) *Science* **248**, 854–857.
Johnson, S.A., Lampert-Etchells, M., Pasinetti, G.M., Rozovsky, I. & Finch, C.E. (1992) *Neurobiol. Aging* **13**, 641–648.
Jucker, M., Walker, L.C., Martin, L.J., Kitt, C.A., Kleinman, H.K., Ingram, D.K. & Price, D.L. (1992) *Science* **255**, 1443–1445.
Kammesheidt, A., Boyce, F.M., Spanoyannis, A.F., Cummings, B.J., Ortegón, M., Cotman, C., Vaught, J.L. & Neve, R.L. (1992) *Proc. Natl Acad. Sci. USA* **89**, 10857–10861.
Kang, J., Lemaire, H.-G., Unterbeck, A., Salbaum, J.M., Masters, C.L., Grzeschik, K.-H., Multhaup, G., Beyreuther, K. & Müller-Hill, B. (1987) *Nature* **325**, 733–736.
Kawabata, S., Higgins, G.A. & Gordon, J.W. (1991) *Nature* **354**, 476–478.
Kawabata, S., Higgins, G.A. & Gordon, J.W. (1992) *Nature* **356**, 23.
Khachaturian, Z. (1985) *Arch. Neurol.* **42**, 1097–1105.
Kitaguchi, N., Takahashi, Y., Tokushima, Y., Shiojiri, S. & Ito, H. (1988) *Nature* **331**, 530–532.
Kitt, C.A., Price, D.L., Struble, R.G., Cork, L.C., Wainer, B.H., Becher, M.W. & Mobley, W.C. (1984) *Science* **226**, 1443–1445.
Koo, E.H. (1993) 52–55.
Koo, E.H. & Squazzo, S.L. (1994) *J. Biol. Chem.* **269**, 17386–17389.
Koo, E.H., Abraham, C.R., Potter, H., Cork, L.C. & Price, D.L. (1991) *Neurobiol. Aging* **12**, 495–501.
Lamb, B.T., Sisodia, S.S., Lawler, A.M., Slunt, H.H., Kitt, C.A., Kearns, W.G., Pearson, P.L., Price, D.L. & Gearhart, J.D. (1993) *Nature Genet.* **5**, 22–30.
Martin, L.J., Sisodia, S.S., Koo, E.H., Cork, L.C., Dellovade, T.L., Weidemann, A., Beyreuther, K., Masters, C. & Price, D.L. (1991) *Proc. Natl Acad. Sci. USA* **88**, 1461–1465.
Masters, C.L., Multhaup, G., Simms, G., Pottgiesser, J., Martins, R.N. & Beyreuther, K. (1985a) *EMBO J.* **4**, 2757–2763.
Masters, C.L., Simms, G., Weinman, N.A., Multhaup, G., McDonald, B.L. & Beyreuther, K. (1985b) *Proc. Natl Acad. Sci. USA* **82**, 4245–4249.
McGeer, P.L., Akiyama, H., Itagaki, S. & McGeer, E.G. (1989) *Neurosci. Lett.* **107**, 341–346.

McKhann, G., Drachman, D., Folstein, M., Katzman, R., Price, D. & Stadlan, E.M. (1984) *Neurology* **34**, 939–944.

Mullan, M., Crawford, F., Axelman, K., Houlden, H., Lillius, L., Winblad, B. & Lannfelt, L. (1992) *Nature Genet.* **1**, 345–347.

Naruse, S., Igarashi, S., Kobayashi, H., Aoki, K., Inuzuka, T., Kaneko, K., Shimizu, T., Iihara, K., Kojima, T., Miyatake, T. & Tsuji, S. (1991) *Lancet* **337**, 978–979.

Oltersdorf, T., Fritz, L.C., Schenk, D.B., Lieberburg, I., Johnson-Wood, K.L., Beattie, E.C., Ward, P.J., Blacher, R.W., Dovey, H.F. & Sinha, S. (1989) *Nature* **341**, 144–147.

Palmert, M.R., Podlisny, M.B., Witker, D.S., Oltersdorf, T., Younkin, L.H., Selkoe, D.J. & Younkin, S.G. (1989a) *Proc. Natl Acad. Sci. USA* **86**, 6338–6342.

Palmert, M.R., Siedlak, S.L., Podlisny, M.B., Greenberg, B., Shelton, E.R., Chan, H.W., Usiak, M., Selkoe, D.J., Perry, G. & Younkin, S.G. (1989b) *Biochem. Biophys. Res. Commun.* **165**, 182–188.

Palmert, M.R., Usiak, M., Mayeux, R., Raskind, M., Tourtellotte, W.W. & Younkin, S.G. (1990) *Neurology* **40**, 1028–1034.

Pasternack, J.M., Abraham, C.R., Van Dyke, B.J., Potter, H. & Younkin, S.G. (1989) *Am. J. Pathol.* **135**, 827–834.

Ponte, P., Gonzalez-DeWhitt, P., Schilling, J., Miller, J., Hsu, D., Greenberg, B., Davis, K., Wallace, W., Lieberburg, I., Fuller, F. & Cordell, B. (1988) *Nature* **331**, 525–532.

Presty, S.K., Bachevalier, J., Walker, L.C., Struble, R.G., Price, D.L., Mishkin, M. & Cork, L.C. (1987) *Neurobiol. Aging* **8**, 435–440.

Probst, A., Langui, D., Ipsen, S., Robakis, N. & Ulrich, J. (1991) *Acta Neuropathol.* **83**, 21–29.

Quon, D., Wang, Y., Catalano, R., Scardina, J.M., Murakami, K. & Cordell, B. (1991) *Nature* **352**, 239–241.

Rapp, P.R. & Amaral, D.G. (1989) *J. Neurosci.* **9**, 3568–3576.

Rapp, P.R. & Amaral, D.G. (1991) *Neurobiol. Aging* **12**, 481–486.

Rumble, B., Retallack, R., Hilbich, C., Simms, G., Multhaup, G., Martins, R., Hockey, A., Montgomery, P., Beyreuther, K. & Masters, C.L. (1989) *N. Engl. J. Med.* **320**, 1446–1452.

Scott, M.R., Köhler, R., Foster, D. & Prusiner, S.B. (1992) *Protein Sci.* **1**, 986–997.

Selkoe, D.J., Bell, D.S., Podlisny, M.B., Price, D.L. & Cork, L.C. (1987) *Science* **235**, 873–877.

Seubert, P., Vigo-Pelfrey, C., Esch, F., Lee, M., Dovey, H., Davis, D., Sinha, S., Schlossmacher, M., Whaley, J., Swindlehurst, C., McCormack, R., Wolfert, R., Selkoe, D., Lieberburg, I. & Schenk, D. (1992) *Nature* **359**, 325–327.

Seubert, P., Oltersdorf, T., Lee, M.G., Barbour, R., Blomquist, C., Davis, D.L., Bryant, K., Fritz, L.C., Galasko, D., Thal, L.J., Lieberburg, I. & Schenk, D.B. (1993) *Nature* **361**, 260–263.

Shoji, M., Golde, T.E., Ghiso, J., Cheung, T.T., Estus, S., Shaffer, L.M., Cai, X.-D., McKay, D.M., Tintner, R., Frangione, B. & Younkin, S.G. (1992) *Science* **258**, 126–129.

Sisodia, S.S. (1992) *Proc. Natl Acad. Sci. USA* **89**, 6075–6079.

Sisodia, S.S., Koo, E.H., Beyreuther, K., Unterbeck, A. & Price, D.L. (1990) *Science* **248**, 492–495.

Strittmatter, W.J., Saunders, A.M., Schmechel, D., Pericak-Vance, M., Enghild, J., Salvesen, G.S. & Roses, A.D. (1993) *Proc. Natl Acad. Sci. USA* **90**, 1977–1981.

Struble, R.G., Cork, L.C., Whitehouse, P.J. & Price, D.L. (1982) *Science* **216**, 413–415.

Struble, R.G., Price, Jr., Cork, L.C. & Price, D.L. (1985) *Brain Res.* **361**, 267–275.

Suzuki, N., Cheung, T.T., Cai, X.-D., Odaka, A., Otvos, L., Jr., Eckman, C., Golde, T.E. & Younkin, S.G. (1994) *Science* **264**, 1336–1340.

Tanzi, R.E., McClatchey, A.I., Lampert, E.D., Villa-Komaroff, L., Gusella, J.F. & Neve, R.L. (1988) *Nature* **331**, 528–530.

Terry, R.D., Masliah, E., Salmon, D.P., Butters, N., DeTeresa, R., Hill, R., Hansen, L.A. & Katzman, R. (1991) *Ann. Neurol.* **30**, 572–580.

Walker, L.C., Kitt, C.A., Schwam, E., Buckwald, B., Garcia, F., Sepinwall, J. & Price, D.L. (1987) *Neurobiol. Aging* **8**, 291–296.

Walker, L.C., Kitt, C.A., Cork, L.C., Struble, R.G., Dellovade, T.L. & Price, D.L. (1988) *Neuropathol. Exp. Neurol.* **47**, 138–144.
Walker, L.C., Masters, C., Beyreuther, K. & Price, D.L. (1990) *Acta Neuropathol.* **80**, 381–387.
Wang, R., Meschia, J.F., Cotter, R.J. & Sisodia, S.S. (1991) *J. Biol. Chem.* **266**, 16960–16964.
Wegiel, J. & Wisniewski, H.M. (1990) *Acta Neuropathol.* **81**, 116–124.
Weidemann, A., König, G., Bunke, D., Fischer, P., Salbaum, J.M., Masters, C.L. & Beyreuther, K. (1989) *Cell* **57**, 115–126.
Whitehouse, P.J., Price, D.L., Struble, R.G., Clark, A.W., Coyle, J.T. & DeLong, M.R. (1982) *Science* **215**, 1237–1239.
Wirak, D.O., Bayney, R., Ramabhadran, T.V., Fracasso, R.P., Hart, J.T., Hauer, P.E., Hsiau, P., Pekar, S.K., Scangos, G.A., Trapp, B.D. & Unterbeck, A.J. (1991) *Science* **253**, 323–325.
Wisniewski, H.M. & Terry, R.D. (1973) In H.M. Zimmerman (Ed.) *Progress in Neuropathology.* New York, Grune & Stratton.
Wisniewski, H.M. & Wegiel, J. (1991) *Neurobiol. Aging* **12**, 593–600.
Wisniewski, H.M., Wegiel, J., Wang, K.C., Kujawa, M. & Lach, B. (1989) *Can. J. Neurol. Sci.* **16**, 535–542.

CHAPTER 11

RELATIONSHIP OF PLAQUES AND TANGLES TO ALZHEIMER'S DISEASE PHENOTYPE

John C. Morris

Washington University School of Medicine, 660 South Euclid Avenue, St Louis, Missouri 63110, USA

Table of Contents

11.1 Introduction

At a psychiatric meeting in southwestern Germany on 3–4 November 1906, Alois Alzheimer reported clinicopathological findings in a woman who developed dementia at age 51 years. Publication of the case report the following year (Alzheimer, 1907) led to lively debates concerning functional versus organic bases for mental illnesses (Torack, 1979) and whether 'senium praecox', later designated as Alzheimer's disease, was an entity distinct from senile dementia (Bick, 1994). Underlying these arguments were uncertainties about the relationship of senile plaques and neurofibrillary tangles, the neuropathological lesions reported by Alzheimer, to the clinical symptoms seen in dementing disease and in ageing. Lacking a clear understanding of the pathogenetic role and disease specificity of these abnormalities, Alzheimer cogently predicted that in mental diseases, '. . . histological examination must be effected to determine the characteristics of each single case. We must reach the stage in which the vast well-known disease groups must be subdivided into many smaller groups, each with its own clinical and anatomical characteristics.' (Alzheimer, 1907; translation in Bick and Amaducci, 1989). Almost 90 years later, the central question encountered by Alzheimer and his contemporaries of how plaques and tangles relate to basic

Pathobiology of Alzheimer's Disease
ISBN 0-12-286965-5

mechanisms of Alzheimer's disease (AD) and its features still is unresolved. After reviewing the clinical phenotype of AD, this chapter will discuss the importance of plaques and tangles for the histopathological diagnosis of the disorder and their correlation with dementia severity and other clinical features of AD.

11.2 Clinical phenotype

The clinical features of AD are pleiomorphic and in the early stages overlap the changes associated with normal ageing. In the absence of a validated biological marker for the disease, clinical diagnosis rests first on establishing the presence of dementia. The probability that dementia is caused by AD then is determined by the exclusion of other known causes of dementia and by the degree to which the clinical presentation and course are characteristic of known cases of AD.

11.2.1 Differentiation of dementia from ageing: clinical distinctions

Dementia is a symptom of brain dysfunction and can be caused by many factors. It is defined as 'the development of multiple cognitive deficits that include memory impairment and at least one of the following cognitive disturbances: aphasia, apraxia, agnosia or a disturbance in executive functioning' (American Psychiatric Association, 1994). Operationally, a demented person performs everyday activities less well than before because of diminished cognitive ability.

Diagnosis of dementia often is greatly facilitated by information provided by a collateral source, generally the spouse or adult child of the patient, who can address changes in the patient's cognitive function. The level of premorbid cognitive performance may vary greatly among individuals because of differences in native intelligence, educational attainment and accustomed cognitive demand (e.g. demand generally is greater for patients who are still employed versus those who have retired). Although clinical and neuropsychological procedures can quantify cognitive function at the time of examination, a collateral source can best provide the historical observations of how function has changed relative to past performance. Collateral source information about these changes can be sensitive to even very mild impairment (Morris *et al.*, 1991; Koss *et al.*, 1993) and can be more reliable than the patient's own memory complaints, which tend to correlate better with personality traits and depression than with dementia (Jorm, 1988; Bolla *et al.*, 1991; Hänninen *et al.*, 1994). It has been stated that 'the patient's self-report of cognitive function bears no relation to cognitive performance whereas the informant's judgement provides a reliable guide' (Evans, 1988).

Quantitative measures of cognitive performance are used to: (i) provide objective documentation of impaired cognition and thus support the clinical judgement that dementia is present; (ii) longitudinally assess cognitive changes; and (iii) identify patterns of selective cognitive deficits for elucidation of brain–behaviour relationships. Brief cognitive tests allow the rapid assessment of a limited number of cognitive domains for screening and follow-up purposes. Commonly used brief measures include the informant-based Dementia Scale of Blessed and colleagues (Blessed *et al.*, 1968), the Information–Memory–Concentration test of the patient in both its original (Blessed *et al.*, 1968) and shortened (Katzman *et al.*, 1983) versions, and the Mini-Mental State (Folstein *et al.*, 1975). Full-scale neuropsychological batteries systematically evaluate the major cognitive functions affected by dementia: primary and secondary memory (verbal and nonverbal), language, visuospatial function, problem solving and sequencing ability, attention and psychomotor performance, including timed tasks (Salmon and Butters, 1992; Schmitt and Sano, 1994). Quantitative cognitive measures, however, are limited by several factors, including the influence of age (Rubin *et al.*, 1993), and race, ethnicity and education (Tombaugh and McIntyre, 1992). Moreover, they may be insensitive to very mild stages of dementia where substantial individual variability in performance overlaps the distributions of normal elderly persons (Storandt and Hill, 1989; Hansen *et al.*, 1988; Morris *et al.*, 1991). The diagnosis of dementia, therefore, is based on the informant-based clinical judgement that cognitive decline is sufficient to interfere with everyday functioning and not simply on cognitive test performance.

As noted above, cognitive changes associated with normal ageing can complicate the diagnosis of dementia, particularly in its mildest forms. The distinction is further compounded by heteroscedasticity, or the wide individual variation among elderly persons in physiological and psychological function (Rubin *et al.*, 1993; Morse, 1993), and by the relative paucity of prospective comprehensive studies of intellectual function with age. Lifestyle influences (e.g. exercise) and cohort effects (secular differences, such as nutrition, in generations) can further blur consideration of what is truly normal for age. A pertinent factor in the interpretation of longitudinal data from control research participants is selective attrition, in which 'at risk' subjects are progressively less likely to be available for evaluation; the residual sample increasingly consists of 'supernormals' (Perls *et al.*, 1993). The terms 'usual' and 'successful' ageing have been suggested to partially address this issue and differentiate persons with little or no loss of function ('successful ageing') from those with clinically evident age-associated changes, such as arthritis or mild cognitive deficits ('usual ageing') (Rowe and Kahn, 1987). The difficulty in defining a 'normal' range of performance in older adults is reflected in the statement that 'to draw a distinction between disease and normal ageing is to attempt to separate the undefined from the undefinable' (Evans, 1988).

With the above caveats in mind, normal ageing generally is characterized by decline in some aspects of cognition but not in others (Craik, 1991). Measures of secondary memory or 'learning' typically show age-related deterioration, particularly when retrieval of the learned material (e.g. a person's name) is self-initiated. Use of environmental guides or other cues can minimize this deficit (Petersen *et al.*, 1992; Verhaeghen *et al.*, 1993), which has been interpreted as reflecting an age-related reduction in cerebral processing resources needed for encoding and retrieval of information (Craik, 1991). This interpretation is consistent with an adverse effect of age on 'working memory', in which certain information must be 'held' in memory during performance of another cognitive activity. Increasing age may be associated with progressive reductions in the pool of processing resources needed to maintain the desired information during the competing task (Salthouse and Skovronek, 1992). Additional deficits with age are seen for divided attention, visuospatial abilities, language and psychomotor speed (Howieson *et al.*, 1993; Rubin *et al.*, 1993), but tasks of primary memory ('immediate' memory, where the material to be assessed remains in conscious awareness), implicit memory (unconscious memory, such as priming) and remote memory (for facts or events of years ago) are relatively unaffected by age.

In intellectually healthy older adults, age-associated decrements in secondary (recall) and working memory correspond to complaints of poor retrieval of names of familiar persons and limited capacity to keep several things in mind simultaneously ('I forgot what I came here to get'). These changes may be more apparent with increased extraneous demands on resources (job, family, social activities); typically, the items to be recalled are trivial and have merited little attention or emotional charge. These age-associated changes generally do not interfere substantively with everyday life, especially when compensatory strategies (e.g. making lists) are used. Self-reported memory complaints of normal older persons usually are not associated with progressive cognitive decline (Flicker *et al.*, 1993). The neurobiological basis for senescent memory changes may be neuronal degeneration in the hippocampus, entorhinal cortex and other medial temporal lobe regions. These structures have been implicated in both normal and disordered memory function (Hyman *et al.*, 1990; Squire and Zola-Morgan, 1991).

11.2.2 Clinical diagnostic criteria for dementia of the Alzheimer type

Standardization of diagnostic criteria for AD has notably improved diagnostic accuracy and allowed meaningful comparison of results from therapeutic trials and other clinical investigations. In the United States, two compatible sets of clinical criteria have been widely used. The Work Group convened by the National Institute of Neurological and Communicative Disorders and Stroke (NINCDS) and the Alzheimer's Disease and Related Disorders Association (ADRDA) published criteria in 1984 that established three levels of confidence for the diagnosis of AD (McKhann *et al.*, 1984). In a person with dementia (defined as documented deficits in at least two cognitive areas in an alert patient), *probable AD* is present 'if there is a typical insidious onset of dementia with progression and if there are no other systemic or brain diseases that could account for the progressive memory and other cognitive deficits' (McKhann *et al.*, 1984). *Possible AD* is diagnosed when there are variations in the presentation or course of dementia, such as early and disproportionate language disturbance, or when another potentially dementing disorder (e.g. stroke) is present but is not believed to be responsible for the dementia. *Definite AD* is reserved for probable AD patients in whom the histopathological lesions of AD are confirmed by biopsy or autopsy. The NINCDS–ADRDA criteria for probable AD are similar to revised criteria for dementia of the Alzheimer type (DAT) as recently published in the *Diagnostic and Statistical Manual of Mental Disorders*, 4th edition (DSM-IV) (American Psychiatric Association, 1994). In patients in whom other causes of dementia have been ruled out, DAT is defined as the gradual onset and progression of cognitive deficits, including memory impairment, sufficiently severe to 'cause significant impairment in social or occupational functioning and represent a decline from a previous level of functioning' (American Psychiatric Association, 1994). In a demented person without known cause, the salient clinical diagnostic feature for AD in both the NINCDS-ADRDA and DSM-IV criteria is the gradual onset and progression of dementia.

With the recognition that the two terms largely are interchangeable, for simplicity and consistency 'DAT' rather than 'probable AD' is chosen here to designate the clinical syndrome of AD.

11.2.3 Phenomenology of DAT

In addition to gradual onset and progression of dementia, other clinical features also are characteristic of DAT. A description of the usual picture must be tempered, however, by acknowledging that the manifestations of the disorder are rich and varied and can result in markedly different presentations and rates of progression in individual patients.

Memory impairment is the cardinal symptom of DAT. Minor forgetfulness begins so insidiously that it is difficult, even in retrospect, to date the onset precisely. It often takes several years for the memory changes to interfere sufficiently with everyday functioning and to be recognized by the family as serious enough to warrant evaluation. Common examples of mild memory impairment are the failure to recall details of recent conversations or events, misplacement of items without independent retrieval, and repetition of statements or questions. These examples indicate that there is an early defect in the acquisition and retrieval of new information, at least for events and facts. Temporal and geographical disorientation also is common in this stage and the patient may need directions to find even familiar locations. As dementia progresses, recent events are rapidly forgotten, sometimes within minutes of their occurrence, and recall of highly learned material (e.g. relationships of spouses or relatives) also is impaired such that not only is there an inability to process new information but stored memories eventually deteriorate as well.

Executive functioning, or the ability to inhibit inappropriate responses and select key information and behaviours for action (Oscar-Berman *et al.*, 1991) often is affected early in DAT, and results in poor judgement and problem solving. Inability to operate appliances (especially those that are complicated or new), increased distractibility, inappropriate or disinhibited behaviours, and reduced or absent insight (including into the presence of dementia) are common manifestations of the dysexecutive syndrome. Related symptoms include personality changes such as passivity, disinterest and withdrawal (Rubin *et al.*, 1987; Bozzola *et al.*, 1992). Impaired ability for calculations is manifested by difficulty in balancing the cheque book, figuring a tip and making change. Word-finding difficulty, hesitancy of speech and diminished verbal output are among the language disturbances associated with the early stages of DAT. Visuospatial dysfunction can also occur and may compound geographical disorientation.

With progression of DAT, day–night disorientation and other troublesome behaviours complicate the clinical picture. Restlessness and wandering, aggression (verbal and physical), suspiciousness, delusions (e.g. belief that misplaced items were stolen; accusations of spousal infidelity), and occasionally visual and auditory hallucinations can be very disruptive (Swearer *et al.*, 1988; Teri *et al.*, 1989; Lachs *et al.*, 1992) and may prompt institutionalization. As dementia progresses, language deteriorates further with impaired comprehension of written and spoken language, paraphasic errors and dysgraphia (Cummings *et al.*, 1985; Faber-Langendoen *et al.*, 1988). It is difficult for the patient to accomplish even simple tasks, such as washing dishes or setting the table, and increasing dependence on others occurs for performance of basic care functions such as dressing, eating, personal hygiene and toileting. With advanced DAT only fragments of memory remain, language declines to near mutism, and urinary and then faecal incontinence is present. Eventually, total dependence on caregivers is the rule. The terminal stage is marked by a bedridden, uncomprehending, vegetative state with swallowing difficulties and weight loss. Death usually results from inanition, aspiration, or complicating illnesses such as pulmonary embolus, pneumonia or urosepsis. Although highly variable, the total duration of DAT from onset until death averages 8–10 years (Walsh *et al.*, 1990).

Many aspects of DAT display a clear stage dependency. For example, disordered language increases in prevalence with dementia severity (Faber-Langendoen *et al.*, 1988), behavioural disturbances most often occur in moderate, as opposed to mild or severe, DAT (Drevets and Rubin, 1989), and many motor complications, including falls (Morris *et al.*, 1987), extrapyramidal dysfunction (Morris *et al.*, 1989; Chui *et al.*, 1994), and generalized seizures (Romanelli *et al.*, 1990), develop with increasing severity of dementia. Several clinical staging instruments provide a global rating of dementia severity and serve as meaningful measures of cognitive disability over the full range of DAT, in part because they are less subject to 'floor' and 'ceiling' effects than are neuropsychological tests (Berg *et al.*, 1992). Some widely used global scales include the Global Deterioration Scale (Reisberg *et al.*, 1982), the Cambridge Mental Disorders of the Elderly Examination (CAMDEX) (Roth *et al.*, 1986) and the Clinical Dementia Rating (Hughes *et al.*, 1982; Morris, 1993).

In summary, the clinical phenotype of AD consists of a symptom complex of slowly progressive cognitive and noncognitive dysfunction. Memory impairment is uniformly present and is combined with deficits in executive functioning, language, visuospatial abilities, personality, behaviour and self-care. Not all areas are equally impaired at the same time and wide interindividual variability exists for the expression of certain behaviours or symptoms.

11.2.4 Heterogeneity in DAT

Variations in the presentation and course of DAT are common. There is increasing support for the possibility that phenotypic variants of DAT reflect different aetiologies that share a final common pathway of cerebral degeneration that in turn produces the general clinical features of gradually progressive impairment of memory and other cognitive functions (Khachaturian, 1992). Selected aspects of the heterogeneity of DAT are presented here.

11.2.4.1 Age at onset

AD generally occurs after age 65 years. Its prevalence is strongly age-associated (Graves and Kukull, 1994) and increases exponentially in successively older age groups such that, by age 85 years, DAT may affect up to 47% of individuals (Evans *et al.*, 1989). However, patients with earlier ages of onset are well documented, including those in the fourth decade of life (Heyman *et al.*, 1983; Schoenberg *et al.*, 1987). Age at onset prior to age 65 years (arbitrarily designated 'early-onset' AD) has been associated with more rapid progression of dementia (Chui *et al.*, 1985; Huff *et al.*, 1987), greater language involvement (Selnes *et al.*, 1988), and increased frequency of seizures and myoclonus (Bird *et al.*, 1989); there also is the perception that early-onset cases more often are familial (Farrer *et al.*, 1990). A recent report, however, found no differences between early- and late-onset DAT patients in rate of cognitive decline (Haupt *et al.*, 1993). It is likely that age at onset modifies the clinical expression of the illness (e.g. dementia likely will be recognized earlier in a corporate executive in her 40s than in a retired person in his 80s). Until aetiological differences can be shown for the two forms, most investigators consider early- and late-onset AD to represent the same disorder.

11.2.4.2 Family history

Several types of inheritance have been proposed for at least some forms of AD. Inheritance patterns include Mendelian (single gene) autosomal dominant, chromosomal (e.g. Down's syndrome), mitochondrial and multifactorial (combination of genetic and environmental factors) (Bird, 1994). Genetic transmission clearly has been established for multigenerational families with autosomal dominant early-onset AD, with point mutations identified in the amyloid β-protein precursor gene on chromosome 21 (Goate *et al.*, 1991) and on another locus on chromosome 14 (Sherrington *et al.*, 1995); a particular form of familial AD has been localized to yet another gene on chromosome 1 (Levy-Lahad *et al.*, 1995). Such mutations account for only a very small number of all AD cases but contribute to evidence that early-onset familial AD is more likely to be autosomal dominant, whereas late-onset familial AD is more likely to demonstrate other inheritance patterns (e.g. shared polygenes; multifactorial) (Farrer *et al.*, 1990). Although multigenerational pedigrees of AD are rare, a familial occurrence (i.e. one or more affected first-degree relatives) can be shown for about 20–40% of all AD cases (Bird, 1994). Numerous epidemiological studies have demonstrated an increased risk for AD in individuals with affected relatives (Graves and Kukull, 1994), with a risk of 2–5 times that of the general population. One genetically determined risk factor for AD, mediated by allelic variants of apolipoprotein E, recently has been characterized in both familial and ostensibly nonfamilial ('sporadic') late onset AD (Saunders *et al.*, 1993; Strittmatter *et al.*, 1993). The apolipoprotein E gene is on chromosome 19; combined with the known loci on chromosomes 1, 14 and 21, it is clear that AD is genetically heterogeneous (Clark and Goate, 1993). Phenotypic heterogeneity also is likely in familial AD, not only for age at onset (i.e. early- versus late-onset) but also for the presence of unusual associated features such as amyotrophy and anterior horn cell loss (Bird *et al.*, 1989). In general, however, familial AD cases are identical to sporadic AD in the spectrum and frequency of clinical symptoms and signs, rate of decline, and mean duration of disease (Edwards *et al.*, 1991; Duara *et al.*, 1993; Haupt *et al.*, 1993).

11.2.4.3 'Focal' presentations

The classical approach to the understanding of brain function has been correlational studies of behavioural deficits in persons with circumscribed areas of damaged brain (Damasio and Damasio, 1989). It is now appreciated that, rather than localization to a single anatomic site, many complex psychological functions are subserved by large-scale, multifocal neural networks (Mesulam, 1990). Nonetheless, insight into brain–behaviour relationships still can be gained from the occasional patient who presents with a seemingly isolated and unexplained cognitive deficit; in many instances, the deficit is associated with corresponding focal cerebral atrophy on neuroimaging studies (Caselli *et al.*, 1992). Specific 'focal' syndromes include progressive aphasia (Mesulam and Weintraub, 1992), apraxia (Crystal *et al.*, 1982), several forms of visual agnosia, including Balint's syndrome (Mendez *et al.*, 1990), and behavioural (frontal lobe) disorders (Gustafson, 1987). These focally presenting cognitive deficits may gradually progress to global DAT and AD has been confirmed pathologically in some cases (Crystal *et al.*, 1982; Green *et al.*, 1990; Levine *et al.*, 1993; Victoroff *et al.*, 1994). Perhaps closely related to these cases are patients who are disproportionately impaired in language, visuospatial or frontal lobe functions in the early stages of DAT. Such occurrences illustrate that occasionally memory disturbance in DAT can be overshadowed by other cognitive dysfunction, at least initially, and predict that, in addition to memory structures, higher order association cortex may demonstrate selective vulnerability to the AD pathogenic process.

11.3 Plaques, tangles and Alzheimer's disease

11.3.1 Histopathology of AD

Macroscopic changes in AD are inconsistent and nondiagnostic. The diagnosis thus is established by histological criteria as described by Alzheimer: '[the examination

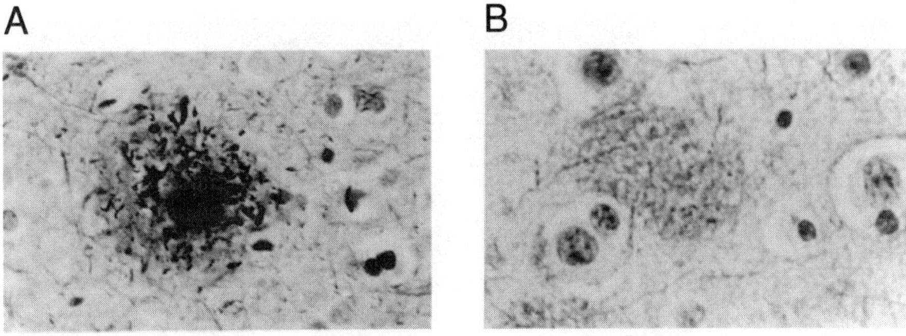

Figure 1 (A) Typical neuritic plaque with thickened neurites surrounding a central amyloid core in the neocortex of a patient with Alzheimer's disease (Bielschowsky stain, ×300). (B) Diffuse plaque without thickened neurites (Bielschowsky stain, ×300). Reproduced with permission from Mirra *et al.* (1993).

showed] an evenly affected atrophic brain without macroscopic foci . . . Numerous neurons, especially in upper cell layers, had totally disappeared. Dispersed over the entire cortex, and in large numbers especially in the upper layers, miliary foci could be found. . . . The Bielschowsky silver preparation showed very characteristic changes in the neurofibrils . . . they accumulated forming dense bundles.' (Alzheimer, 1907; translation in Bick and Amaducci, 1989). The 'miliary foci', which had been described in 1892 by Blocq and Marinesco, now are termed 'senile plaques' and the bundled neurofibrils are known as 'neurofibrillary tangles'. Both are argyrophilic and silver stains such as Bielschowsky's remain the principal method for their detection.

Senile plaques are extracellular and are found in two major forms – diffuse and neuritic (Figure 1). The amorphous diffuse plaques vary from 20 μm to 80 μm in diameter. Not infrequently, they are clustered and it can be difficult, particularly when small, to distinguish individual lesions. Diffuse plaques are immunoreactive for the AD-associated amyloid protein (Aβ) but contain little or no neuritic pathology (Schmidt *et al.*, 1994). In contrast, neuritic plaques are characterized by deposits of amyloid fibrils, often coalesced into a central amyloid core about which abnormally thickened axons or dendrites (neurites) are radially arranged to form a discrete spherical lesion averaging about 30 μm in diameter. Variations in size and in the presence of an amyloid core are common (Mirra and Gearing, 1994). The dystrophic neurites contain paired helical filaments (PHF) and various organelles, including lysosomes, mitochondria, and synaptic vesicles. Neuritic plaques typically are surrounded by astrocytes and microglia (Kalaria, 1993). In addition to Aβ, neuritic plaques are immunoreactive with many molecules, including tau, growth associated protein (GAP43), α-1-antichymotrypsin, apolipoprotein E, glycosaminoglycans and several neurotransmitters, including acetylcholine (Terry *et al.*, 1994). Both diffuse and neuritic plaques are distributed in varying proportions throughout AD neocortex and total plaque burden can be extensive (Figure 2).

Neurofibrillary tangles are intracellular aggregates of PHF that occupy the cell

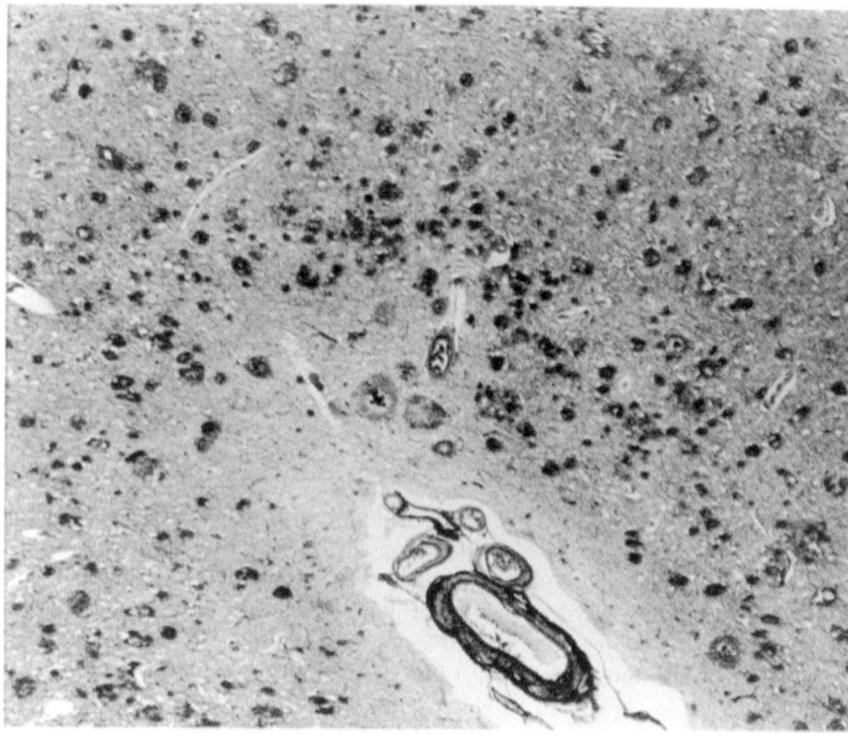

Figure 2 Bielschowsky stain (×34) of neocortex of Alzheimer's disease brain with many neuritic and diffuse plaques. Reproduced with permission from Mirra *et al.* (1993).

body and apical dendrite and generally are observed in large pyramidal neurons. Both globoid and flame-shaped forms are seen (Figure 3). Although primarily intraneuronal, in some regions (e.g. the entorhinal cortex) extraneuronal or 'ghost' tangles can be recognized and apparently represent the insoluble residue of neurons that have died. The major constituent of PHF is abnormally phosphorylated tau, a microtubule-associated protein. Tau normally polymerizes tubulin but in states of abnormal phosphorylation is unable to bind to microtubules with consequent destabilization of the neuronal cytoskeleton and disruption of axonal transport (Trojanowski *et al.*, 1993). Abnormally phosphorylated tau and PHF also are found in dystrophic neurites, both those associated with senile plaques and in individual neurites in the cortex between plaques. These latter forms are termed 'neuropil threads' and are believed to originate in axonal or dendritic processes of tangle-bearing neurons (Braak *et al.*, 1986).

In addition to these classic markers of AD, other lesions often are present. In the hippocampus, these include granulovacuolar degeneration, in which a small vacuole containing a dense central granule is present in the cytoplasm of pyramidal neurons, and the Hirano body, a rod-like filamentous structure generally adjacent to pyrami-

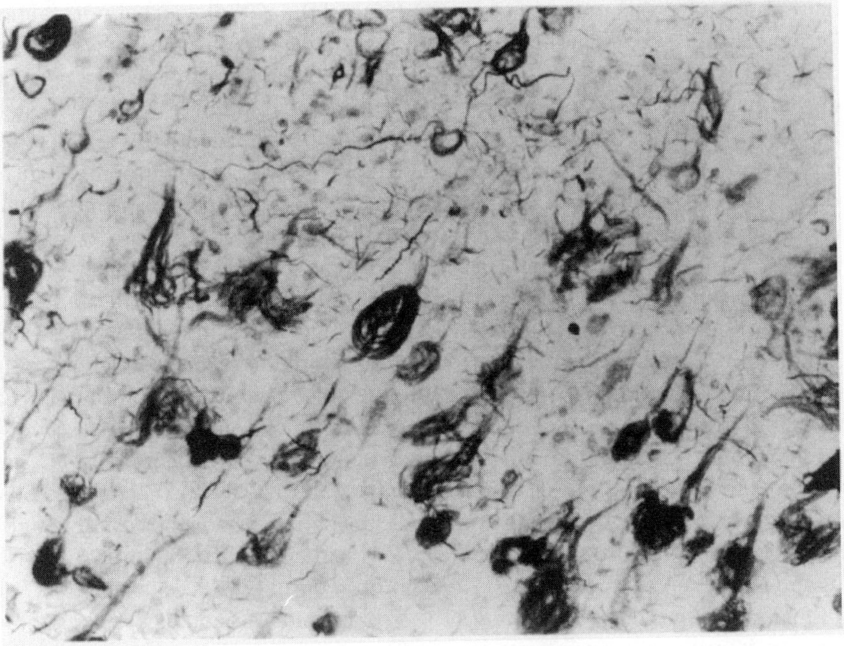

Figure 3 Neurofibrillary tangles as demonstrated by Gallyas silver impregnation of subiculum in the hippocampal formation (scale bar = 20 μm). Reproduced with permission from Vallet *et al.* (1992).

dal neurons. Amyloid, indistinguishable from the peptide found in cores of senile plaques, not infrequently is deposited in the walls of leptomeningeal and small cortical vessels. Neuronal loss in the hippocampus (Ball, 1977), entorhinal cortex (Hyman *et al.*, 1984), nucleus basalis of Meynert (Whitehouse *et al.*, 1982; Oyanagi *et al.*, 1989), brainstem (Bondareff *et al.*, 1982; Aletrino *et al.*, 1992), and neocortex has been demonstrated in AD. As compared with age-matched controls, 30–60% (younger patients with greater decline) of large neurons in midfrontal, superior temporal and inferior parietal cortex are lost in AD (Hansen *et al.*, 1990). Extensive synaptic loss also occurs in AD. Quantitative ultrastructural (Scheff *et al.*, 1990) and immunocytochemical (Hamos *et al.*, 1989; Terry *et al.*, 1991; Zhan *et al.*, 1993) studies reveal that densities of presynaptic and total synaptic elements in AD patients are reduced by 25–50% from control levels.

11.3.2 Neuropathological criteria for the diagnosis of AD

The seminal studies of Tomlinson, Blessed and Roth in Newcastle, UK (Blessed *et al.*, 1968; Tomlinson *et al.*, 1968; Tomlinson *et al.*, 1970) underscored that quantitative measures of plaques and tangles distinguished AD from vascular and other causes of dementia, and could serve as histological correlates of the cognitive deficits observed in AD patients. However, none of the pathological abnormalities associated with AD, including plaques and tangles (Wisniewski *et al.*, 1979; Katzman *et al.*, 1988), are pathognomonic. In particular, most of the histological features of AD are found to a lesser extent in brains of apparently nondemented older adults (Katzman *et al.*, 1988; Crystal *et al.*, 1993; Price, 1993). A further complication is that there is considerable heterogeneity in the neuropathological findings of AD. Not only can plaques and tangles vary in their distribution and density, but occasionally tangles may be virtually absent from neocortex (Terry *et al.*, 1987) or in unusual cases plaques may be lacking (Sumi *et al.*, 1992). Finally, variations in sampling, staining techniques and interpretation can lead to marked inter-rater differences in the quantitative assessment of AD. These differences were highlighted in a study (Mirra *et al.*, 1994) of 24 neuropathologists, independently using their routine staining methods and interpretative criteria, for quantitative and semiquantitative rankings of plaques and tangles in contiguous frontal lobe sections from the same AD brain; although good inter-rater reliability was found for tangles ($r = 0.66$), diffuse plaque counts ranged from $0/\text{mm}^2$ to $204/\text{mm}^2$ (Figure 4). This range suggests that the same case could be classified from normal to fully established AD, depending on the methodology. Reproducibility can be improved with consensus discussions (Paulus *et al.*, 1992), rank ordering (McKeel *et al.*, 1993) or semiquantitative methods (Mirra *et al.*, 1994), but what is accepted as AD and how to define it remain major issues in need of resolution for neuropathologists (Tomlinson, 1989).

In an attempt to establish minimum microscopic standards for AD, a panel of experienced neuropathologists, convened by the National Institutes of Health (NIH) and the American Association of Retired Persons (AARP), recommended diagnostic criteria based on age-adjusted plaque densities in the neocortex (Khachaturian, 1985). In the absence of other obvious dementing illnesses, these criteria suggest for a diagnosis of 'pure' AD that the number of senile plaques in a 1 mm^2 field should exceed 2–5 in a patient less than 50 years of age, eight in a patient 50–65 years of age, ten in a patient 66–75 years of age, or 15 in a patient greater than 75 years of age (Khachaturian, 1985). Downward revision of the criteria, perhaps as much as 50%, was acceptable in the presence of a clinical history of dementia. Reservations about the application of these criteria were revealed several years following their publication when only 21% of neuropathologists specifically used them for the histopathological assessment of AD (Wisniewski *et al.*, 1979). To encourage adoption of standardized diagnostic criteria, the Consortium to Establish a Registry for Alzheimer's Disease (CERAD) proposed a uniform methodology and a simple protocol for AD diagnosis based on semiquantitative assessment (sparse, moderate, frequent) of age-related neocortical neuritic plaque density in an area of maximum involvement (Mirra *et al.*, 1991). Although some neuropathologists have adopted, at least in modified form, the NIH/AARP or the CERAD criteria (Joachim *et al.*, 1988; Jellinger *et al.*, 1990; Mendez *et al.*, 1992a; Berg *et al.*, 1993), a substantial number continue to use other quantitative or qualitative criteria to diagnose AD (Wisniewski *et al.*, 1979; Alafuzoff *et al.*, 1987; Wade *et al.*, 1987; Tierney *et al.*, 1988; Förstl *et al.*,

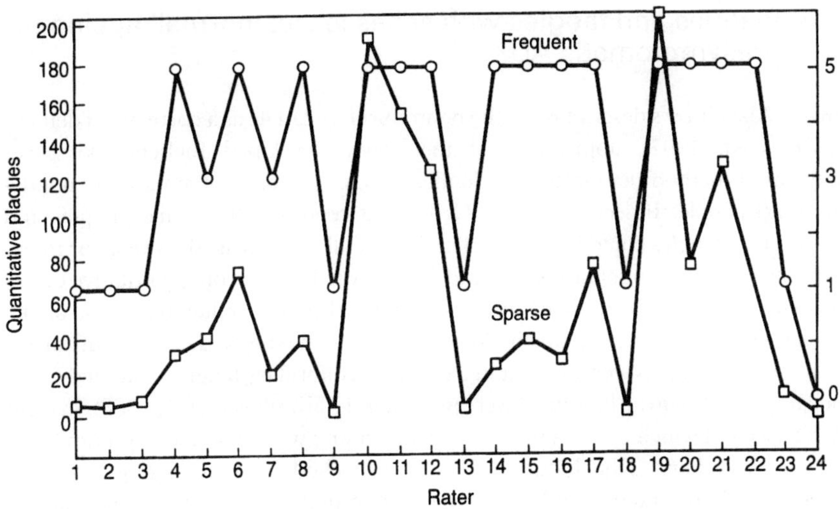

Figure 4 Range of quantitative (□) and semiquantitative (○) counts of diffuse plaques in frontal cortex of a single AD case as generated by individual neuropathologists relying on their routine staining methods. The marked variability associated with nonuniform methods and criteria is evident. Courtesy of Suzanne Mirra, MD, based on data from Mirra *et al.* (1994).

1992; Crystal *et al.*, 1993). The lack of agreement resulting from the differing neuropathological methodologies and criteria for AD severely limits comparison of data across centres.

Part of the dissatisfaction regarding recommended neuropathological diagnostic criteria (Khachaturian, 1985; Mirra *et al.*, 1991) derives from their reliance on plaque densities; tangle burden is de-emphasized. Although remarkable advances in the understanding of the molecular neurobiology of amyloid suggest to some that plaque formation (abnormal deposition of Aβ) represents the critical event in the pathogenesis of AD (Rosenberg, 1993), tangles remain as a characteristic lesion of AD. Several classification systems for AD have been proposed on the basis of neurofibrillary degeneration alone (Bondareff *et al.*, 1993; Braak *et al.*, 1993) and the consistent involvement by tangles of hippocampal structures has contributed to the proposal that AD can be defined as a hippocampal disorder (Ball *et al.*, 1985). Even 'plaque-only' cases of AD in which tangles are virtually absent from the neocortex have substantial numbers of tangles in the hippocampus (Terry *et al.*, 1987). It remains uncertain, however, that either plaques or tangles are appropriate markers for AD as they may represent downstream events from more fundamental aetiopathogenic mechanisms. In this view, they may only be secondary and relatively nonspecific results of neuronal degeneration that may be produced by several different aetiologies, at least in the case of tangles. Of particular importance for the diagnostic utility of plaques and tangles for AD is whether neuronal changes occurring with age alone can produce these lesions.

11.3.3 Plaques and tangles with senescence: 'normal' ageing or presymptomatic AD?

Following Alzheimer's description of the neurofibrillary tangle as a correlate of dementia (Alzheimer, 1907), application of the newly developed Bielschowsky silver impregnation method demonstrated identical lesions in brains of apparently nondemented old people. Indeed, in 1938, Gellerstedt reported that senile plaques and neurofibrillary tangles could be detected in 80% of all persons aged 65 years or more. The distinction between senile change and AD grew to be based on quantitative rather than qualitative criteria. This concept is embodied in the influential work of the Newcastle group, who showed that in 28 hospitalized nondemented patients (mean age 75 years), senile plaques were present in 22 and neurofibrillary tangles (confined to the hippocampal region in all but three) were present in 17 (Tomlinson *et al.*, 1968). In eight of the 22 cases in which plaques were detected, plaques were described as 'numerous' and by today's criteria (Khachaturian, 1985) approached densities diagnostic for AD. Tomlinson and coworkers (1968) interpreted these findings to indicate that plaques and tangles increasingly appeared with age and stated that their presence 'is not synonymous with the diagnosis of senile dementia, for all these changes may be found in nondemented old people'. In their studies of 50 demented patients (mean age = 76 years), all hospitalized in a mental institution, plaques were present in 42; plaque densities in 26 were greater than any of the nondemented brains (Tomlinson *et al.*, 1970). Plaque distribution, however, was similar in both demented and nondemented patient groups. Tangles were present in 35 of the 50 demented cases and involved the neocortex in 31. Thus, tangles occurred in the hippocampus in both normal and demented elderly individuals but 'severe generalized neurofibrillary change' (i.e. tangles in neocortex) was observed only in demented persons (Tomlinson *et al.*, 1970). These findings led to the conclusion that 'no qualitative histologic feature of senile . . . type was encountered in the dements that was not found in the controls; only the degree to which the change occurred or the distribution of the change differed in the two groups' (Tomlinson *et al.*, 1970). The current assumption that individuals without cognitive impairment accumulate plaques and tangles as part of normal ageing derives largely from these studies and forms the basis of the age-adjusted histological criteria found in the NIH/AARP and CERAD protocols.

By current standards, the 'control' group used by Tomlinson *et al.* was inappropriate: not only were all putatively nondemented patients hospitalized with terminal physical illnesses, encephalopathies and psychiatric disorders, the criterion for cognitive normality was a 'cutoff' score of less than four on the Dementia Scale (Blessed *et al.*, 1968). It is now recognized that pathologically confirmed AD can be clinically detected in individuals scoring in the 'normal' range (0–3) on this measure (Morris *et al.*, 1991). As noted, over a third of the 'normal' cases of Tomlinson *et al.* would satisfy current histological criteria for AD. It is highly probable that the conclusions of Tomlinson and colleagues regarding a quantitative rather than qualitative distinction between AD and normal ageing may have been flawed by the contamination of the control group with unrecognized cases of AD.

Clearly, the crucial issue in distinguishing AD from senescence is the definition of 'normal'. It may not be possible, given the limited current knowledge about AD pathogenesis and the absence of a surrogate marker for AD, to choose between the traditional model of an age-disease dichotomy, in which biological ageing is separate from pathological ageing, versus a unimodal process in which AD merely is an accentuation of normal ageing phenomena (Von Dras and Blumenthal, 1992). To resolve this dilemma, carefully assessed control subjects, evaluated by informant-based clinical methods in addition to neuropsychological measures, must be studied longitudinally to determine the natural history of normal ageing and at autopsy for clinicopathological correlation. This approach can identify intellectually healthy patients up to age 89 years of age who on post-mortem examination have no neocortical plaques or tangles (Morris *et al.*, 1991; Kazee *et al.*, 1993). Although these findings support (but do not prove) the concept that ageing and AD are qualitatively distinct, they also can be interpreted to indicate that the individuals simply did not survive to the age where they would naturally manifest AD lesions and thus cannot refute the notion that plaques and tangles are an inevitable consequence of ageing. Although virtually all centenarians studied to date have amyloid (Aβ) plaques (Delaére *et al.*, 1993), there are exceptions (Giannakopoulos *et al.*, 1993) and thus the dilemma is unresolved.

It is increasingly evident that accumulation of neurofibrillary tangles restricted to the hippocampus and other limbic regions is an inevitable accompaniment of age. Ulrich (1985) assessed 51 brains from putatively nondemented individuals who died between the ages of 55 and 64 years, and found tangles in 35, almost always localized in the entorhinal cortex with less frequent involvement of the hippocampus, olfactory bulb, amygdala and nucleus basalis of Meynert. Because no clinical data were available, the clinical correlates of Ulrich's findings cannot be assessed but the frequency of tangle detection and the confined topography in these brains is very similar to the findings of Tomlinson *et al.* (1968) in nondemented patients. Studies at Washington University's Alzheimer's Disease Research Center have focused on detecting the earliest clinical manifestations of AD, as staged by the Clinical Dementia Rating (CDR) (Hughes *et al.*, 1982; Morris, 1993). This global staging instrument uses an informant interview and clinical examination of the patient to determine the presence or absence of dementia and, when present, to rate its severity on a five-point scale: CDR = 0 indicates no cognitive impairment, and CDR = 0.5, 1, 2 or 3 indicates questionable, mild, moderate or severe dementia. The longitudinal study design has permitted recognition of several cases just on the threshold of cognitive impairment (e.g. a subject who consistently was CDR 0 over several years but at last assessment merited a CDR 0.5 rating). Experience has shown that these 'questionably' impaired patients have fully developed AD on histopathological examination (Morris *et al.*, 1991) and thus the CDR 0.5 assignment in this selected population may characterize very mild AD rather than 'questionable' dementia.

Many CDR 0.5 subjects perform in the normal range on brief cognitive tests and neuropsychological measures (Storandt and Hill, 1989; Morris *et al.*, 1991); the designation rests on clinical judgement based heavily on the informant report of subtle cognitive decline relative to past abilities. Reliance on neuropsychological performance alone to identify 'normal' cases almost certainly would contaminate control samples with these very mild AD patients. In spite of minimal or no quantitative cognitive deficit, these cases already have very high plaque concentrations in the neocortex (Morris *et al.*, 1991; Price *et al.*, 1991); other workers have reported similar findings (Hof *et al.*, 1992). True control brains (CDR 0) have few if any neocortical lesions. In controls over the age of 54 years, studies mapping the distribution of plaques and tangles in Bielschowsky-stained sections of ventral forebrain with a computer-assisted microscope digitizer (Price *et al.*, 1991) or with manual quantitative neuropathology (Morris *et al.*, 1991; Berg *et al.*, 1993), reveal that tangles invariably are present in hippocampus (CA1) and parahippocampal gyrus (entorhinal and perirhinal cortices). Moreover, tangle density in these cytoarchitectural areas increases as a function of age (Price *et al.*, 1991). The topography of tangles in normal ageing is qualitatively identical to that seen in mild and severe AD, and follows a hierarchical pattern of involvement from limbic areas, especially those close to the hippocampal formation, to neocortex with a gradient in density from inferior temporal cortex to superior temporal cortex (Price, 1995). Similar results have been reported by others (Arriagada *et al.*, 1992a). The appearance of neocortical tangles outside of medial temporal cortex essentially does not occur with normal ageing and thus the presence of tangles in other cortical regions serves as a reliable indicator of AD (Wilcock and Esiri, 1982; Morris *et al.*, 1991; Price *et al.*, 1991).

Tangle distribution also follows a laminar pattern in certain regions. Cells in layer II of the entorhinal cortex give rise to projections to the granule cells of the dentate gyrus of the hippocampus; the hippocampus in turn projects back to the entorhinal cortex from cells of origin in CA1/subicular zone. Layer II entorhinal and CA1 and subicular neurons are most vulnerable to tangle formation in nondemented elderly persons; progressively severe involvement of these cells is seen with AD (Hyman *et al.*, 1990; Arriagada *et al.*, 1992b; Braak *et al.*, 1993). Processing of information from visual, auditory, olfactory and somatosensory areas involves the convergence of input from multimodal association cortex on the entorhinal cortex, which relays the information via the perforant pathway to the dentate gyrus; information is returned to the neocortex through the subiculum and entorhinal cortex (Zola-Morgan and Squire, 1993). Nearly all cortical input to the hippocampus thus is carried by the perforant pathway. The progressive neurofibrillary involvement of perforant pathway cells of origin in layer II of the entorhinal cortex in ageing and in AD is postulated effectively to disconnect the hippocampus from neocortex, and interfere with the acquisition and learning of new facts and events (Hyman *et al.*, 1986).

Unlike tangles, plaques are not invariably associated with ageing. Although the frequency of detection of plaques increases with advancing age, sensitive silver stains and amyloid immunohistochemistry fail to reveal plaques in some nondemented individuals up to age 100 years (Giannakopoulos *et al.*, 1993; Price, 1995), although in another series plaques were detected in all 15 nondemented patients aged 100 years or more (Delaére *et al.*, 1993). When plaques are present in nondemented ageing, typically they are fewer in number and are proportionately greater for diffuse plaques

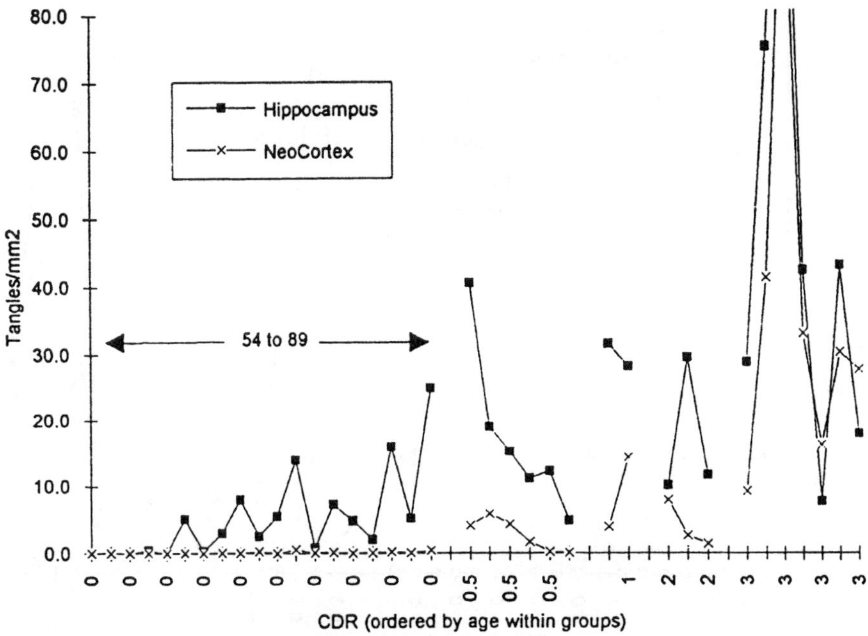

Figure 5 The relation between tangle density, dementia severity, and age in the hippocampus (average of field CA1 and the entorhinal cortex) and neocortex (average of temporal, insular and orbital cortex). Each dot represents one case. The cases are grouped according to their Clinical Dementia Rating (CDR); within each CDR group the cases are arranged by age (the nondemented cases vary from 54 to 89 years of age). Within the CDR 0 group (see text) the Spearman's rank correlation coefficient of age vs tangle density in the hippocampus is 0.69, $p < 0.01$. Reproduced with permission from Price (1993).

than is typical for AD (Price, 1995). Although the increasingly frequent appearance of at least some diffuse plaques with advancing age has been interpreted to indicate that, like tangles, Aβ deposition represents an age-related process, plaque density does not correlate with age (Price, 1995). The failure to find a correlation of increasing numbers of plaques with age is inconsistent with an age-associated phenomenon.

In ageing, tangles are present in limbic structures in the absence of detectable plaques, and in AD follow a hierarchical pattern of progression with initial and uniform involvement of the hippocampal formation and entorhinal cortex and, ultimately, the neocortex (Braak *et al.*, 1993). Plaques, however, typically affect the neocortex in AD to a much greater extent than the hippocampal formation (Price *et al.*, 1991; Arriagada *et al.*, 1992b) and do not demonstrate a regular relationship with age. These findings imply that plaques and tangles develop by independent processes, at least initially (Price, 1993). The different relationships of plaques and tangles to age and to the presence and severity of AD are shown in Figures 5 and 6.

Although unanswerable at present, the question arises whether the brains of indi-

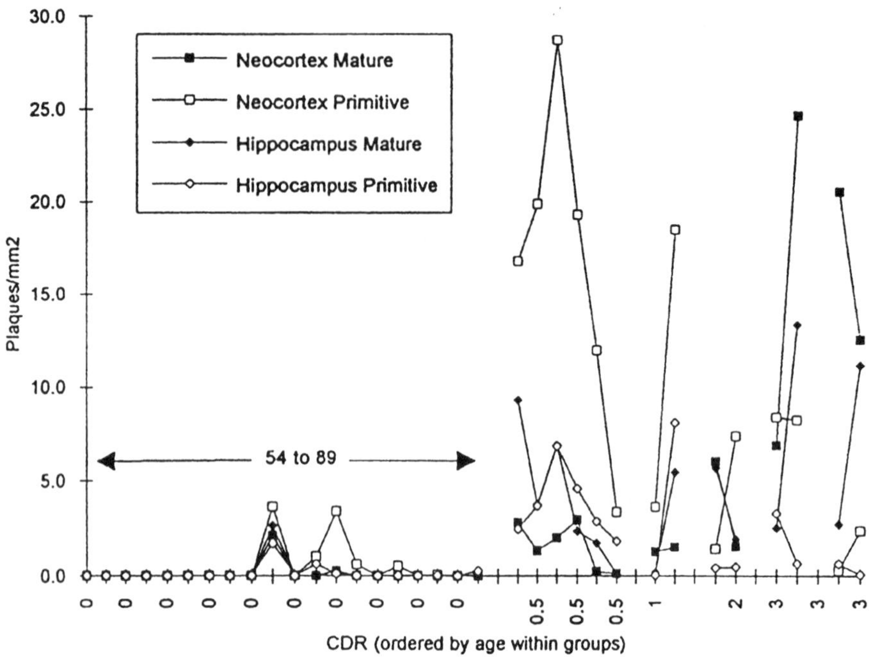

Figure 6 The relation between diffuse ('primitive') and neuritic ('mature') plaque density, dementia severity, and age in the hippocampus (average of field CA1 and the entorhinal cortex) and neocortex (average of temporal, insular and orbital cortex). The cases are ordered as in Figure 5. Note that there is no clear relation between age and plaque density in the nondemented (CDR 0) group (see text). Reproduced with permission from Price (1993).

viduals known to be nondemented during life but displaying appreciable numbers of neocortical plaques at autopsy represent presymptomatic AD. This concept is suggested by the remarkably high densities of plaques (preponderantly but not exclusively of the diffuse type) in cortical regions of patients just at the threshold of detectable dementia, i.e. CDR 0.5 individuals (Morris *et al.*, 1991; Berg *et al.*, 1993; Price, 1993) (see Figure 6). Intuitively, years or even decades must be required for the accumulation of such large numbers of plaques. That is, plaque development must be ongoing during the years of cognitive normality prior to CDR 0.5 designation. Evidence in support of this hypothesis is found in Down's syndrome patients, in whom neocortical plaques and tangles in densities and neuroanatomical distribution commensurate with AD occur uniformly past the age of 30–40 years (Wisniewski *et al.*, 1985; Mann *et al.*, 1990). Dementia develops at a later age, with mean age of onset after 50 years (Lai and Williams, 1989; Evenhuis, 1990). The incubation period of 10 years or more between the onset of neuropathological lesions and expression of DAT in Down's syndrome patients is consistent with a threshold effect in which a critical level of pathology must develop before symptoms appear (Mann *et al.*, 1990). A threshold effect also has been postulated to explain the apparent protective influence of education on the development of AD in elderly populations (Stern *et al.*, 1994), perhaps because an increased 'brain reserve' (e.g. greater number of neurons, strengthened synaptic connections) presumably associated with higher education requires a greater pathological burden before symptoms of dementia appear (Zhang *et al.*, 1990; Katzman, 1993).

A presymptomatic stage in which histological AD is present years before dementia is recognized appears increasingly likely (Hof *et al.*, 1992; Berg *et al.*, 1993; Coria *et al.*, 1993; Kazee *et al.*, 1993). Reports of cases who potentially represent presymptomatic AD but are interpreted as 'pathological ageing' have relied on neuropsychological measures and cognitive test scores to define cognitive normality, and thus it is uncertain whether very mild dementia as detected by collateral sources may have been present (Crystal *et al.*, 1988, 1993; Katzman *et al.*, 1988; Delaére *et al.*, 1990). It is also possible that pathological ageing and presymptomatic AD are two sides of the same coin; that is, given appropriate survival, all pathological ageing cases will manifest DAT. What is needed are more studies of normal ageing in which cases of very mild dementia are excluded so that the neuropathology of truly normal cognitive ageing can be defined and the frequency of presymptomatic AD determined, if indeed it can be shown to exist.

11.3.4 Correlation of plaque and tangle densities with dementia severity

Neither plaque nor tangle densities correlate particularly well with dementia severity. Although Blessed and coworkers (1968) reported significant correlations of mean cortical plaque counts with scores on brief cognitive tests, they pointed out that the correlations may have been skewed by inclusion of a substantial number of control patients with normal test scores and few or no plaques. Similar interpretative problems apply to the significant correlations between test scores and plaque counts reported by the Katzman group, who included not only normal subjects but patients with multiple infarcts in their analyses (Katzman *et al.*, 1983), and the Charles Foix Longitudinal Study (Duyckaerts *et al.*, 1988). Although AD may occur in the presence of other potentially dementing conditions (Hansen *et al.*, 1990), elimination of non-AD case material is necessary to determine the true correlation of plaque counts with dementia severity. With this restriction, numerous studies find weak or no correlation between neocortical plaque densities and clinical severity measures (Morimatsu *et al.*, 1975; Wilcock and Esiri, 1982; Neary *et al.*, 1986; McKee *et al.*, 1991; Morris *et al.*, 1991; Terry *et al.*, 1991; Arriagada *et al.*, 1992b; Berg *et al.*, 1993; McKeel *et al.*, 1993).

Tangle densities in the hippocampus and entorhinal cortex generally (but see McKee *et al.*, 1991) do not correlate with dementia severity (Arriagada *et al.*, 1992b; Berg *et al.*, 1993; Price, 1993). The lack of correlation of hippocampal tangles with presence or severity of AD is consistent with the interpretation that neurofibrillary degeneration in these regions reflects ageing rather than disease (Price, 1995). However, in the neocortex tangles are significantly related to the presence and severity of dementia (Wilcock and Esiri, 1982; McKee *et al.*, 1991; Morris *et al.*, 1991;

Arriagada *et al.*, 1992b; Berg *et al.*, 1993). A significant association between dystrophic neurites (neuropil threads) with clinical measures of dementia also has been noted (McKee *et al.*, 1991). Neuropil threads are dense intraneuronal (probably intra-axonal) accumulations of PHF and, in conjunction with tangles and neuritic plaques, form part of the spectrum of neurofibrillary changes observed in AD. Along with neurofibrillary tangles, they are useful markers of the pathological stages of AD (Braak *et al.*, 1993).

In general, correlations between neocortical tangle (and neuropil thread) densities and clinical measures of dementia severity are greater than for neocortical plaques and dementia severity. However, neither tangles nor plaques are wholly adequate markers for AD. It is unresolved whether plaques, especially the diffuse type, can be a phenomenon of normal ageing. Tangles are uniformly present in limbic regions in AD but often in densities comparable to those found in these areas with age alone. As discussed, tangles characterize other diseases (Wisniewski *et al.*, 1979) and thus are less specific than plaques for AD. There is also the consideration of cases of 'plaque only' AD in which neocortical tangles are few or absent (Terry *et al.*, 1987). Finally, neither plaques nor tangles increase consistently with disease duration: comparison of quantitative assessments of cortical biopsy material and adjacent cortical sections at autopsy from 3 to 7 years later in five AD patients found no accumulation of these lesions in spite of documented cognitive decline (Mann *et al.*, 1987).

To summarize, the distribution of plaques and tangles may be as meaningful as their densities in the assessment of AD. Current evidence suggests that tangle deposition in limbic regions is a ubiquitous function of age. Plaques, on the other hand, appear to develop first in the neocortex where their presence establishes the diagnosis of AD. This last statement assumes that plaques in nondemented individuals indicate presymptomatic AD rather than normal or pathological ageing. Although the appearance of tangles in the neocortex is associated with the presence of AD, plaques are more reliable markers for AD diagnosis because tangles may be relatively absent from neocortex in some cases (Terry *et al.*, 1987). When present, neocortical tangles are better correlates than plaques of dementia severity. Both of these classical lesions remain useful pathological hallmarks for AD in spite of their limitations (Table 1). It is appropriate, however, to consider whether other alterations are better, more relevant markers of AD.

11.3.5 Nonplaque, nontangle structural correlates of dementia severity

Neuronal counts, synaptic densities and vascular amyloid burden have been investigated for associations with clinical status in AD (Terry *et al.*, 1994). Glial markers and neurotransmitter indices also have been studied but will not be discussed here.

11.3.5.1 Neuronal loss

Although the mechanism of cell death in AD is unknown even with conventional light microscopy, neuronal loss has been appreciated since the time of Alzheimer's case report (Alzheimer, 1907). The degree of loss is variable and can be difficult to detect

Table 1 Mean values for neuropathological measures in control and Alzheimer's disease (AD) groups

	Controls ($n = 5$)	All AD ($n = 36$)	Pure AD ($n = 25$)
Brain weight (g)	1346	1133†	1132†
Regional atrophy			
Surface	1.58	2.27‡	2.30‡
Coronal	1.46	1.96	1.96
Ventricular dilatation	1.25	1.91‡	1.85‡
CAAM, region			
1–3	0.87	1.52	1.43
5A	0	0.26	0.29
5B	0.60	0.63	0.38
CAAP, region			
1–3	0.33	0.85	0.80
5A	0	0.47	0.67
5B	0	0.39	0.42
NFT, region			
1–3	0.12	9.93‡	11.91‡
5A	8.46	41.43‡	46.23‡
5B	12.50	20.56	22.70
TSP, region			
1–3	1.33	80.10†	85.79†
5A	0.02	11.07†	12.57†
5B	1.36	30.77†	35.40†
CSP, region			
1–3	0.16	7.39†	8.01†
5A	0.02	2.28†	2.52‡
5B	0	4.43†	4.44†

NFT, neurofibrillary tangles; TSP and CSP, total and amyloid cored (neuritic) senile plaques, all expressed as counts/mm^2. CAAM and CAAP, meningeal and parenchymal cerebral amyloid angiopathy. Regions 1–3 indicate frontal, temporal, and parietal cortex; region 5A, CA1/subiculum; region 5B, entorhinal cortex.
Significant differences between control and AD groups: †$p < 0.01$ and ‡$p < 0.05$. Reproduced with permission from Berg *et al.* (1993).

in some cases; when present, there often is a microcystic appearance of the involved areas, especially in the superficial layers of neocortex. Several reports have shown significant relationships between loss of large neurons (greater than 90 μm^2 in cross-sectional area) in frontal, temporal and parietal cortex, and cognitive measures (Neary *et al.*, 1986), age (Hansen *et al.*, 1988) and duration of AD (Mann *et al.*, 1987). However, morphometric studies can be cumbersome and confounded by methodological difficulties (West, 1993), and the correlations of dementia with neuronal loss have not been sufficiently compelling to favour this method over traditional quantification of plaques and tangles.

11.3.5.2 Synaptic density

Because cognition is dependent on large-scale distributed neural networks (Mesulam, 1990) and hence on neuronal connectivity, synaptic integrity is of clear biological relevance for dementia. Ultrastructural studies and the development of immunohistochemical methods for the measurement of synaptic markers, including synapsin and synaptophysin, have yielded convincing data to show that extensive synaptic loss is characteristic of AD (Hamos *et al.*, 1989; Zhan *et al.*, 1993). In frontal cortex, synaptic loss in AD approaches 40% of normal and overcomes a compensatory increase in the size of remaining synapses such that total synaptic contact area is reduced (Scheff *et al.*, 1990). Synaptic loss in the frontal lobe correlates powerfully with measures of dementia severity (Dekosky and Scheff, 1990; Terry *et al.*, 1991) and occurs to a greater degree than the loss of large neurons in the same region (Terry *et al.*, 1991), suggesting that the synapse is an earlier target of the AD degenerative process. This process has been hypothesized to involve neurotoxic effects, perhaps precipitated by Aβ deposition and plaque-associated reactive microglia with consequent cytokine production and complement activation (McGeer and Rogers, 1992; Kalaria, 1993). Alternatively, synaptic injury may precede and contribute to Aβ deposition (Masliah *et al.*, 1993). Disrupted axonal transport (Trojanowski *et al.*, 1993) is another possible mechanism underlying synaptic loss.

11.3.5.3 Cerebrovascular amyloid

Amyloid angiopathy commonly, but not always, is present in AD and is variable in intensity and distribution. Small leptomeningeal and cortical vessels typically are involved, particularly in the occipital and cerebellar regions. Because plaques and tangles generally are sparse in occipital cortex, this dissociated topographical predilection suggests that parenchymal and vascular amyloid deposition are not closely related (Terry *et al.*, 1994). No significant relationships of meningeal or parenchymal amyloid angiopathy with the presence of dementia or its severity have been found (Berg *et al.*, 1993). Particular mutations in the amyloid precursor protein gene have been associated in families in the Netherlands with hereditary cerebral haemorrhages caused by marked Aβ deposition in cerebral blood vessels (Vinters, 1992).

11.4 Clinicopathological correlations in AD

11.4.1 Diagnostic accuracy

As discussed above, plaques and tangles are not entirely adequate as markers for AD. Combined with varying methodologies for their detection and differences in the interpretation of their role in the assessment of AD, it is apparent that a neuropathological 'gold standard' for the diagnosis of AD does not currently exist. In the absence of an established biological marker for the disease, however, validation of the clinical diagnosis of AD rests on tissue confirmation, generally through autopsy studies, of the presence of plaques and tangles in the neocortex. When informant-based methods and adherence to specific clinical criteria are lacking, accuracy rates range only from 45% to 58% (Alafuzoff *et al.*, 1987; Homer *et al.*, 1988). The development of standardized clinical diagnostic criteria for DAT, however, has improved diagnostic accuracy in recent years, and rates of 80% or greater now are the rule (Joachim *et al.*, 1988; Mirra *et al.*, 1991; Mendez *et al.*, 1992a; Becker *et al.*, 1994). Two studies with small samples of highly selected DAT cases report 100% accuracy rates (Martin *et al.*, 1988; Morris *et al.*, 1988). The high accuracy of clinical diagnostic methods alone must be equalled or surpassed by proposed surrogate markers for AD before they can be accepted as diagnostic.

11.4.2 Cognitive features

11.4.2.1 Memory

The hierarchical neuroanatomical involvement of limbic structures by neurofibrillary degeneration (Hyman *et al.*, 1990; Price *et al.*, 1991) and neuronal loss (West, 1993) in ageing and AD can be considered as the morphological correlates of the characteristic memory changes seen in these conditions. These changes selectively damage components of the perforant pathway (Vogels *et al.*, 1990) and can be postulated to isolate the entorhinal cortex and hippocampus from their cortical and subcortical connections. The medial temporal lobe memory components (hippocampus; entorhinal, perirhinal, and parahippocampal cortices) subserve a fast, limited-capacity system for establishing long-term memories, i.e. the ability to learn facts and events and consciously recall them (Squire and Zola-Morgan, 1991; Hamos *et al.*, 1989). The pathological involvement of these structures predicts that acquisition of new knowledge ('learning') is adversely affected, as is observed with normal ageing (Petersen *et al.*, 1992) and to a greater extent in very mild DAT (Petersen *et al.*, 1994). In contrast, immediate recall, remote memory for events stored (presumably) in the neocortex, and nondeclarative memory (skill and habit learning) are independent of the medial temporal lobe memory system and can remain intact even in fully amnestic patients (Zola-Morgan and Squire, 1993; Victor and Agamonolis, 1990).

In AD, lesions in other cortical and subcortical structures almost certainly contribute to memory deficits such that they are more severe than those seen with age alone. Cholinergic neurons in the nucleus basalis of Meynert are notably reduced in number in AD and residual cells commonly contain tangles. In addition to providing cortical cholinergic innervation, the nucleus basalis is closely related to medial temporal lobe structures and AD pathology in the cholinergic system may contribute substantially to memory dysfunction (Whitehouse *et al.*, 1981).

11.4.2.2 Cortical disconnection syndrome

Consistent with the prevalent intellectual and personality deficits in DAT, AD is predominantly a neocortical disease and specifically displays regional and laminar predilections that subserve the cognitive and behavioural functions most affected in the disorder. Tangles are found principally in pyramidal neurons in layers III and V of cerebral cortex. These neurons have been implicated as cells of origin for cortico-cortical pathways with three patterns of connectivity: feedforward (from a primary sensory area to an association area), feedback (from an association area to a primary sensory area), and lateral (integration of cortical regions of similar functional levels) (Felleman and Van Essen, 1991). Moreover, tangle density is approximately 20 times greater in association cortex compared with primary sensory cortex (Lewis *et al.*, 1987). Plaques are more diffusely distributed than tangles throughout the neocortex but may have slightly higher densities in layer IV (Lewis *et al.*, 1987), the termination area of feedforward corticocortical pathways. Taken together, the preferential regional and laminar distribution patterns of AD histopathology are consistent with the concept of AD as a cortical disconnection syndrome. The degeneration of corticocortical circuits may be the distinguishing feature of AD as compared with normal ageing, which shares with AD neurofibrillary alterations within limbic areas, and is an appropriate pathoanatomic correlate of the multiple cognitive dysfunctions that characterize the clinical syndrome (Hof and Morrison, 1994). Investigations of the selective vulnerability of the specific neuronal classes to the AD process may yield important insights into the pathogenesis of the disorder.

11.4.2.3 Heterogeneous aspects

The presence or absence of a family history of AD in a clinicopathological study of 407 AD patients did not influence the neuropathological findings; that is, the neuropathology of cases with and without a family history were not different (Mendez *et al.*, 1992b). Another investigation found no correlation of age at dementia onset with the number of hippocampal tangles (Bondareff *et al.*, 1993). Although these studies are limited, the findings are consistent with the prevailing consensus that early- and late-onset, and familial and nonfamilial AD all represent the same general clinocopathological disorder (this does not suggest that these forms all share the same aetiology). Disproportionate, regionally appropriate AD pathology has been documented in cases that present with focal cognitive deficits, although only a very few have been examined histopathologically. In patients with primary progressive aphasia, significantly greater plaque density was found at post-mortem examination in the left (mean = 58 plaques/mm^2) versus right (mean = 21 plaques/mm^2) inferior parietal cortex (Green *et al.*, 1990). Selective left hemispheric involvement by AD pathology was reported in a similar case (Pogacar and Williams, 1984). DAT patients presenting with marked visuospatial disturbances have been reported (Hof *et al.*, 1993; Levine *et al.*, 1993) who at autopsy showed widespread AD lesions with very high densities of plaques and tangles in primary, secondary and association visual cortex. These cases indicate that AD can selectively involve specific cortical systems and that primary as well as association cortex can be vulnerable to plaques and tangles. The pathogenic basis for these findings is not understood and is made more complex by reports in which AD lesions do not predominate in those cortical regions corresponding to focal cognitive syndromes (Victoroff *et al.*, 1994), and by cases notable for extensive focal neuronal loss and gliosis but lacking plaques, tangles or hallmarks of other neurodegenerative diseases (Green *et al.*, 1990; Neary *et al.*, 1993; Scheltens *et al.*, 1994; Victoroff *et al.*, 1994).

11.4.3 Noncognitive features

Tangles can be found in large numbers in several subcortical structures that project to the cerebral cortex, including the amygdala, ventral tegmental area, dorsal raphe, locus ceruleus, and certain thalamic and hypothalamic nuclei. Relatively little attention has been given, however, to neuropathological correlates of the behavioural disturbances associated with AD. Major depression occurring during the course of DAT is related at autopsy to degenerative changes in the substantia nigra and locus ceruleus (Zubenko and Moossy, 1988; Zweig *et al.*, 1988). Degeneration of the ventral tegmental area, the medial-most region of dopaminergic cell bodies in the mesencephalon particularly may be associated with depression as a prominent feature of dementia, although the degenerative changes consist of neuronal loss and gliosis rather than neurofibrillary change (Torack and Morris, 1988). In addition to the association with depression, tangle density in the locus ceruleus is reported to correlate with duration and severity of dementia (Bondareff *et al.*, 1993). Tangles frequently involve the dorsal raphe nucleus in AD. The clinical significance of tangle-bearing raphe neurons is not apparent, although only a handful of cases have been studied systematically (Halliday *et al.*, 1992).

11.5 Conclusions

Essential for investigations of AD is the definition of normal ageing in both clinical and pathological terms. Evidence is accumulating to suggest that: (i) tangle deposition in the hippocampal formation is an invariable accompaniment of age; (ii) the presence of tangles in neocortex indicates AD; and (iii) the appearance of plaques, including the diffuse form, in putatively nondemented elderly persons represents presymptomatic or unrecognized very mild AD. The regional and laminar distribution of tangles in AD disconnects neural networks important for cognitive processing, and corresponds well with the memory impairment and other intellectual deficits characteristic of the clinical disorder.

The fundamental relationship, however, of plaques and tangles to AD phenotype is only incompletely understood. The presence of these lesions in the neocortex remains the cardinal feature for both the qualitative and quantitative assessment of AD, but neither lesion is pathognomonic. The growing view that these lesions mark a final common pathway of cerebral degeneration as secondary rather than primary changes cannot be proven until the causes of AD are elucidated. Both plaques (e.g. Aβ deposition) and tangles (e.g. abnormally phosphorylated tau) provide important clues in the search for aetiopathogenic mechanisms of AD, and thus validate the critical significance of Alzheimer's original observations of nearly a century ago.

Acknowledgements

Supported in part by National Institute on Aging grants AG 03991, AG 05682 and AG 06790. Many thanks are due to colleagues in the Memory and Aging Project of Washington University's Alzheimer's Disease Research Center. In particular, Drs Leonard Berg, Daniel McKeel, Joseph Price, Martha Storandt and Eugene Rubin have contributed in many ways to important aspects of the manuscript.

Linda Vaughn expertly and patiently typed the manuscript. Dr Virginia Buckles and Jennifer Zieleskiewicz provided invaluable editorial assistance.

References

Alafuzoff, I., Iqbal, K., Friden, H., Adolfsson, R. & Winblad, B. (1987) *Acta Neuropathol.* **74**, 209–225.

Aletrino, M.A., Vogels, O.J.M., Van Domberg, P.H.M.F. & Ten Donneklaar, H.J. (1992) *Neurobiol Aging* **13**, 461–468.

Alzheimer, A. (1907) *Allg. Z. Psychiatr. Psych. Geriat. Med.* **64**, 146–148.

American Psychiatric Association (1994) *Diagnostic and Statistical Manual of Mental Disorders*, 4th edn. Washington, DC, American Psychiatric Association.

Arriagada, P.V., Marzloff, K. & Hyman, B.T. (1992a) *Neurology* **42**, 1681–1688.

Arriagada, P.V., Growdon, J.H., Hedley-Whyte, E.T. & Hyman, B.T. (1992b) *Neurology* **42**, 631–639.

Ball, M.J. (1977) *Acta Neuropathol.* **37**, 111–118.

Ball, M.J., Hachinski, V., Fox, A. *et al.* (1985) *Lancet* **1**(8419), 14–16.

Becker, J.T., Boller, F., Lopez, O., Saxton, J. & McGonigle, K.L. (1994) *Arch Neurol.* **51**, 585–594.

Berg, L., Miller, J.P., Baty, J., Rubin, E.H., Morris, J.C. & Figiel, G. (1992) *Ann. Neurol.* **31**, 242–249.

Berg, L., McKeel, D.W., Miller, J.P., Baty, J. & Morris, J.C. (1993) *Arch. Neurol.* **50**, 349–358.

Bick, K.L. (1994) In *Alzheimer Disease* (eds Terry, R.D., Katzman, R. & Bick, K.L.), pp. 1–8. New York, Raven Press.

Bick, K.L. & Amaducci, L. (1989) *The Early Story of Alzheimer's Disease.* Padova, Liviana Press.
Bird, T.D. (1994) In *Alzheimer Disease* (eds Terry, R.D., Katzman, R. & Bick, K.L.), pp. 65–74. New York, Raven Press.
Bird, T.D., Sumi, S.M., Nemens, E.J. *et al.* (1989) *Ann. Neurol.* **25**, 12–25.
Blessed, G., Tomlinson, B.E. & Roth, M. (1968) *Br. J. Psychiatry* **114**, 797–811.
Bolla, K.I., Lindgren, K.N., Bonaccorsy, C. & Bleecker, M.L. (1991) *Arch. Neurol.* **48**, 61–64.
Bondareff, W., Mountjoy, C.Q. & Roth, M. (1982) *Neurology* **32**, 164–168.
Bondareff, W., Mountjoy, C.Q., Wischik, C.M., Hauser, D.L., LaBree, L.D. & Roth, M. (1993) *Arch. Gen. Psychiatry.* **50**, 350–356.
Bozzola, F., Gorelick, P. & Freels, S. (1992) *Arch. Neurol.* **49**, 297–300.
Braak, H., Braak, E., Grundke-Iqbal, I. & Iqbal, K. (1986) *Neurosci. Lett.* **65**, 351–355.
Braak, H., Duyckaerts, C., Braak, E. & Piette, F. (1993) In *Alzheimer's Disease: Advances in Clinical and Basic Research* (eds Corain, B., Iqbal, K., Nicolini, M., Winbald, H. & Zatta, P.), pp. 131–137. Chichester, John Wiley & Sons.
Caselli, R.J., Jack, C.R., Petersen, R.C., Wahner, H.W. & Yanagihara, T. (1992) *Neurology* **42**, 1462–1468.
Chui, H.C., Teng, E.L., Henderson, V.W. & Moy, A.C. (1985) *Neurology* **35**, 1544–1550.
Chui, H.C., Lyness, S.A., Sobel, E. & Schneider, L.S. (1994) *Arch. Neurol.* **51**, 676–681.
Clark, R.F. & Goate, A.M. (1993) *Arch. Neurol.* **50**, 1164–1172.
Coria, F., Moreno, A., Rubio, I., Garciá, M.A., Morato, E. & Mayor, F. (1993) *Neuropathol. Appl. Neurobiol.* **19**, 261–268.
Craik, F.I.M. (1991) In *Memory Disorders, Research and Clinical Practice* (eds Yanagihara, T. & Petersen, R.C.), pp. 347–367. New York, Marcel Dekker.
Crystal, H.A., Horoupian, D.S., Katzman, R. & Jotkowitz, S. (1982) *Ann. Neurol.* **12**, 186–188.
Crystal, H., Dickson, D., Fuld, P. *et al.* (1988) *Neurology* **38**, 1682–1687.
Crystal, H.A., Dickson, D.W., Sliwinski, M.J. *et al.* (1993) *Ann. Neurol.* **34**, 566–573.
Cummings, J.L., Benson, D.F., Hill, M.A. & Read, S. (1985) *Neurology* **35**, 394–397.
Damasio, H. & Damasio, A.R. (1989) *Lesion Analysis in Neuropsychology.* New York, Oxford University Press.
Dekosky, S.T. & Scheff, S.W. (1990) *Ann. Neurol.* **27**, 457–464.
Delaére, P., Duyckaerts, C., Masters, C., Beyreuther, K., Piette, F. & Hauw, J.J. (1990) *Neurosci. Lett.* **116**, 87–93.
Delaére, P., Fayet, G., Duyckaerts, C. & Hauw, J.J. (1993) *Neurobiol. Aging* **14**, 191–194.
Drevets, W.C. & Rubin, E.H. (1989) *Biol. Psychiatry* **25**, 39–48.
Duara, R., Lopez-Alberola, R.F., Barker, W.W. *et al.* (1993) *Neurology* **43**, 1377–1384.
Duyckaerts, C., Delaére, P., Poulain, P., Brion, J.P. & Hauw, J.J. (1988) *Neurosci. Lett.* **91**, 354–359.
Edwards, J.K., Larson, E.B., Hughes, J.P. & Kukall, W.A. (1991) *J. Am. Geriatr. Soc.* **39**, 477–483.
Evans, D.A., Funkenstein, H., Albert, M.S. *et al.* (1989) *JAMA* **262**, 2551–2556.
Evans, J.G. (1988) In *Research and the Aging Population. Ciba Foundation Symposium* (eds Evered, D. & Whalen, J.), pp. 38–57. Chichester, John Wiley.
Evenhuis, H.M. (1990) *Arch. Neurol.* **47**, 263–267.
Faber-Langendoen, K., Morris, J.C., Knesevich, J., LaBarge, E. & Berg, L. (1988) *Ann. Neurol.* **23**, 365–370.
Farrer, L.A., Myers, R.H., Cupples, L.A. *et al.* (1990) *Neurology* **40**, 395–403.
Felleman, D.J. & Van Essen, D.C. (1991) *Cereb. Cortex* **1**, 1–47.
Flicker, C., Ferris, S.H. & Reisberg, B. (1993) *J. Am. Geriatr. Soc.* **41**, 1029–1032.
Folstein, M.F., Folstein, S.E. & McHugh, P.R. (1975) *J. Psychiatr. Res.* **12**, 189–198.
Förstl, H., Burns, A., Levy, R., Cairns, N., Luthert, P. & Lantos, P. (1992) *Arch. Neurol.* **49**, 1038–1042.
Gellerstedt, N. (1938) *Ups. LaKaref. Forh.* **38**, 193–408.
Giannakopoulos, P., Hof, P.R., Surini, M., Michel, J.P. & Bouras, C. (1993) *Acta Neuropathol.* **85**, 602–610.

Goate, A., Chartier-Harlin, M.C., Mullan, M. *et al.* (1991) *Nature* **349**, 704–706.
Graves, A.B. & Kukull, W.A. (1994) In *Handbook of Dementing Illnesses* (ed. Morris, J.C.), pp. 23–69. New York, Marcel Dekker.
Green, J., Morris, J.C., Sandson, J., McKeel, D.W. & Miller, J.W. (1990) *Neurology* **40**, 423–429.
Gustafson, L. (1987) *Arch. Gerontol. Geriatr.* **6**, 209–223.
Halliday, G.M., McCann, H.L., Pamphlett, R. *et al.* (1992) *Acta Neuropathol.* **84**, 638–650.
Hamos, J.E., DeGennaro, L.J. & Drachman, D.A. (1989) *Neurology* **39**, 355–361.
Hansen, L.A., DeTeresa, R., Davies, P. & Terry, R.D. (1988) *Neurology* **38**, 48–54.
Hansen, L., Salmon, D., Galasko, D. *et al.* (1990) *Neurology* **40**, 1–8.
Haupt, M., Pollmann, S. & Kurz, A. (1993) *Acta Neurol. Scand.* **88**, 349–353.
Hänninen, T., Reinikainen, K.J., Helkala, E.L. *et al.* (1994) *J. Am. Geriatr. Soc.* **42**, 1–4.
Heyman, A., Wilkinson, W.E. & Hurwitz, B.J. (1983) *Ann. Neurol.* **14**, 507–515.
Hof, P.R. & Morrison, J.H. (1994) In *Alzheimer Disease* (eds Terry, R.D., Katzman, R. & Bick, K.L.), pp. 197–229. New York, Raven Press.
Hof, P.R., Bierer, L.M., Perl, D.P. *et al.* (1992) *Arch. Neurol.* **49**, 956–953.
Hof, P.R., Archin, N., Psmand, A.P. *et al.* (1993) *Acta Neuropathol.* **86**, 215–223.
Homer, A.C., Honavar, M., Lantos, P.L. *et al.* (1988) *Br. Med. J.* **297**, 894–896.
Howieson, D.B., Holm, L.A., Kaye, J.A., Oken, B.S. & Howieson, J. (1993) *Neurology* **43**, 1882–1886.
Huff, F.J., Growdon, J.H., Corkin, S. & Rosen, T.J. (1987) *J. Am. Geriatr. Soc.* **35**, 27–30.
Hughes, C.P., Berg, L., Danziger, W.L., Coben, L.A. & Martin, R.L. (1982) *Br. J. Psychiatry* **140**, 566–572.
Hyman, B.T., Van, H.G.W., Damasio, A.R. & Barnes, C.L. (1984) *Science* **225**, 1168–1170.
Hyman, B.T., Van Hoesen, G.W., Kromer, L.J. & Damasio, A.R. (1986) *Ann. Neurol.* **20**, 472–481.
Hyman, B.T., Van Hoesen, G.W. & Damasio, A.R. (1990) *Neurology* **40**, 1721–1730.
Jellinger, K., Danielczyk, W., Fischer, P. & Gabriel, E. (1990) *J. Neurol. Sci.* **95**, 239–258.
Joachim, C.L., Morris, J.H. & Selkoe, D.J. (1988) *Ann. Neurol.* **24**, 50–56.
Jorm, A.F. (1988) *Br. J. Psychiatry* **152**, 209–213.
Kalaria, R.N. (1993) *Brain Pathol.* **3**, 333–347.
Katzman, R. (1993) *Neurology* **43**, 13–20.
Katzman, R., Brown, T., Fuld, P., Peck, A., Scechler, R. & Schimmel, H. (1983) *Am. J. Psychiatry* **140**, 734–739.
Katzman, R., Terry, R., DeTeresa, R. *et al.* (1988) *Ann. Neurol.* **23**, 138–144.
Kazee, A., Eskin, T., Lapham, L., Gabriel, K., McDaniel, K. & Hamill, R. (1993) *Alzheimer Dis. Assoc. Disord.* **7**, 152–164.
Khachaturian, Z.S. (1985) *Arch. Neurol.* **42**, 1097–1105.
Khachaturian, Z.S. (1992) In *Heterogeneity of Alzheimer's Disease* (eds Boller, F., Forette, F., Khachaturian, Z.S., Poncet, M. & Christen, Y.), pp. 1–3. Berlin, Springer-Verlag.
Koss, E., Patterson, M.B. & Ownby, R. (1993) *Arch. Neurol.* **50**, 92–97.
Lachs, M.S., Becker, M., Siegal, A.P., Miller, R.L. & Tinetti, M.E. (1992) *J. Am. Geriatr. Soc.* **40**, 768–773.
Lai, F. & Williams, R.S. (1989) *Arch. Neurol.* **46**, 849–853.
Levine, D.N., Lee, M.J. & Fisher, C.M. (1993) *Neurology* **43**, 305–313.
Levy-Lahad, E., Wasco, W., Poorkaj, P. *et al.* (1995) *Science* **269**, 973–977.
Lewis, D.A., Campbell, M.J., Terry, R.D. & Morrison, J.H. (1987) *J. Neurosci.* **7**, 1799–1808.
Mann, D.M.A., Marcyniuk, B., Yates, P.O., Neary, D. & Snowden, J.S. (1987) *Neuropathol. Appl. Neurobiol.* **14**, 177–195.
Mann, D.M.A., Jones, D., Prinja, D. & Purkiss, M.S. (1990) *Acta Neuropathol.* **80**, 318–327.
Martin, E.M., Wilson, R.E., Penn, R.D. *et al.* (1988) *Neurology* **37**, 1201–1207.
Masliah, E., Mallory, M., Hansen, L., DeTeresa, R. & Terry, R.D. (1993) *Neurology* **43**, 192–197.
McGeer, P.L. & Rogers, J. (1992) *Neurology* **42**, 447–449.
McKee, A.C., Kosik, K.S. & Kowall, N.W. (1991) *Ann. Neurol.* **30**, 156–165.

McKeel, D.W., Ball, M.J., Price, J.L. *et al.* (1993) *Alzheimer Dis. Assoc. Disord.* **7**, 136–151.
McKhann, G., Drachman, D., Folstein, M., Katzman, R., Price, D. & Stadlan, E.M. (1984) *Neurology* **34**, 939–944.
Mendez, M.F., Mendez, M.A., Martin, R. & Smyth, K.A. (1990) *Neurology* **40**, 439–443.
Mendez, M.F., Mastri, A.R., Sung, J.H. & Frey, W.H. (1992a) *Alzheimer Dis. Assoc Disord.* **6**, 35–43.
Mendez, M.F., Underwood, K.L., Zander, B.A., Mastri, A.R., Sung, J.H. & Frey, W.H. (1992b) *Neurology* **42**, 770–775.
Mesulam, M.M. (1990) *Ann. Neurol.* **28**, 587–613.
Mesulam, M.M. & Weintraub, S. (1992) In *Heterogeneity of Alzheimer's Disease* (eds Boller, F., Forette, F., Khachaturian, Z.S., Poncet, M. & Christen, Y.), pp. 43–66. Berlin, Springer-Verlag.
Mirra, S.S. & Gearing, M. (1994) In *Handbook of Dementing Illnesses* (ed. Morris, J.C.), pp. 189–226. New York, Marcel Dekker.
Mirra, S.S., Heyman, A., McKeel, D.W. *et al.* (1991) *Neurology* **41**, 479–486.
Mirra, S.S., Hart, M.N. & Terry, R.D. (1993) *Arch. Pathol. Lab. Med.* **117**, 132–144.
Mirra, S.S., Gearing, M., McKeel, D.W. *et al.* (1994) *J. Neuropathol. Exp. Neurol.* **53**, 303–315.
Morimatsu, M., Hirai, S., Muramatsu, A. & Yoshikawa, M. (1975) *J. Am. Geriat. Soc.* **23**, 390–406.
Morris, J.C. (1993) *Neurology* **43**, 2412–2414.
Morris, J.C., Rubin, E.H., Morris, E.J. & Mandel, S. (1987) *J. Gerontol.* **42**, 412–417.
Morris, J.C., McKeel, D.W., Fulling, K., Torack, R.M. & Berg, L. (1988) *Ann. Neurol.* **24**, 17–22.
Morris, J.C., Drazner, M., Fulling, K., Grant, E.A. & Goldring, J. (1989) *Arch. Neurol.* **46**, 651–657.
Morris, J.C., McKeel, D.W., Jr, Storandt, M. *et al.* (1991) *Neurology* **41**, 469–478.
Neary, D., Snowden, J.S., Mann, D.W.A. *et al.* (1986) *J. Neurol. Neurosurg. Psychiatry* **49**, 229–237.
Neary, D., Snowden, J.S. & Mann, D.M.A. (1993) *Dementia* **4**, 154–159.
Oscar-Berman, M., McNamara, P., Freedman, M. (1991) In *Frontal Lobe Function and Dysfunction* (eds Levin, H.S., Eisenberg, H.M. & Benton, A.L.), p. 238. New York, Oxford University Press.
Oyanagi, K., Takahashi, H., Wakabayashi, K. & Ikuta, F. (1989) *Brain Res.* **504**, 354–357.
Paulus, W., Bancher, C. & Jellinger, K. (1992) *Neurology* **42**, 329–332.
Perls, T.T., Morris, J.N., Ovi, W.L. & Lipsitz, L.A. (1993) *J. Am. Geriatr. Soc.* **41**, 1193–1201.
Petersen, R.C., Smith, G., Kokmen, E., Ivnik, R.J. & Tangalos, E.G. (1992) *Neurology* **42**, 396–401.
Petersen, R.C., Smith, G.E., Ivnik, R.J., Kokmen, E. & Tangalos, E.G. (1994) *Neurology* **44**, 867–872.
Pogacar, S. & Williams, R.S. (1994) *R. I. Med J.* **67**, 181–185.
Price, J.L. (1993) *Neurobiol. Aging* **14**, 661–663.
Price, J.L. (1995) *Semin. Neurosci.* **6**, 305–402.
Price, J.L., Davis, P.B., Morris, J.C. & White, D.L. (1991) *Neurobiol. Aging* **12**, 295–312.
Reisberg, B., Ferris, S.H. & deLeon, M.J. (1982) *Am. J. Psychiat.* **139**, 1136–1139.
Romanelli, M.F., Morris, J.C., Ashkin, K. & Coben, L.A. (1990) *Arch. Neurol.* **47**, 847–850.
Rosenberg, R.N. (1993) *Neurology* **43**, 851–856.
Roth, M., Tym, E., Mountjoy, C.Q. *et al.* (1986) *Br. J. Psychiatry* **149**, 698–709.
Rowe, J.W. & Kahn, R.L. (1987) *Science* **237**, 143–149.
Rubin, E.H., Morris, J.C. & Berg, L. (1987) *J. Am. Geriatr. Soc.* **35**, 721–725.
Rubin, E.H., Storandt, M., Miller, J.P. *et al.* (1993) *Arch. Neurol.* **50**, 380–383.
Salmon, D.P. & Butters, N.M. (1992) In *Principles of Geriatric Neurology* (eds Katzman, R. & Rowe, J.W.), pp. 144–163. Philadelphia, F.A. Davis.
Salthouse, T.A. & Skovronek, E. (1992) *J. Gerontol.* **47**, P110–P120.
Saunders, A.M., Strittmatter, W.J., Schmechel, D. *et al.* (1993) *Neurology* **43**, 1467–1472.
Scheff, S.W., Dekosky, S.T. & Price, D.A. (1990) *Neurobiol. Aging* **11**, 29–37.

Schellenberg, G.D., Bird, T.D., Wijsman, E.M. *et al.* (1992) *Science* **258**, 668–671.
Scheltens, P., Ravid, R. & Kamphorst, W. (1994) *Neurology* **44**, 279–282.
Schmidt, M.L., DiDario, A.G., Lee, V.M-Y., Trojanowski, J.Q. (1994) *FEBS Lett*, **344**, 69–73.
Schmitt, F.A. & Sano, M. (1994) In *Handbook of Dementing Illnesses* (ed. Morris, J.C.), pp. 89–123. New York, Marcel Dekker.
Schoenberg, B.S., Kokmen, E. & Ozakazi, H. (1987) *Ann. Neurol.* **22**, 724–729.
Selnes, O.A., Carson, K., Rovner, B. & Gordon, B. (1988) *Neurology* **38**, 1053–1056.
Sherrington, R., Rogaev, E.I., Liang, Y. *et al.* (1995) *Nature* **375**, 754–760.
Squire, L.R. & Zola-Morgan, S. (1991) *Science* **253**, 1380–1386.
Stern, Y., Gurland, B., Tatemichi, T.K., Tang, M.X., Wilder, D. & Mayeux, R. (1994) *JAMA* **271**, 1004–1010.
Storandt, M. & Hill, R.D. (1989) *Arch. Neurol.* **46**, 383–386.
Strittmatter, W.J., Saunders, A.M., Schmechel, D. *et al.* (1993) *Proc. Natl Acad. Sci. USA* **90**, 1977–1981.
Sumi, S.M., Bird, T.D., Nochlin, D. & Raskind, M.A. (1992) *Neurology* **42**, 120–127.
Swearer, J.M., Drachman, D.A., O'Donnell, B.F. & Mitchell, A.L. (1988) *J. Am. Geriatr. Soc.* **36**, 784–790.
Teri, L., Borson, S., Kiyak, A. & Yamagishi, M. (1989) *J. Am. Geriatr. Soc.* **37**, 109–116.
Terry, R.D., Hansen, L.A., DeTeresa, R., Davies, P., Tobias, H. & Katzman, R. (1987) *J. Neuropathol. Exp. Neurol.* **46**, 262–268.
Terry, R.D., Masliah, E., Salmon, D.P. *et al.* (1991) *Ann. Neurol.* **30**, 572–580.
Terry, R.D., Masliah, E. & Hansen, L.A. (1994) In *Alzheimer Disease* (eds Terry, R.D., Katzman, R. & Bick, K.L.), pp. 179–196. New York, Raven Press.
Tierney, M.C., Fisher, R.H. & Lewis, A.J. *et al.* (1988) *Neurology* **38**, 359–364.
Tombaugh, T.N. & McIntyre, N.J. (1992) *J. Am. Geriatr. Soc.* **40**, 922–935.
Tomlinson, B.E. (1989) *Neuropathol. Appl. Neurobiol.* **15**, 491–512.
Tomlinson, B.E., Blessed, G. & Roth, M. (1968) *J. Neurol. Sci.* **7**, 331–356.
Tomlinson, B.E., Blessed, G. & Roth, M. (1970) *J. Neurol. Sci.* **11**, 205–242.
Torack, R.M. (1979) *Neurosurgery* **4**, 432–434.
Torack, R.M. & Morris, J.C. (1988) *Arch. Neurol.* **45**, 497–501.
Trojanowski, J.Q., Schmidt, M.L., Shin, R.W., Bramblett, G.T., Goedert, M. & Lee, V.M.-Y. (1993) *Clin. Neurosci.* **1**, 184–191.
Ulrich, J. (1985) *Ann. Neurol.* **17**, 273–277.
Vallet, P.G., Guntern, R., Hof, P.R., Golaz, J., Delacourte, A., Robakis, N.K. & Bouras, C. (1992) *Acta Neuropathol.* **83**, 170–178.
Verhaeghen, P., Marcoen, A. & Goossens, L. (1993) *J. Gerontol.* **48**, 157–171.
Victor, M. & Agamonolis, D. (1990) *J. Cog. Neurosci.* **2**, 246–257.
Victoroff, J., Ross, W.G., Benson, D.F., Verity, M.A. & Vinters, H.V. (1994) *Arch. Neurol.* **51**, 269–274.
Vinters, H. (1992) *J. Neurol. Sci.*, **112**, 1–3.
Vogels, O.J.M., Broere, A.J., Ter Laak, H.J., Ten Donneklaar, H.J., Nieuwenhuys, R. & Schulte, B.P.M. (1990) *Neurobiol. Aging* **11**, 3–13.
Von Dras, D.D. & Blumenthal, H.T. (1992) *J. Am. Geriatr. Soc.* **40**, 285–294.
Wade, J.P.H., Mirsen, T.R., Hachinski, V.C. *et al.* (1987) *Arch. Neurol.* **44**, 24–29.
Walsh, J.S., Welch, H.G. & Larson, E.B. (1990) *Ann. Intern. Med.* **113**, 429–434.
West, M.J. (1993) *Neurobiol. Aging* **14**, 287–293.
Whitehouse, P.W., Price, D.L., Clark, A.W., Coyle, J.T. & DeLong, M.R. (1981) *Ann. Neurol.* **10**, 122–126.
Whitehouse, P.J., Price, D.L., Struble, R.G., Clark, A.W., Coyle, J.T. & DeLong, M.R. (1982) *Science* **215**, 1237–1239.
Wilcock, G.K. & Esiri, M.M. (1982) *J. Neurol. Sci.* **56**, 343–356.
Wisniewski, K., Jervis, G.A., Moretz, R.C. & Wischik, H.M. (1979) *Ann. Neurol.* **5**, 288–294.

Wisniewski, K.E., Wisniewski, H.M. & Wen, G.Y. (1985) *Ann. Neurol.* **17**, 278–282.
Zhan, S.S., Beyreuther, K. & Schmitt, H.P. (1993) *Dementia* **4**, 66–74.
Zhang, M., Katzman, R., Salmon, D. *et al.* (1990) *Ann. Neurol.* **27**, 428–437.
Zola-Morgan, S. & Squire, L.R. (1993) In *Annual Review of Neuroscience*, Vol. 16 (eds Cowan, W.M., Shooter, E.M., Stevens, C.F. & Thompson, R.F.), pp. 547–563. Palo Alto, CA, Annual Reviews, Inc.
Zubenko, G.S. & Moossy, J. (1988) *Arch. Neurol.* **45**, 1182–1186.
Zweig, R.M., Ross, C.A., Hedreen, J.C. *et al.* (1988) *Ann. Neurol.* **24**, 233–242.

CHAPTER 12

APOLIPOPROTEIN E AND ALZHEIMER'S DISEASE

Judes Poirier, Josephine Nalbantoglu, Daniel Guillaume and Philippe Bertrand

McGill Centre for Studies in Aging, Departments of Psychiatry and Neurology and Neurosurgery, and the Douglas Hospital Research Centre, Verdun, Québec, Canada, H4H 1R3

Table of Contents

12.1 Apolipoprotein E

Apolipoprotein E (ApoE) is a polymorphic protein associated with plasma lipoproteins (Mahley, 1988). It interacts with the 'remnant receptor' (ApoE receptor) and the LDL receptor (ApoE/B receptor) of liver and other organs, to modulate the catabolism of triglyceride-rich lipoprotein particles. It has been extensively studied in nervous and non-nervous tissues as one of several apolipoproteins that direct lipid metabolism. Because ApoE facilitates cholesterol movement both in and out of cells, and from cell to cell, it has been well characterized in studies dealing with abnormal cholesterol and phospholipid metabolism (Mahley, 1988).

Human ApoE is a 37 kDa protein encoded by a four-exon gene (3.6 kb in length) on the long arm of chromosome 19. Common ApoE polymorphisms at residues 112 and 158 are determined by three different alleles at the ApoE locus, designated E4, E3 and E2 (for a review, see Davignon *et al.*, 1988) (Figure 1). The ancestral isoform, ApoE3, has a cysteine at residue 112 and an arginine at residue 158, whereas argi-

Pathobiology of Alzheimer's Disease
ISBN 0-12-286965-5

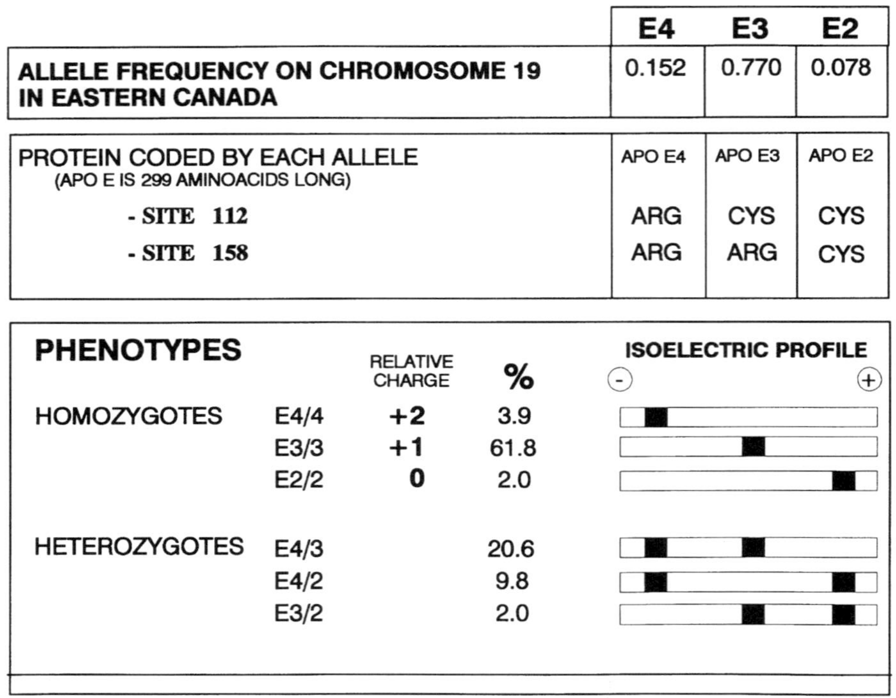

Figure 1 Apolipoprotein E polymorphisms, population distribution and biochemical properties.

nine in ApoE4 and cysteine in ApoE2 are present at both sites. This polymorphism results in six ApoE phenotypes in the population: E2/2, E3/3 and E4/4 in homozygotes, and E3/2, E4/2 and E4/3 in heterozygotes. Functionally, ApoE2 has a lower affinity for the LDL receptor than E3 and E4 (Wesgraber *et al.*, 1982). Lipoproteins associated with ApoE4 are cleared more efficiently than the ones containing ApoE3 and ApoE2. It has been estimated that 60% of the variation in cholesterol is genetically determined and that ApoE polymorphisms account for 15% of this genetic variation (Sing and Davignon, 1985).

ApoE-containing lipoproteins such as VLDL and LDL, which contain apolipoprotein B (ApoB), bind to the same cell surface receptor, namely the LDL receptor (Wesgraber and Mahley, 1986). The LDL receptor has been extensively characterized. It is a glycoprotein of an apparent molecular weight of 160 kDa, which is encoded by a 45 kb gene made up of 18 exons and 17 introns (Sudhof *et al.*, 1985). A number of lines of evidence suggest that positively charged amino acids of ApoE located between residues 140 and 150 interact with negatively charged amino acids of the LDL receptor. The binding of lipoproteins to the LDL receptor is calcium dependent and inhibited by proteases (Brown and Goldstein, 1986). The LDL receptor pathway consists of cell surface binding of ApoE- or Apo B-containing lipoproteins followed by internalization and degradation of the LDL particle via a lysosomal pathway (Goldstein *et al.*, 1983). As the incoming cholesterol concentration increases, gene expression and activity of the rate-controlling enzyme in cholesterol synthesis, 3,3′-hydroxymethyl-glutaryl coenzyme A reductase (HMG-CoA reductase), becomes progressively repressed (Brown *et al.*, 1973). The intracellular accumulation of cholesterol also reduces the synthesis of LDL receptor and favours the esterification of the cholesterol excess via activation of acyl-CoA : cholesterol acyltransferase for storage.

12.2 ApoE and LDL receptor: a role in PNS regeneration and CNS reinnervation

ApoE is unique among apolipoproteins in that it has a special relevance to nervous tissue where it is involved in the mobilization and redistribution of cholesterol in repair, growth and maintenance of myelin and neuronal membranes during development or after injury in the PNS (Boyle *et al.*, 1985, 1989; Pitas *et al.*, 1987; Mahley, 1988; Leblanc and Poduslo, 1991). In the central nervous system (CNS), ApoE plays a pivotal role in the mobilization and redistribution of cholesterol and phospholipid during membrane remodelling associated with synaptic plasticity (Poirier *et al.*, 1991a, b, c, 1993a). The importance of ApoE in the brain is further underscored by the absence of other key plasma apolipoproteins such as ApoA1 and ApoB (Roheim *et al.*, 1979) in this tissue.

After a sciatic nerve injury in the PNS, large amounts of lipids are released from degenerating axon membranes and myelin. In response, macrophages synthesize and release ApoE within the peripheral lesion in order to scavenge cholesterol from both cellular and myelin debris (Boyle *et al.*, 1985, 1989; Leblanc and Poduslo, 1991).

During this critical phase, cholesterol synthesis is markedly depressed, suggesting a receptor-mediated down-regulation of cholesterol synthesis by the internalization of lipoprotein (Goodrum, 1990). Much of the cholesterol generated during this phase is stored in macrophages and appears to be reused during axonal regeneration (Rawlins *et al.*, 1970).

In the CNS, the ability of neurons to regenerate is very limited. However, the brain has the ability to induce proliferation of presynaptic extensions from axons or terminals derived from undamaged neurons in order to compensate for the loss of specific input. A classical example of reactive synaptogenesis is illustrated by the compensatory response of the hippocampal formation to entorhinal cortex lesions (ECL). ECL was shown to cause the loss of nearly 60% of the synaptic input to the granule cell layer of the hippocampus. The loss of synapses, however, is transient. Beginning a few days after denervation, new synapses are formed, virtually replacing the lost inputs within a few months (Matthews *et al.*, 1976). These new synapses originate from the cholinergic septal neurons (Lynch *et al.*, 1972), glutamatergic commissural-associational pyramidal cells of the CA3/Hilus areas (Lynch *et al.*, 1976; Scheff *et al.*, 1980) and, to a lesser extent, from neurons of the contralateral entorhinal cortex (Steward and Scoville, 1976).

This sequence of compensatory changes in the CNS was found to correlate with the increased expression of ApoE in the deafferented zone of the molecular layer (Poirier *et al.*, 1991b, c, 1993a). In the injured CNS, cells that synthesize ApoE have been identified as astrocytes (Poirier *et al.*, 1991c) rather than macrophages.

Ultrastructural studies of the molecular layer of the dentate following ECL showed that throughout the 2–11 days post-lesion, astrocytes progressively engulf both presynaptic terminals and preterminal axons (Lee *et al.*, 1977). It has been postulated that once metabolized, these neuron-derived particles could generate a large astroglial store of lipids, providing a convenient and readily retrievable pool for membrane synthesis of precursors used in the formation of neuronal sprouts and in the reorganization of the dendritic field of granule cell neurons in the hippocampus (Figure 2, #1 and #2). The accumulation of large amounts of sudanophilic lipids and deposits (cholesterol esters and phospholipids) in the molecular layer of the dentate gyrus during the early phase of the reinnervation process is compatible with such a concept (Poirier *et al.*, 1993a).

Figure 2 Role of astrocytes in the catabolism of degenerating terminals following hippocampal deafferentation. Degenerating terminals are initially internalized and degraded. The non-esterified cholesterol (FC) (1) is used (2) as free cholesterol for the assembly of an ApoE/cholesterol/lipoprotein complex or it is converted into cholesterol esters (CE) for storage purposes. The newly formed ApoE/cholesterol/lipoprotein complexes are then directed toward the circulation and to specific brain cells requiring lipids. ApoE complexes are apparently internalized by the neuronal LDL receptor pathway (4) and the cholesterol released for dendritic proliferation and/or synaptogenesis (5). As a consequence of the internalization process, cholesterol synthesis in neurons (via the HMG-CoA reductase pathway) becomes progressively repressed.

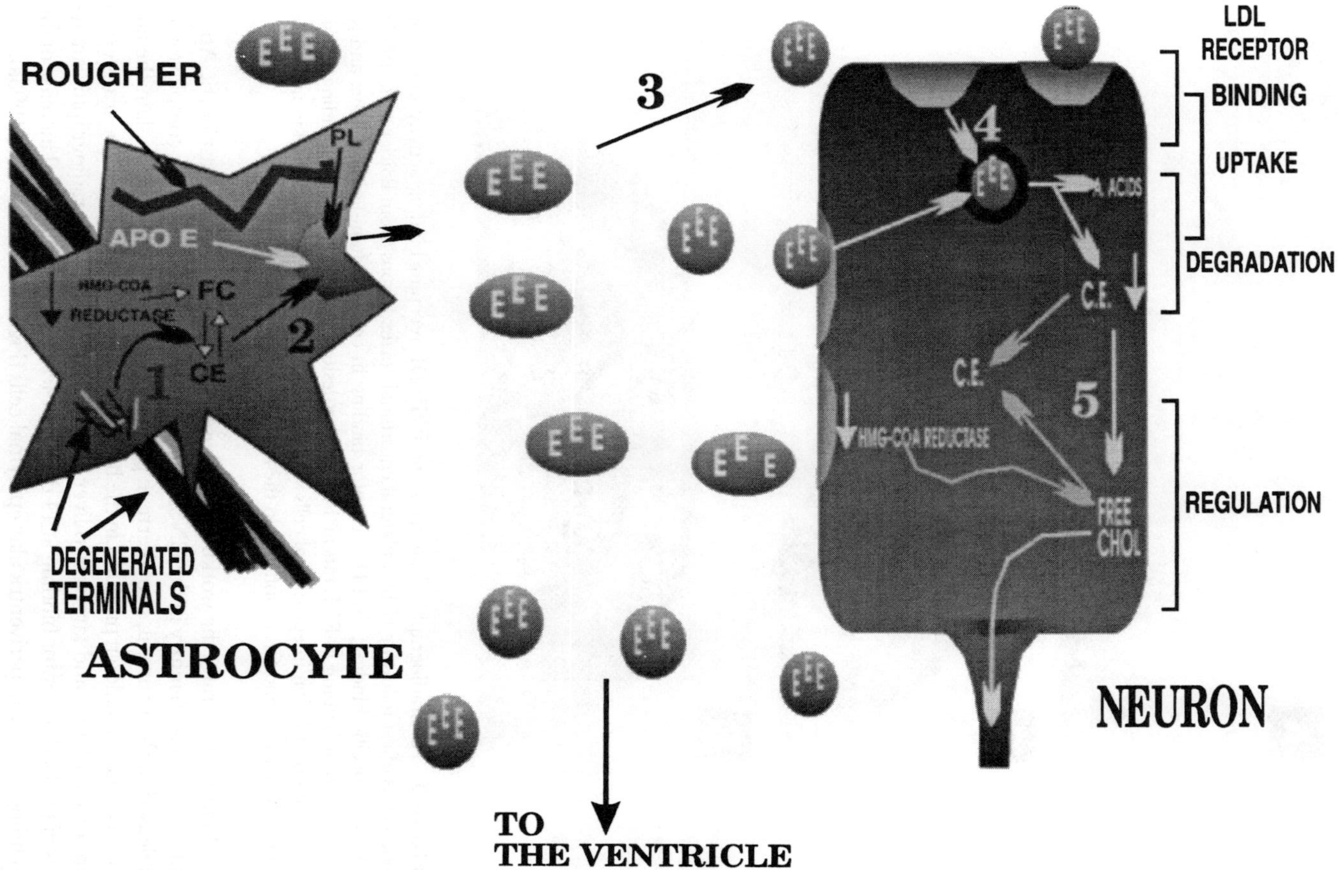
ROUGH ER
PL
APO E
HMG-COA
FC
REDUCTASE
CE
1
2
DEGENERATED
TERMINALS
ASTROCYTE
3
TO
THE VENTRICLE
4
A. ACIDS
C.E.
C.E.
5
HMG-COA REDUCTASE
FREE
CHOL
LDL
RECEPTOR
BINDING
UPTAKE
DEGRADATION
REGULATION
NEURON

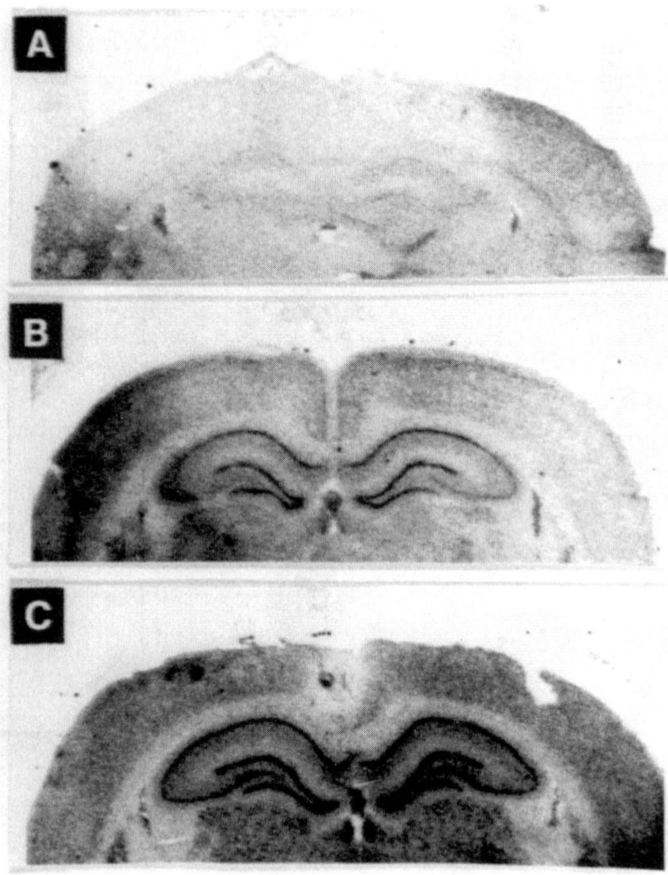

Figure 3 Autoradiographic analysis of the [^{125}I]LDL receptor binding activity in the hippocampus of rat following right unilateral entorhinal cortex lesions (from Poirier *et al.*, 1993a). (A) Non-specific binding; (B) LDL receptor binding in control (sham-lesioned) rats; and (C) LDL receptor binding in ECL rats at 14 days post-lesion. Note the increased binding activity at 14 days post-lesion in ependymal cells surrounding the ventricular space and in neurons of the dentate gyrus and CA1 subfields on the right (lesioned) side.

Figure 2 illustrates the working model used to examine the role of astrocytic ApoE in the response to injury in the CNS. Once lipid-rich ApoE-containing complexes are released in the neuropil, two alternate pathways can be proposed to explain the fate of terminal breakdown products in the hippocampus. First, lipids could be transferred to the circulation and targeted to liver for disposal. Indeed, autoradiographic analysis of the LDL receptor binding at 14 days post-lesion reveals a marked increase of the binding activity in periventricular epidymal cells (Figure 3; Poirier *et al.*, 1993a).

Alternatively, ApoE/lipoprotein complexes could be taken up by local cells requiring cholesterol and phospholipids for membrane synthesis. The time-dependent, cell-specific induction of [^{125}I]LDL binding in granule cells (Figure 3) in ECL rats is consistent with such a concept. Figure 4 illustrates the concomitant alterations in ApoE protein levels, LDL receptor binding activity in granule neurons of the dentate and total hippocampal HMG-CoA reductase activity of rat following ECL (Poirier *et al.*, 1993a). The accumulation of cholesterol in granule cell neurons via the ApoE/LDL receptor pathway is certainly compatible with the down-regulation of the hippocampal HMG-CoA reductase activity at 6–8 days post-lesion, and is apparently the result of a cholesterol-mediated suppression of the HMG-CoA reductase transcription in reinnervated neurons.

Interestingly, the induction of the LDL receptor binding activity in granule cell neurons correlates very well with the extent of the compensatory cholinergic plasticity in the outer portion of molecular layer of the dentate gyrus (Poirier *et al.*, 1993a). On account of the role of ApoE–lipoprotein complexes in the transport of phospholipids in blood, it is tempting to postulate that ApoE could play a similar role in the CNS. Membrane phospholipids such as phosphatidyl-ethanolamine and phosphatidylcholine are well-known precursors of acetylcholine. It is thus quite conceivable that, in addition to the transport of cholesterol required for synaptogenesis, ApoE could also affect cholinergic activity of the hippocampus through the availability of precursor lipid in the synthesis of acetylcholine. This possibility will be discussed later in relation to the recent ApoE4 association with sporadic Alzheimer's disease.

Together, these results indicate that the CNS has evolved a powerful mechanism allowing lipid redistribution during neuronal reinnervation, despite the fact that it lacks some critical apolipoproteins used in PNS regeneration. Moreover, since the adult brain does not accumulate cholesterol (Pardridge and Mietus, 1980), the role played by ApoE and its receptor during reinnervation becomes even more critical for the redistribution of available cholesterol and phospholipids to sprouting neurons and glial cells. The cascade of cholesterol/phospholipid transport and re-uptake may ultimately prove to be essential in the ability of the CNS to compensate for the neuronal loss associated with normal brain ageing and neurodegenerative conditions such as Alzheimer's disease. Alternatively, impairment of these compensatory mechanisms might likewise contribute to the emergence of CNS dysfunctions.

12.3 Synaptic remodelling and Alzheimer's disease: a role for ApoE?

During Alzheimer's disease (AD), the hippocampus undergoes selective neuronal degeneration, particularly in the CA1 and subiculum area (Brun and Englund, 1981). Neurofibrillary tangles accumulate in the subiculum and CA1 areas, whereas plaques are more prominent in the dentate gyrus than in the CA1 and CA3 regions. Deficits in hippocampal function are attributed to the loss of intrinsic neuronal connections as well as to degeneration of the entorhinal cortical neurons of layers II and III, which project to the hippocampus via the perforant pathway (Hyman *et al.*, 1984).

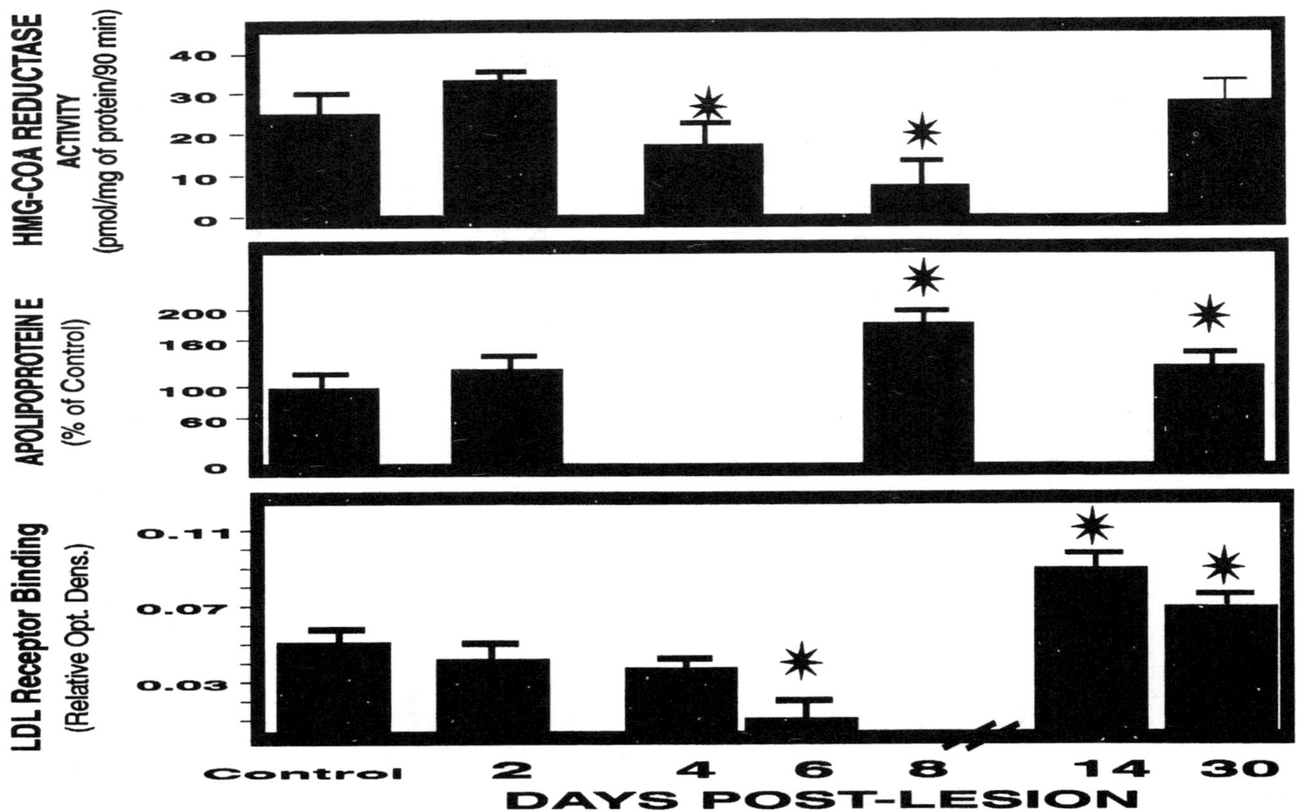

Figure 4 Time-course analysis of ApoE, HMG-CoA activity (cholesterol synthesis) and LDL receptor binding sites density (cholesterol internalization) in the hippocampus of rats following entorhinal cortex lesions. *$p < 0.05$. Adapted from Poirier *et al.* (1993a).

This functional deafferentation elicits in a small group of patients regenerative changes in the dentate gyrus (Geddes *et al.*, 1985; Hyman *et al.*, 1987), which resemble the hippocampal response of rats submitted to entorhinal cortex lesions. However, this observation has been challenged by a number of independent investigators.

Geddes *et al.* (1985) reported an expansion of the glutamatergic receptor field (kainate type) in the molecular layer of the dentate, whereas Greenamyre *et al.* (1987) and Represa *et al.* (1988) showed severe reduction of the [^{3}H]glutamate and [^{3}H]-kainate binding sites in the dentate gyrus of AD brains. Hyman *et al.* (1987) reported enhanced acetylcholine esterase (AChE) staining in the molecular layer of the dentate in a subgroup of AD subjects, which was interpreted as evidence of compensatory sprouting of septohippocampal cholinergic neurons. A decrease rather than an increase in cholinergic sprouting was recently reported by Ransmayr *et al.* (1989) using choline acetyltransferase (ChAT) immunoreactivity. Electron microscopy studies failed to detect the expected compensatory growth associated with the entorhinal innervation deficits. Flood and Coleman (1987) demonstrated a loss of proliferating dendrites in the hippocampus of AD while de Ruiter and Uyling (1987) showed a loss of dendritic spine and length in the dentate of AD patients.

These results suggest that the usual compensatory growth of fibres associated with hippocampal deafferentation is lost in most AD patients. These observations in the hippocampus add up to recent reports on cerebral cortex integrity in AD that showed a 45% decrease in presynaptic terminal density in the neocortex (Masliah *et al.*, 1989) and a significant synaptic loss in layer V of the frontal cortex (–27%) and temporal cortex (–36%) (Davies *et al.*, 1987).

On account of the critical role played by ApoE and the LDL receptor in compensatory reactive synaptogenesis in the injured rat hippocampus following ECL, it is quite conceivable that the poor reinnervation capacity observed in most AD subjects could be caused by a selective impairment of the ApoE/LDL receptor axis.

Previous studies have shown that ApoE mRNA levels are increased (Deitrich *et al.*, 1991) or unchanged (Poirier *et al.*, 1991a) in post-mortem brains of AD patients. These results were obtained from patients whose ApoE genotypes were undetermined. Since ApoE expression (mRNA and protein levels) is markedly increased in the hippocampus of entorhinal cortex lesioned rats (Poirier *et al.*, 1991b, c, 1993a) and during the demyelination phase in humans with multiple sclerosis (Rifai *et al.*, 1987), we examined ApoE mRNA levels in the hippocampus of AD patients whose genotypes have previously been established. Figure 5 illustrates Northern blot analyses of ApoE levels in the autopsied hippocampus of control and AD subjects as a function of their respective ApoE genotype. Autoradiographic analyses of the band corresponding to ApoE revealed no differences between control and AD hippocampal tissues, irrespective of their ApoE genotypes. These results are consistent with a preliminary report (Poirier *et al.*, 1991a) showing that despite astrocytic hyper-reactivity and increases in astrocytic glial fibrillary acidic protein mRNA levels in the hippocampus of AD, ApoE gene expression remained relatively unchanged in the hippocampus.

At least three different hypotheses can be proposed to explain the lack of changes in ApoE gene expression in AD.

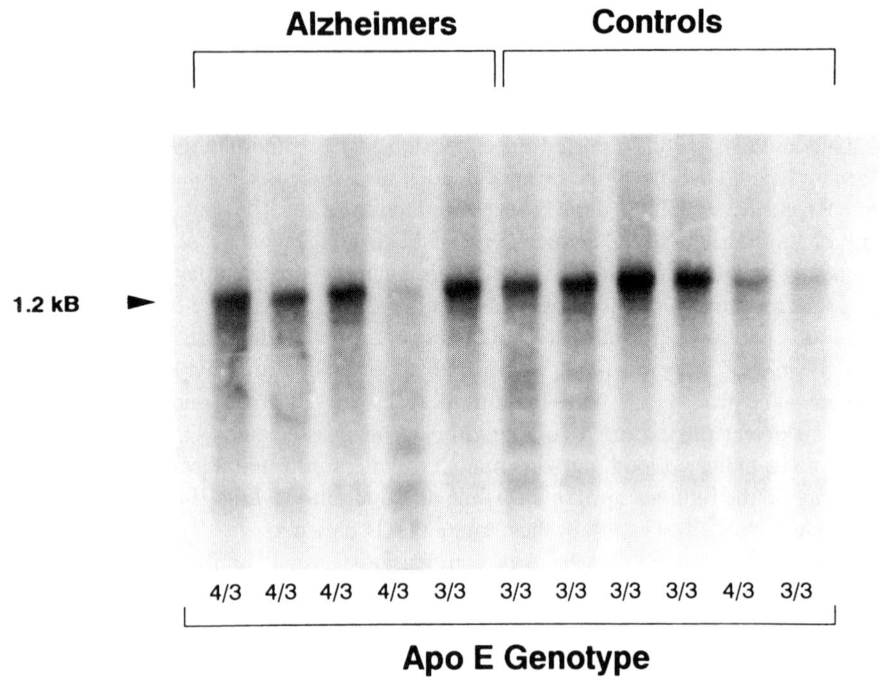

Figure 5 Northern blot analysis of ApoE mRNA prevalence in the post-mortem hippocampus of AD and age-matched control subjects whose ApoE genotypes were determined as described in Nalbantoglu *et al.* (1994).

First, in contrast to animal models, terminal sprouting and reactive synaptogenesis in humans may not require the presence of ApoE for the transport of cholesterol and phospholipids. However, reports that patients with multiple sclerosis show increased cerebrospinal fluid (CSF) levels of ApoE during remission phases (active remyelination) but not during the exacerbation phases (myelin breakdown) (Rifai *et al.*, 1987) suggest a role for ApoE in lipid homeostasis associated with CNS remodelling in humans.

A second possibility could be a selective alteration of the ApoE-synthesizing glial cell population in AD. Although there is little information regarding the number of ApoE-immunopositive astrocytes in AD versus control subjects, there is a marked increase in the number of glial fibrillary acidic protein (GFAP)-positive astrocytes and in GFAP mRNA prevalence in the hippocampus of AD (Finch, 1987).

A third possible explanation is that the unidentified aetiological factor(s) causing AD could modulate directly or indirectly ApoE gene expression. Alternatively, the presence of specific mutation(s) in the ApoE gene could result in the poor regulation of its own expression. Because of the widespread expression of ApoE in the human body, one has to assume that a disruption of ApoE gene regulation should have a generalized impact on lipid homeostasis. Indeed, several previous studies have reported high blood cholesterol levels in AD versus control and stroke subjects (Giubilei *et al.*, 1990), and in AD versus multiple-infarct dementia subjects (Muckle and Roy, 1985).

Differences in amino-acid sequences corresponding to the major ApoE isoforms were shown to be major determinants of the differences in plasma cholesterol levels within the general population, with ApoE playing a key role in the clearance of cholesterol from plasma (Davignon *et al.*, 1988). The synthesis of ApoE is under the control of three independent alleles, located at a single gene locus, coding for the major isoforms E2, E3 and E4 with respective frequencies of 8, 77 and 15% for the Eastern Canada population (Sing and Davignon, 1985). The homozygote E3/E3 is the most common phenotype encountered and E2/E2 is the least common. As a result, the avidity of ApoE-containing lipoproteins for lipoprotein receptors increases from ApoE2 to ApoE3 to ApoE4. The gradient in serum cholesterol is associated with a gradient in ApoE phenotype, E2 being associated with lower serum lipoproteins and total cholesterol levels than E3 and E4 (Davignon *et al.*, 1988). Thus, if an impaired regulation of lipid homeostasis was to be part of the aetiological process associated with AD, then subjects with the E4/E4 or E3/E4 phenotype should be at an increased risk of having AD.

This concept would also agree with the reported genetic association between some families with late-onset AD. The 3.5 kb allele of a Taq1 RFLP on chromosome 19 is less than 50 kb from the ApoE locus on the 5′ end of the gene (Schellenberg *et al.*, 1992).

12.4 Alzheimer's disease and ApoE4

Recently, several independent research groups have examined the frequency distribution of ApoE isoforms in AD subjects in the USA, Canada, France, Japan and Germany. The frequency of the ApoE4 allele was shown to be markedly increased in sporadic (Lucotte *et al.*, 1993; Noguchi *et al.*, 1993; Poirier *et al.*, 1993b; Saunders *et al.*, 1993) as well as in late-onset familial AD (Corder *et al.*, 1993; Payami *et al.*, 1993; Saunders *et al.*, 1993). A gene dosage effect was observed in both familial (Corder *et al.*, 1993) and sporadic (Poirier *et al.*, 1993b) cases (i.e. as age of onset increases, E4 allele copy number decreases).

Figure 6 represents the ApoE4 allele frequency and prevalence as a function of age distribution in living patients with probable AD (Poirier *et al.*, 1993a) and autopsied-confirmed AD subjects (Nalbantoglu *et al.*, 1994) from Eastern Canada. As many as 80% of all clinical AD with age 65–75 carry at least one copy of the ApoE4 allele. Interestingly, there is a sharp decline in the prevalence of ApoE4 in AD at very old age; this suggests the presence of a late-late onset form of AD. A similar distribution profile was observed in the post-mortem AD cases. Comparable results were obtained by Rebeck *et al.* (1994) with a large clinical cohort of AD patients from the Boston area.

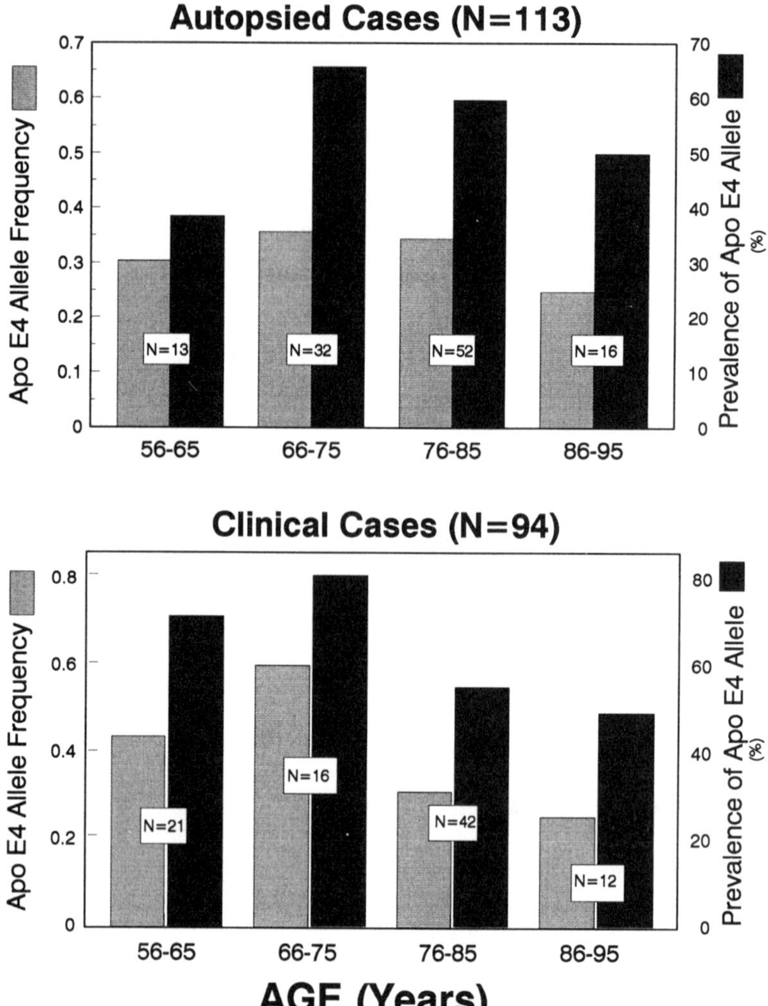

Figure 6 ApoE4 allele frequency and prevalence in autopsied (top) and clinical (bottom) sporadic Alzheimer's disease subjects from Eastern Canada. Note the age-dependent reduction of the ApoE4 allele frequency in both living and autopsied AD cases.

Control cases represent an interesting challenge. It was shown that living control subjects with mini-mental state scores greater than 28 (n = 74; mean age 75.5 years) exhibit an ApoE4 allele frequency of 12.2% (Poirier *et al.*, 1993b). Post-mortem control subjects with plaque and tangle counts below AD threshold (n = 78; mean age 70.2) exhibit an ApoE4 allele frequency of 5% (Nalbantoglu *et al.*, 1994). Finally, living French centenarians (n = 325; mean age 100.7) showing no signs of dementing illnesses exhibit an ApoE4 allele frequency of 5% (Schachter *et al.*, 1994). There is an important drop in ApoE4 allele frequency between cognitively normal octogenarians (Poirier *et al.*, 1993b) and cognitively normal nonagerians (Rebeck *et al.*, 1994) living in North America.

These studies suggest that more than 60% of cognitively normal individuals aged 65–80 carrying a copy of ApoE4 allele are at risk of developing AD before they reach the age of 90. Canadian women, who are at a greater risk of developing sporadic AD (women/men ratio: 2/1; from the Canadian Study of Health and Aging Working Group, 1994), showed a marked increase in ApoE4 allele frequency when compared to age-matched men (Poirier *et al.*, 1993b).

The inheritance of late-onset AD is clearly associated with the inheritance of the E4 allele in individuals who live long enough to be at risk. Thus, the population frequencies of the E4 allele may contribute to the prevalence of AD in different geographical locations and in genetically distinct populations. Analysis of ApoE allele frequencies in 45 populations shows that there are differences in ApoE allele frequencies between Caucasians, Japanese and Chinese (Gerdes *et al.*, 1992). Some differences in ApoE allele frequencies between Caucasian populations were noted and discussed by Davignon *et al.* (1988).

Although the definition of dementia or the methods of assessment may vary from centre to centre, there is a noticeable variation in prevalence and incidence of AD around the world (Breteler *et al.*, 1992). Interestingly, a recent study by Mayeux *et al.* (1993) showed that the association between ApoE4 and AD varied significantly across three different ethnic groups (black, hispanic and white) from New York City.

A careful examination of the published ApoE phenotypes and genotypes for several different populations worldwide (Gerdes *et al.*, 1992) and the age-adjusted prevalence of AD (Breteler *et al.*, 1992) reveals an interesting relationship (Figure 7). There is a highly significant correlation between the ApoE4 allele frequency and the AD prevalence in six different countries (and seven regions) for which allele frequencies and AD prevalence values are currently available. In contrast, there is no correlation between the ApoE3 and ApoE2 allele frequencies and AD prevalence in those same regions. This observation suggests that the well-established association between ApoE4 and AD is not population specific but represents a worldwide effect of genetic aetiology.

12.5 ApoE4 and neuropathological features

It has been shown that ApoE-like immunoreactivity was found to be associated with amyloid in senile plaques, cerebral vessels and neurofibrillary tangles (Namba *et al.*, 1991). *In vitro*, amyloid β (Aβ) binds more avidly to ApoE4 than to ApoE3 (Strittmatter *et al.*, 1993; Wisniewski *et al.*, 1993). Increased peptide deposition was recently reported in the cerebral cortex (Schmechel *et al.*, 1993) and hippocampus (Poirier *et al.*, 1994) of late-onset AD carrying one or two copies of the ApoE4 allele. Analyses of the effect of ApoE4 allele copy number on the prevalence of senile plaques in the

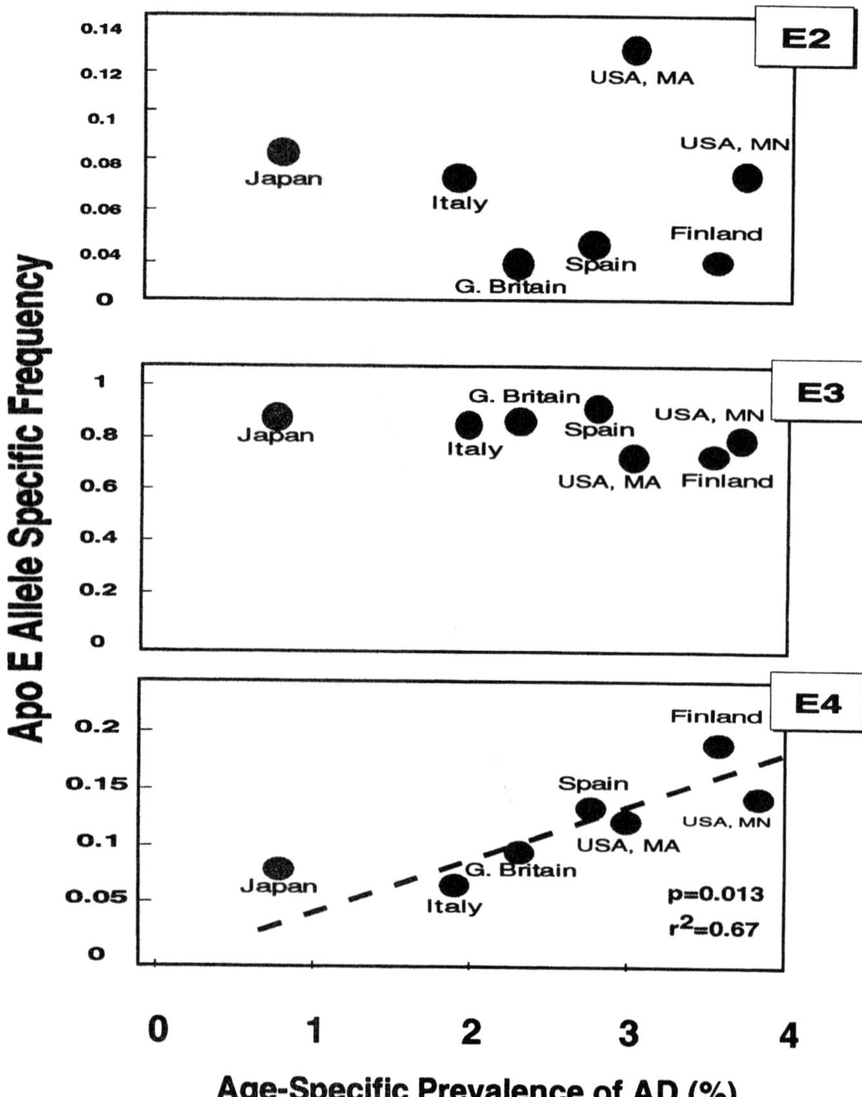

Figure 7 Correlational analysis of the incidence of the different ApoE polymorphisms as a function of the age-adjusted prevalence for AD in six different countries and seven distinct regions. Adapted from Gerdes *et al.* (1992) and Breteler *et al.* (1992).

hippocampus of 59 AD patients (Figure 8) reveal highly significant correlations between those two markers in the CA1 subfield, the subiculum and the parasubiculum (Poirier *et al.*, 1994). This study also revealed significant correlations between the ApoE4 allele copy number and the density of neurofibrillary tangles in the CA1 subfield and the subiculum but not in the parasubiculum of sporadic AD subjects.

12.6 ApoE4 and cholinergic deficits

As losses of cholinergic neurons and/or ChAT activity are well-known classical hallmarks of AD (Davies and Maloney, 1976; Perry *et al.*, 1977; McGeer, 1984), a possible relationship between the ApoE4 genotype and a cholinergic marker was deemed as highly relevant. On the basis of its well-recognized post-mortem stability, ChAT activity was examined in relation to ApoE genotype. Figure 9 illustrates that the reduction in ChAT activity [as measured by the method of Tucek (1978)] in the hippocampus and temporal cortex of AD cases is proportional to the ApoE4 allele copy number (i.e. as ApoE4 allele copy number increased, ChAT activity decreased) (Poirier *et al.*, 1994). Results obtained in our laboratory clearly indicate the existence of distinct genetic entities in sporadic AD, which show a differential degree of alteration of cholinergic innervation, at least as revealed by ChAT activity.

The ApoE4 allele copy number-related reduction in ChAT activity could be caused by at least two distinct phenomena. First, phospholipids such as phosphatidylcholine (PC) and phosphatidyl-ethanolamine (PE), which serve as precursors to choline in the synthesis of acetylcholine (Ach), could be transported into neurons via the classical ApoE–LDL receptor pathway. An isoform-dependent impaired regulation of the transport of phospholipids in the brain of ApoE4 carriers could explain the reduced levels of PC, PE and choline reported in the AD brain (Pettigrew, 1989; Nitsch *et al.*, 1992); this would then lead to decreased Ach synthetic capacities.

Alternatively, the reduction in neuronal ChAT activities and choline levels could be secondary to losses of cholinergic neurons in the basal forebrain, including the nucleus basalis of Meynert (NBM), diagonal band of Broca (DBB) and the septal area. A preliminary analysis of the number of acetylcholinesterase-positive neurons in the NBM and DBB of a small number of AD patients ($n = 7$) revealed marked losses of cholinergic neurons in ApoE4 versus ApoE3 homozygous AD cases (↓70% in NBM and ↓45% in DBB; Figure 10).

Taken together, these results clearly suggest that cholinergic function in AD-E3/3 subjects may at least be partially spared when compared to AD-E4/3 and AD-E4/4 carriers. Most importantly, this genetic susceptibility could result in subgroups of AD patients that may respond differently to cholinomimetics, ApoE4 subjects having lost most of their Ach synthetic capacities.

The cholinergic hypothesis of geriatric memory dysfunction as discussed by Bartus more than a decade ago (Bartus *et al.*, 1982) raised some fundamental questions regarding the observed heterogeneity of clinical responses toward various choli-

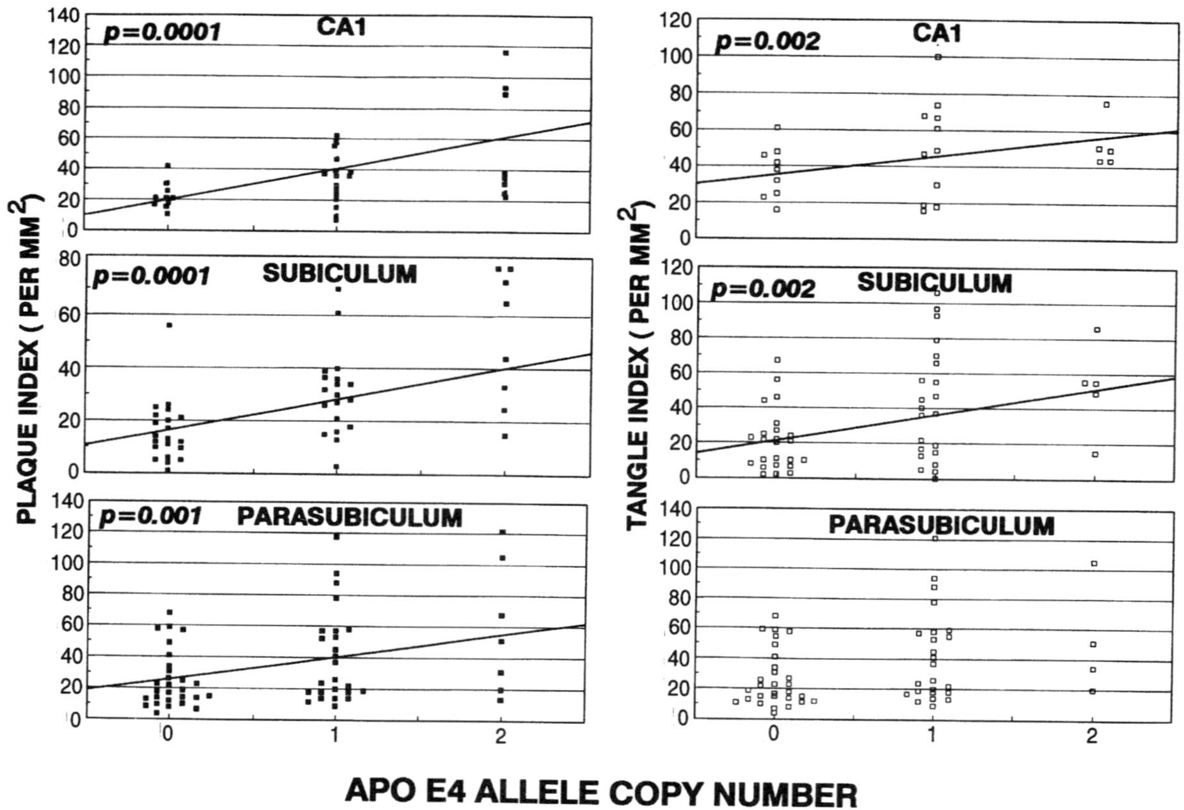

Figure 8 ApoE4 allele copy number, senile plaque (left) and neurofibrillary tangle (right) densities in different subfields of the hippocampus in Alzheimer's disease. Each dot refers to one AD subject. Statistically significant correlations are indicated by the p values in the figure.

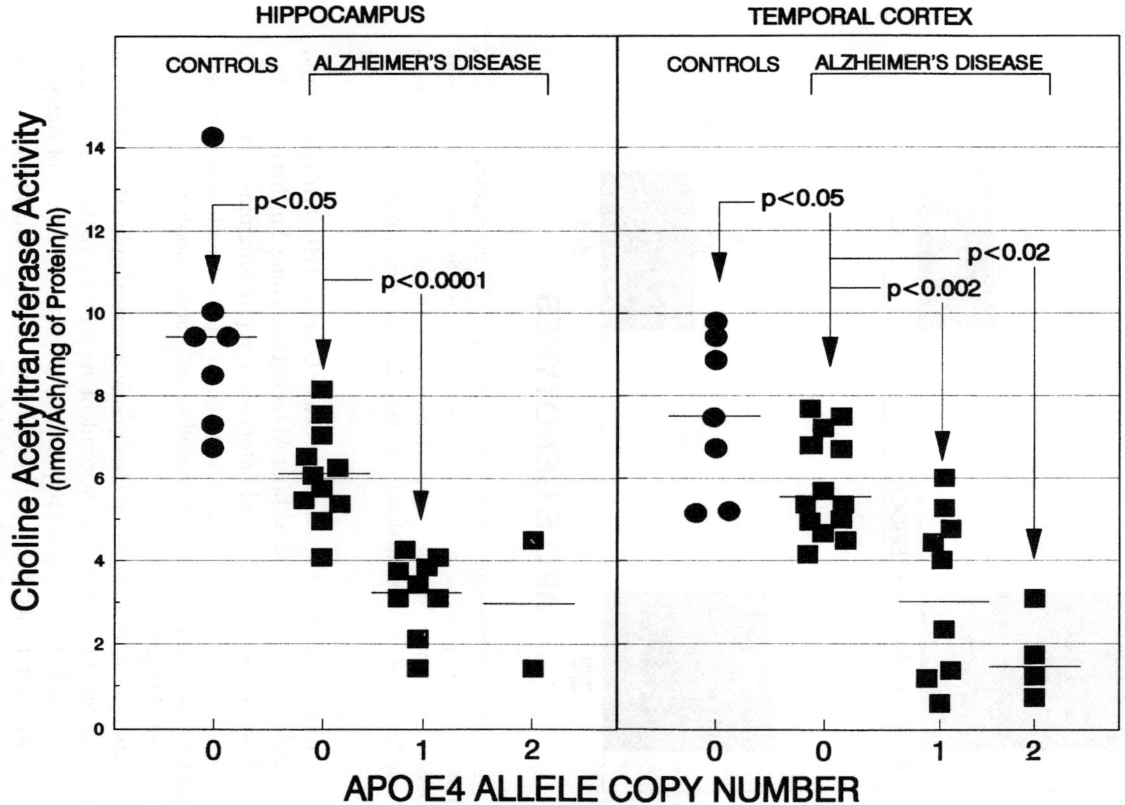

Figure 9 ApoE4 allele copy number and choline acetyltransferase activity in post-mortem control and Alzheimer's disease brains in hippocampal formation and temporal cortex. Each dot refers to one subject: control subjects are represented by solid circles; AD subjects are represented by solid squares. Significant differences between group are indicated by the *p* values on the figure.

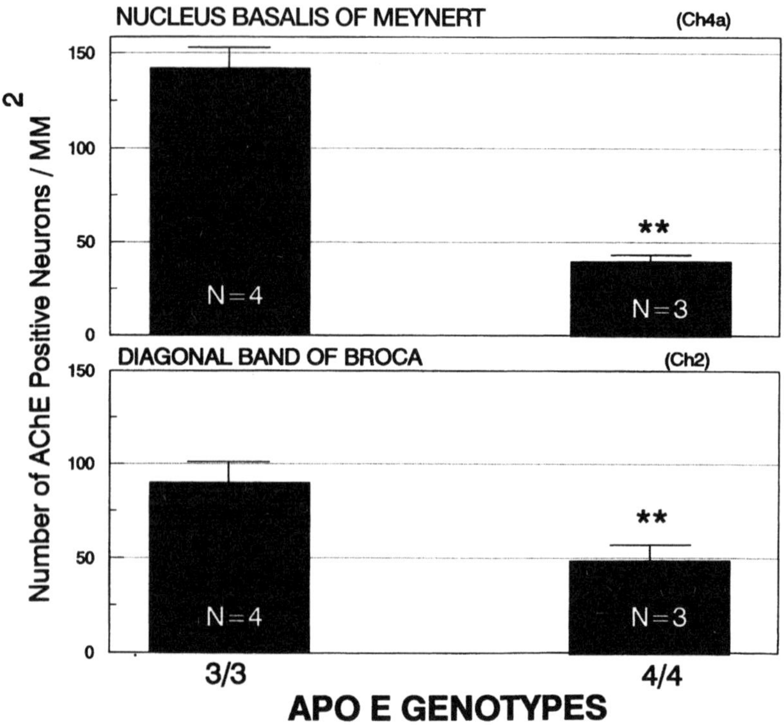

Figure 10 Acetylcholinesterase-positive neuronal cell counts in the nucleus basalis of Meynert (top panel) and diagonal band of Broca (bottom panel) as a function of ApoE genotype in homozygous Alzheimer's disease subjects. Bars represent mean ± S.D. The methods used in this study have been described in detail elsewhere (Etienne *et al.*, 1986).

nomimetics in patients with AD. The absence of clear beneficial effects of choline and lecithin on geriatric patients with and without AD is still intriguing. Furthermore, clinical trials based on the use of esterase inhibitors such as physostigmine (Davis *et al.*, 1979) have shown that in contrast to young subjects, the optimal acute dose necessary to facilitate performance on memory tasks varied considerably among individual aged control subjects and AD patients.

Individual variations in residual brain cholinergic innervation in AD and clinical responses to cholinergic agents could be the result of the presence of the ApoE4 allele; ApoE4 carriers are unlikely to be good responders, at least with the use of Ach precursor and esterase-based therapies. This could have a significant impact on the design of future cholinomimetic-based trials in AD. Prospective-retrospective analyses of genotyped AD patients, which were either good or poor responders to cholinomimetics, are currently underway to examine the effect of the presence of ApoE4 allele on the clinical response of these patients.

12.7 Conclusion

Results gathered around the world indicate a significant over-representation of ApoE4 in sporadic and familial late-onset AD patients compared to controls. These reports indicate that the inheritance of the ApoE4 allele represents a very strong risk factor for AD. Although it is not possible at the moment definitively to assess the aetiological role of ApoE4 in AD without large-scale epidemiological studies, it is clear that ApoE4 has a direct impact on: (i) senile plaque accumulation; (ii) neurofibrillary tangle accumulation in selected brain areas, and (iii) cholinergic deficits (ChAT activity and cholinergic neuronal cell loss) – three neuropathological landmarks of AD. Whether ApoE4 is an aetiologically relevant gene or an amplifying gene remains to be established. Preliminary data suggest that the interactions between ApoE4 and Aβ appear to be closely related to the formation of amyloid plaques and could play a role in the pathophysiological process leading to AD.

References

Bartus, R.T., Dean, R.L., Beer, B. & Lippa, A.S. (1982) *Science* **217**, 408–417.
Boyle, J.K., Pitas, R.E., Wilson, E., Mahley, R.W. & Taylor, J.M. (1985) *J. Clin. Invest.* **76**, 1501–1513.
Boyle, J.K., Zoellner, C.D., Anderson, L.J., Kosick, L.M., Pitas, R.E., Weisgraber, K.H., Hui, D.Y., Mahley, R.W., Gebicke-Haeter, P.J., Ignatius, M.J. & Shooter, E.M. (1989) *J. Clin. Invest.* **83**, 1015–1031.
Breteler, M.M.B., Claus, J.J., Van Duijn, C.M., Launere, L.J. & Hofman, A. (1992) *Epid. Rev.* **14**, 59–93.
Brown, M.S. & Goldstein, J.L. (1986) *Science* **232**, 34–47.
Brown, M.S., Dana, S.E. & Goldstein, J.L. (1973) *Proc. Natl Acad. Sci. USA* **70**, 2162–2166.
Brun, A. & Englund, E. (1981) *Histopathology* **5**, 549–564.
Canadian Study of Health and Aging Working Group (1994) *Can. Med. Assoc. J.* **150**, 899–906.
Corder, E.H., Saunders, A.M., Strittmatter, W.J., Schmechel, D.E., Gaskell, P.C., Small, G.W., Roses, A.D., Pericak-Vance, M.A. (1993) *Science* **261**, 921–923.
Davies, P. & Maloney, A.J.F. (1976) *Lancet* **2**, 1403.
Davies, C.A., Mann, D.M.A., Sumpter, P.Q. & Yates, P.O. (1987) *J. Neurol. Sci.* **78**, 151–164.
Davignon, J., Gregg, R.E. & Sing, C.F. (1988) *Arteriosclerosis* **8**, 1–21.
Davis, K.L., Mohs, J.R. & Tinklenberg, N. (1979) *N. Engl. J. Med.* **301**, 946–951.
Deitrich, J.F., Minnigan, H., Carp, R.I. & Haase, A.T. (1991) *J. Virol.* **65**, 4759–4768.
de Ruiter, J.P. & Uyling, H.B.M. (1987) *Brain Res.* **402**, 217–229.
Etienne, P., Robitaille, Y., Wood, P., Gauthier, S., Nair, N.P.V. & Quirion, R. (1986) *Neuroscience* **4**, 1279–1291.
Finch, C.E. (1987) mRNA alterations in Alzheimer's disease. Banbury Report 27, pp. 143–158. Cold Spring Harbour, Cold Spring Harbor Press.
Flood, D.G. & Coleman, P.D. (1987) *Can. J. Neurol. Sci.* **13**, 475–479.
Geddes, J.W., Monaghan, D.T., Cotman, C.W., Lott, I.T., Kim, R.C. & Chui, H.C. (1985) *Science* **230**, 1179–1181.
Gerdes, L.U., Klausen, I.C. & Faergeman, O. (1992) *Gen. Epid.* **9**, 155–179.
Giubilei, F., D'Antona, R., Antonini, R., Lenzi, G.L. & Fiesci, C. (1990) *Acta Neurol. Scand.* **81**, 84–86.

Goldstein, J.L., Basu, S.K. & Brown, M.S. (1983) *Meth. Enzym.* **98**, 241–260.
Goodrum, J.F. (1990) *J. Neurochem.* **54**, 1709–1715.
Greenamyre, J.T., Penney, J.B., D'Amato, C.B. & Young, A.B. (1987) *J. Neurochem.* **48**, 543–551.
Hyman, B.T., Van Hoesen, G.W., Damasion, A.R. & Barnes, C.L. (1984) *Science* **225**, 1168–1170.
Hyman, B.T., Kromer, L.J. & Van Hoesen, G.W. (1987) *Ann. Neurol.* **21**, 259–267.
Leblanc, A.C. & Poduslo, J.F. (1990) *J. Neurosci. Res.* **25**, 162–171.
Lee, K.S., Standford, E.J., Cotman, C.W. & Lynch, G.S. (1977) *Exp. Brain Res.* **29**, 475–485.
Lucotte, G., David, F., Babron, M.C. & Couderc, R. (1993) *Lancet* (letter) **342**, 1309.
Lynch, G.S., Matthews, D.A., Mosko, S., Parks, T. & Cotman, C.W. (1972) *Brain Res.* **42**, 311–318.
Lynch, G.S., Gall, C., Rose, G. & Cotman, C.W. (1976) *Brain Res.* **110**, 57–71.
Mahley, R.W. (1988) *Science* **40**, 622–630.
Masliah, E., Terry, R.D., deTeresa, R.M. & Hansen, L.A. (1989) *Neurosci. Lett.* **103**, 234–239.
Matthews, D.A., Cotman, C.W. & Lynch, G. (1976) *Brain Res.* **115**, 1–41.
Mayeux, R., Stern, Y., Ottman, R., Tatemichi, T.K., Tang, M.X., Maestre, G., Ngai, C. & Ginsberg, H. (1993) *Ann. Neurol.* **34**, 752–754.
McGeer, P.L. (1984) *Can. J. Physiol. Pharmacol.* **62**, 741–754.
Muckle, T.J. & Roy, J.R. (1985) *Lancet* **1**, 1191–1192.
Nalbantoglu, J., Gilfix, B.M., Bertrand, P., Robitaille, Y., Gauthier, S., Rosenblatt, D.S. & Poirier, J. (1994) *Ann. Neurol.* **36**, 889–895.
Namba, Y., Tomonaga, M. & Kawasaki, H. (1991) *Brain Res.* **541**, 163–166.
Nitsch, R.M., Blusztajn, J.K., Pittas, A.G., Slack, B.E., Growdon, J.H. & Wurtman, R.J. (1992) *Proc. Natl Acad. Sci. USA* **89**, 1671–1675.
Noguchi, S., Murakami, K. & Yamada, N. (1993) *Lancet* **342**, 737.
Pardridge, W.M. & Mietus, L.J. (1980) *Neurochem.* **34**, 463–466.
Payami, H., Kaye, J., Heston, L.L. & Schellenberg, G.D. (1993) *Lancet* (letter) **342**, 738.
Pettegrew, J.W. (1989) *Ann. N. York Acad. Sci.* **568**, 5–28.
Perry, E.K., Gibson, P.H. & Blessed, G. (1977) *J. Neurol. Sci.* **34**, 247–265.
Pitas, R.E., Boyle, J.K., Lee, S.H., Foss, D. & Mahley, R.W. (1987) *Biochem. Biophys. Acta* **917**, 148–161.
Poirier, J., Hess, M., May, P.C., Pasinetti, G. & Finch, C.E. (1990) In *Basic and Therapeutic Strategies in Alzheimer's and Parkinson's diseases* (eds. Nagatsu, T. & Fisher, A.), pp. 191–194. New York, Plenum Press.
Poirier, J., Hess, M., May, P.C., Pasinetti, G. & Finch, C.E. (1991a) In *Basic and Therapeutic Strategies in Alzheimer's and Parkinson's Diseases.* (eds Nagatsu, T. & Fisher, A.), pp. 191–194. New York, Plenum Press.
Poirier, J., Hess, M., May, P.C. & Finch, C.E. (1991b) *Mol. Brain Res.* **9**, 191–195.
Poirier, J., Hess, M., May, P.C. & Finch, C.E. (1991c) *Mol. Brain Res.* **11**, 97–106.
Poirier, J., Baccichet, A., Dea, D. & Gauther, S. (1993a) *Neuroscience* **55**, 81–90.
Poirier, J., Davignon, J., Bouthillier, D., Bertrand, P. & Gauthier, S. (1993b) *Lancet* **342**, 697–699.
Poirier, J., Aubert, I., Bertrand, P., Quirion, R., Gauthier, S. & Nalbantoglu, J. (1994) In *Alzheimer's Disease: Therapeutic Strategies* (eds Giacobini, E. & Becker, R.F.), pp. 72–76. Boston, Birkhauser.
Ransmayr, G., Cervera, P., Hirsch, E., Ruberg, M., Hershi, L.B. & Aging, Y. (1989) *Neuroscience* **32**, 701–714.
Rawlins, F.A., Hedley-White, E.T., Villegas, G. & Uzman, B.G. (1970) *Lab. Invest.* **22**, 237–240.
Rebeck, G.A., Perls, T.T., West, H.L., Sodhi, P., Lipsitz, L.A. & Hyman, B.T. (1994) (Submitted)
Represa, A., Duyckaerts, C., Tremblay, E., Hauw, J.J. & Ben-Hari, Y. (1988) *Brain Res.* **457**, 355–359.
Rifai, N., Christenson, R.H., Gelman, B.B. & Silverman, L.M. (1987) *Clin. Chem.* **33**, 1155–1157.
Roheim, P.S., Carey, M., Forte, T. & Vega, G.L. (1979) *Proc. Natl Acad. Sci. USA* **76**, 4646–4649.

Saunders, A.M., Strittmatter, W.J., Schmechel, D., St George-Hyslop, P.H., Pericak-Vance, M.A., Joo, S.H., Rosi, B.A., Gusella, J.F., McClaclans, D.R., Alberts, M.J. & Roses, A.D. (1993) *Neurology* **43**, 1467–1472.
Schachter, F., Faure-Delanef, L., Guénot, F., Rouger, H., Froguel, P. & Cohen, D. (1994) *Nature Genet.* **6**, 29–31.
Scheff, S.W., Bernardo, L.S. & Cotman, C.W. (1980) *Brain Res.* **199**, 21–38.
Schellenberg, G.D., Boehnke, M., Wijsman, E.M., Moore, D.K. & Martin, G. (1992) *Ann. Neurol.* **31**, 223–227.
Schmechel, D., Saunders, A.M., Strittmatter, W.J., Crain, B.J., Hulette, C.M., Joo, S.H. & Roses, A.D. (1993) *Proc. Natl Acad. Sci. USA* **90**, 9649–9653.
Sing, C.F. & Davignon, J. (1985) *Am. J. Hum. Genet.* **37**, 268–285.
Steward, O. & Scoville, S.A. (1976) *J. Comp. Neurol.* **169**, 347–370.
Strittmatter, W.J., Weisgraber, K.H., Huang, D.Y., Dong, L.M., Salvesen, G.S., Pericak-Vance, M., Schmechel, D. & Roses, A.D. (1993) *Proc. Natl Acad. Sci. USA* **90**, 8098–8102.
Sudhof, T.C., Goldstein, J.L., Brown, M.S. & Russell, D.W. (1985) *Science* **228**, 815–822.
Tucek, S. (1978) In *Acetylcholine Synthesis in Neurons*, pp. 29–42. London, Chapman & Hall.
Wesgraber, K.H. & Mahley, R.W. (1986) *Meth. Enzymol.* **128**, 273–287.
Wesgraber, K.H., Innerarity, T.L. & Mahley, R.W. (1982) *J. Biol. Chem.* **257**, 2518–2521.
Wisniewski, T., Golabek, A., Ghiso, J. & Frangione, B. (1993) *Biochem. Biophys. Res. Commun.* **192**, 359–365.

Index